MICROBIAL EVOLUTION AND CO-ADAPTATION

A Tribute to the Life and Scientific Legacies of Joshua Lederberg

Workshop Summary

Rapporteurs: David A. Relman, Margaret A. Hamburg, Eileen R. Choffnes, and Alison Mack

Forum on Microbial Threats
Board on Global Health

INSTITUTE OF MEDICINE
OF THE NATIONAL ACADEMIES

THE NATIONAL ACADEMIES PRESS
Washington, D.C.
www.nap.edu

THE NATIONAL ACADEMIES PRESS   500 Fifth Street, N.W.   Washington, DC 20001

NOTICE: The project that is the subject of this report was approved by the Governing Board of the National Research Council, whose members are drawn from the councils of the National Academy of Sciences, the National Academy of Engineering, and the Institute of Medicine.

This project was supported by contracts between the National Academy of Sciences and the U.S. Department of Health and Human Services: National Institutes of Health, National Institute of Allergy and Infectious Diseases, Centers for Disease Control and Prevention, and Food and Drug Administration; U.S. Department of Defense, Department of the Army: Global Emerging Infections Surveillance and Response System, Medical Research and Materiel Command, and Defense Threat Reduction Agency; U.S. Department of Veterans Affairs; U.S. Department of Homeland Security; U.S. Agency for International Development; Department of Energy: Lawrence Livermore National Laboratory; American Society for Microbiology; Sanofi Pasteur; Burroughs Wellcome Fund; Pfizer; GlaxoSmithKline; Infectious Diseases Society of America; and the Merck Company Foundation. Any opinions, findings, conclusions, or recommendations expressed in this publication are those of the author(s) and do not necessarily reflect the view of the organizations or agencies that provided support for this project.

International Standard Book Number-13: 978-0-309-13121-6
International Standard Book Number-10: 0-309-13121-9

Additional copies of this report are available from the National Academies Press, 500 Fifth Street, N.W., Lockbox 285, Washington, DC 20055; (800) 624-6242 or (202) 334-3313 (in the Washington metropolitan area); Internet, http://www.nap.edu.

For more information about the Institute of Medicine, visit the IOM home page at: **www. iom.edu.**

The serpent has been a symbol of long life, healing, and knowledge among almost all cultures and religions since the beginning of recorded history. The serpent adopted as a logotype by the Institute of Medicine is a relief carving from ancient Greece, now held by the Staatliche Museen in Berlin.

COVER: *E. coli* is a gram-negative, facultatively anaerobic, rod prokaryote. The strains are undergoing conjugation via a pilus (one strain has fimbriae) under ×3,645 magnification. Bacterial conjugation is the ability to transfer DNA between strains of bacteria (via a pilus). It allows a new mutation to spread through an existing population. It is believed that this process led to the spread of toxin synthesis from *Shigella* to *E. coli* (O157:H7). *E. coli* can cause urinary tract infections, traveler's diarrhea, nosocomial infections, meningitis, peritonitis, mastitis, septicemia, and gram-negative pneumonia. It causes a variety of skin and wound infections such as scalded skin syndrome, scarlet fever, erysipelas, and impetigo. SOURCE: Cover art was provided by Dennis Kunkel Microscopy, Inc. The photo of Joshua Lederberg on the spine is courtesy of The Rockefeller University.

Suggested citation: IOM (Institute of Medicine). 2009. *Microbial evolution and co-adaptation: a tribute to the life and scientific legacies of Joshua Lederberg*. Washington, DC: The National Academies Press.

*"Knowing is not enough; we must apply.
Willing is not enough; we must do."*
—Goethe

INSTITUTE OF MEDICINE
OF THE NATIONAL ACADEMIES

Advising the Nation. Improving Health.

THE NATIONAL ACADEMIES
Advisers to the Nation on Science, Engineering, and Medicine

The **National Academy of Sciences** is a private, nonprofit, self-perpetuating society of distinguished scholars engaged in scientific and engineering research, dedicated to the furtherance of science and technology and to their use for the general welfare. Upon the authority of the charter granted to it by the Congress in 1863, the Academy has a mandate that requires it to advise the federal government on scientific and technical matters. Dr. Ralph J. Cicerone is president of the National Academy of Sciences.

The **National Academy of Engineering** was established in 1964, under the charter of the National Academy of Sciences, as a parallel organization of outstanding engineers. It is autonomous in its administration and in the selection of its members, sharing with the National Academy of Sciences the responsibility for advising the federal government. The National Academy of Engineering also sponsors engineering programs aimed at meeting national needs, encourages education and research, and recognizes the superior achievements of engineers. Dr. Charles M. Vest is president of the National Academy of Engineering.

The **Institute of Medicine** was established in 1970 by the National Academy of Sciences to secure the services of eminent members of appropriate professions in the examination of policy matters pertaining to the health of the public. The Institute acts under the responsibility given to the National Academy of Sciences by its congressional charter to be an adviser to the federal government and, upon its own initiative, to identify issues of medical care, research, and education. Dr. Harvey V. Fineberg is president of the Institute of Medicine.

The **National Research Council** was organized by the National Academy of Sciences in 1916 to associate the broad community of science and technology with the Academy's purposes of furthering knowledge and advising the federal government. Functioning in accordance with general policies determined by the Academy, the Council has become the principal operating agency of both the National Academy of Sciences and the National Academy of Engineering in providing services to the government, the public, and the scientific and engineering communities. The Council is administered jointly by both Academies and the Institute of Medicine. Dr. Ralph J. Cicerone and Dr. Charles M. Vest are chair and vice chair, respectively, of the National Research Council.

www.national-academies.org

FORUM ON MICROBIAL THREATS

IOM Forums and Roundtables do not issue, review, or approve individual documents. The responsibility for the published workshop summary rests with the workshop rapporteur(s) and the institution.

JAMES M. HUGHES, Global Infectious Diseases Program, Emory University, Atlanta, Georgia

STEPHEN A. JOHNSTON, Arizona BioDesign Institute, Arizona State University, Tempe

GERALD T. KEUSCH, Boston University School of Medicine and Boston University School of Public Health, Massachusetts

RIMA F. KHABBAZ, National Center for Preparedness, Detection, and Control of Infectious Diseases, Centers for Disease Control and Prevention, Atlanta, Georgia

LONNIE J. KING, Center for Zoonotic, Vectorborne, and Enteric Diseases, Centers for Disease Control and Prevention, Atlanta, Georgia

GEORGE W. KORCH, U.S. Army Medical Research Institute for Infectious Diseases, Fort Detrick, Maryland

STANLEY M. LEMON, School of Medicine, University of Texas Medical Branch, Galveston

EDWARD McSWEEGAN, National Institute of Allergy and Infectious Diseases, National Institutes of Health, Bethesda, Maryland

STEPHEN S. MORSE, Center for Public Health Preparedness, Columbia University, New York

MICHAEL T. OSTERHOLM, Center for Infectious Disease Research and Policy, School of Public Health, University of Minnesota, Minneapolis

GEORGE POSTE, Arizona BioDesign Institute, Arizona State University, Tempe

JOHN C. POTTAGE, JR., GlaxoSmithKline, Collegeville, Pennsylvania

GARY A. ROSELLE, Central Office, Veterans Health Administration, Department of Veterans Affairs, Washington, DC

JANET SHOEMAKER, Office of Public Affairs, American Society for Microbiology, Washington, DC

P. FREDERICK SPARLING, University of North Carolina, Chapel Hill

BRIAN J. STASKAWICZ, Department of Plant and Microbial Biology, University of California, Berkeley

TERENCE TAYLOR, International Council for the Life Sciences, Washington, DC

MURRAY TROSTLE, U.S. Agency for International Development, Washington, DC

Staff

EILEEN CHOFFNES, Director
KATE SKOCZDOPOLE, Senior Program Associate
SARAH BRONKO, Research Associate
KENISHA PETERS, Senior Program Assistant
ALISON MACK, Science Writer

vii

Reviewers

This report has been reviewed in draft form by individuals chosen for their diverse perspectives and technical expertise, in accordance with procedures approved by the National Research Council's Report Review Committee. The purpose of this independent review is to provide candid and critical comments that will assist the institution in making its published report as sound as possible and to ensure that the report meets institutional standards for objectivity, evidence, and responsiveness to the study charge. The review comments and draft manuscript remain confidential to protect the integrity of the deliberative process. We wish to thank the following individuals for their review of this report:

Martin J. Blaser, New York University School of Medicine
Steven J. Brickner, Pfizer Global Research and Development, Pfizer, Inc.
James M. Hughes, Hubert Department of Global Health and Rollins
    School of Public Health, Emory University

Although the reviewers listed above have provided many constructive comments and suggestions, they were not asked to endorse the final draft of the report before its release. The review of this report was overseen by **Dr. Melvin Worth.** Appointed by the Institute of Medicine, he was responsible for making certain that an independent examination of this report was carried out in accordance with institutional procedures and that all review comments were carefully considered. Responsibility for the final content of this report rests entirely with the authoring committee and the institution.

Preface

The Forum on Emerging Infections was created by the Institute of Medicine (IOM) in 1996 in response to a request from the Centers for Disease Control and Prevention (CDC) and the National Institutes of Health (NIH). The purpose of the Forum is to provide opportunities for leaders from government, academia, and industry to meet and examine issues of shared concern regarding research, prevention, detection, and management of emerging or reemerging infectious diseases. In pursuing this task, the Forum provides a venue to foster the exchange of information and ideas, identify areas in need of greater attention, clarify policy issues by enhancing knowledge and identifying points of agreement, and inform decision makers about science and policy issues. The Forum seeks to illuminate issues rather than resolve them; for this reason, it does not provide advice or recommendations on any specific policy initiative pending before any agency or organization. Its value derives instead from the diversity of its membership and from the contributions that individual members make throughout the activities of the Forum. In September 2003, the Forum changed its name to the Forum on Microbial Threats.

ABOUT THE WORKSHOP

To a great extent, the Forum on Microbial Threats (hereinafter, the Forum) owes its very existence to the life and legacies of the late Dr. Joshua Lederberg. Along with the late Robert Shope and Stanley C. Oaks, Jr., Lederberg organized and co-chaired the 1992 Institute of Medicine study, *Emerging Infections: Microbial Threats to Health in the United States*. The *Emerging Infections* report helped to define the factors and dynamic relationships that lead to the emergence of infectious diseases. Its recommendations addressed both the recognition of and interventions against emerging infections as well as identified major unmet challenges in responding to infectious disease outbreaks and monitoring the prevalence of endemic diseases. This report ultimately led to the Forum's creation in 1996. As the first chair of the Forum, 1996-2001, Lederberg was instrumental in establishing it as a venue for the discussion and scrutiny of critical—and sometimes contentious—scientific and policy issues of shared concern related to research on and the prevention, detection, and management of infectious diseases and dangerous pathogens.

Lederberg's long shadow may readily be appreciated in the Forum's 2005 workshop, *Ending the War Metaphor: The Changing Agenda for Unraveling the Host-Microbe Relationship*. Its central theme was derived from a comprehensive essay that he published several years earlier in *Science* entitled "Infectious History." Under the heading, "Evolving Metaphors of Infection: Teach War No More," Lederberg argued that "[w]e should think of each host and its parasites as a superorganism with the respective genomes yoked into a chimera of sorts." Thus began a discussion that developed the concept of the *microbiome*—a term Lederberg coined to denote the collective genome of an indigenous microbial community—as a forefront of scientific inquiry.

Having reviewed the shortcomings and consequences of the war metaphor of infection Lederberg suggested, in the same essay, a "paradigm shift" in the way we collectively identify and think about the microbial world around us, replacing notions of aggression and conflict with a more ecologically—and evolutionarily—informed view of the dynamic relationships among and between microbes, hosts, and their environments. This perspective recognized the participation of every eukaryotic organism—moreover, every eukaryotic cell—in partnerships with microbes and microbial communities, and acknowledged that microbes and their hosts are ultimately dependent upon one another for survival. It also encouraged the exploration and exploitation of these ecological relationships in order to increase agricultural productivity and to improve animal, human, and environmental health.

More than a century of research, sparked by the germ theory of disease, underlies our current appreciation of microbe-host-environment interactions. Our "war" on infectious microbes has restricted the spread of several pathogens and drastically reduced the burden of human disease, but the tide of the human conquest shows many signs of turning. Over the past 30 years, 37 new human pathogens have been identified as disease threats, and an estimated 12 percent of known human pathogens have been recognized as either emerging or reemerging.[1] Due in large part to the HIV/AIDS pandemic, the number of deaths in the

---

[1]Merell, D. S., and S. Falkow. 2004. Frontal and stealth attack strategies in microbial pathogenesis. *Nature* 430(6996):250-256.

United States attributable to infection, having fallen steadily since the turn of the century, began to increase in the early 1980s. Infectious diseases continue to cause high morbidity and mortality throughout the world, particularly in developing countries. In 2001, infectious diseases accounted for an estimated 26 percent of deaths worldwide.

Clearly, a reconsideration of our interactions with pathogenic microbes is warranted, and it must be based on a better understanding of host-microbe relationships in general. Estimates indicate that 90 to 99 percent of the approximately 10^{14} cells that comprise a healthy human body belong to the complex microbiota that share our space. Only a small fraction of the roughly several thousand bacterial species that inhabit our bodies cause illness; very little is known about the other nonpathogenic bacteria, or even about microbes that in most cases cause chronic, subclinical disease in humans, and that only occasionally produce illness and death. Research into our own microbial ecology and that of our fellow eukaryotes, including plants, appears certain to reveal new strategies for preventing and treating a broad spectrum of infectious disease.

Dr. Lederberg's death on February 2, 2008, marked the departure of a central figure of modern science. It is in his honor that the Forum convened this public workshop on May 20 and 21, 2008, to examine Dr. Lederberg's scientific and policy contributions to the marketplace of ideas in the life sciences, medicine, and public policy. The agenda for this workshop demonstrates the extent to which conceptual and technological developments have, within a few short years, advanced our collective understanding of microbial genetics, microbial communities, and microbe-host-environment interactions. Through invited presentations and discussions, participants explored a range of topics related to microbial evolution and co-adaptation, including methods for characterizing microbial diversity; model systems for investigating the ecology of host-microbe interactions and microbial communities at the molecular level; microbial evolution and the emergence of virulence; the phenomenon of antibiotic resistance and opportunities for mitigating its public health impact; and an exploration of current trends in infectious disease emergence as a means to anticipate the appearance of future novel pathogens.

As Adel Mahmoud, first co-chair of the Forum observed, "Joshua believed very strongly in the work of this Forum. He had great confidence in the ability of scientists and researchers to continue to solve some of the riddles that still confront science in the fight against infectious diseases. By remembering him with this tribute, we are also remembering the many things that his life and career can teach all of us. I hope that every time we meet at this Forum, Joshua Lederberg will be an inspiration and a reminder that our work can truly change the world, just as his life and career certainly did."

It is to Josh's life and living legacies that we dedicate this volume.

ACKNOWLEDGMENTS

The Forum on Microbial Threats and the IOM wish to express their gratitude to the individuals and organizations who, through their participation in this workshop, provided invaluable information and advice to the Forum. A full list of presenters may be found in Appendix A.

The Forum is deeply indebted to the IOM staff who contributed during the course of the workshop and the production of this workshop summary. On behalf of the Forum, we gratefully acknowledge the efforts led by Dr. Eileen Choffnes, director of the Forum; Kate Skoczdopole, senior program associate; Sarah Bronko, research associate; and Kenisha Peters, senior program assistant, for dedicating much effort and time to developing this workshop's agenda and for their thoughtful and insightful approach and skill in planning for the workshop and in translating the workshop's proceedings and discussion into this workshop summary. We would also like to thank the following IOM staff and consultants for their valuable contributions to this activity: Alison Mack, Bronwyn Schrecker Jamrok, Jackie Turner, Michael Hayes, and Florence Poillon.

Finally, the Forum wishes to recognize the sponsors that supported this activity: U.S. Department of Health and Human Services: National Institutes of Health, National Institute of Allergy and Infectious Diseases, Centers for Disease Control and Prevention, and Food and Drug Administration; U.S. Department of Defense: Global Emerging Infections Surveillance and Response System, Walter Reed Army Institute of Research, U.S. Army Medical and Materiel Command, and the Defense Threat Reduction Agency; U.S. Department of Veterans Affairs; U.S. Department of Homeland Security; U.S. Agency for International Development; U.S. Department of Energy: Lawrence Livermore National Laboratory; American Society for Microbiology; Sanofi Pasteur; Burroughs Wellcome Fund; Pfizer; GlaxoSmithKline; Infectious Diseases Society of America; and the Merck Company Foundation.

David A. Relman, *Chair*
Margaret A. Hamburg, *Vice Chair*
Forum on Microbial Threats

Contents

Tables, Figures, and Boxes

TABLES

FIGURES

BOXES

Workshop Overview[1]

MICROBIAL EVOLUTION AND CO-ADAPTATION:
A WORKSHOP IN HONOR OF JOSHUA LEDERBERG

Prologue

To a great extent, the Forum on Microbial Threats (hereinafter, the Forum) owes its very existence to the life and legacies of the late Dr. Joshua Lederberg. Dr. Lederberg's death on February 2, 2008, marked the departure of a central figure of modern science. It is in his honor that the Forum hosted this public workshop on "microbial evolution and co-adaptation" on May 20 and 21, 2008.

Along with the late Robert Shope and Stanley C. Oaks, Jr., Lederberg organized and co-chaired the 1992 Institute of Medicine (IOM) study, *Emerging Infections: Microbial Threats to Health in the United States* (IOM, 1992). The *Emerging Infections* report helped to define the factors and dynamic relationships that lead to the emergence of infectious diseases. The recommendations of this report (IOM, 1992) addressed both the recognition of and interventions against emerging infections. This IOM report identified major unmet challenges in responding to infectious disease outbreaks and monitoring the prevalence of endemic diseases, and ultimately led to the Forum's creation in 1996 (Morse, 2008). As the first chair of the Forum, 1996-2001, Dr. Lederberg was instrumental in establishing it as a venue for the discussion and scrutiny of critical—and sometimes contentious—scientific and policy issues of shared concern related to

---

[1]The Forum's role was limited to planning the workshop, and this workshop summary has been prepared by the workshop rapporteurs as a factual summary of what occurred at the workshop.

research on and the prevention, detection, and management of infectious diseases and dangerous pathogens.

Lederberg's influence may readily be appreciated in the 2005 Forum workshop *Ending the War Metaphor: The Changing Agenda for Unraveling the Host-Microbe Relationship* (IOM, 2006a). Its central theme was derived from a comprehensive essay entitled "Infectious History" that he published several years earlier in *Science* (Lederberg, 2000; reprinted as Appendix WO-1). Under the heading, "Evolving Metaphors of Infection: Teach War No More," Lederberg argued that "[w]e should think of each host and its parasites as a superorganism with the respective genomes yoked into a chimera of sorts." Thus began a discussion that developed the concept of the *microbiome*—a term Lederberg coined to denote the collective genome of an indigenous microbial community—as a forefront of scientific inquiry (Hooper and Gordon, 2001; Relman and Falkow, 2001).

Having reviewed the shortcomings and consequences of the war metaphor of infection, Lederberg suggested, in the same essay, a "paradigm shift" in the way we collectively identify and think about the microbial world around us, replacing notions of aggression and conflict with a more ecologically—and evolutionarily—informed view of the dynamic relationships among and between microbes, hosts, and their environments (Lederberg, 2000). This perspective recognizes the participation of every eukaryotic organism—moreover, every eukaryotic cell—in partnerships with microbes and microbial communities, and acknowledges that microbes and their hosts are ultimately interdependent upon one another for survival. It also encourages the exploration and exploitation of these ecological relationships in order to increase agricultural productivity and to improve animal, human, and environmental health.

The agenda of the present workshop demonstrates the extent to which conceptual and technological developments have, within a few short years, advanced our collective understanding of microbial genetics, microbial communities, and microbe-host-environment relationships. Through invited presentations and discussions, participants explored a range of topics related to microbial evolution and co-adaptation, including: methods for characterizing microbial diversity; model systems for investigating the ecology of host-microbe interactions and microbial communities at the molecular level; microbial evolution and the emergence of virulence; the phenomenon of antibiotic resistance and opportunities for mitigating its public health impact; and an exploration of current trends in infectious disease emergence as a means to anticipate the appearance of future novel pathogens.

Organization of the Workshop Summary

This workshop summary was prepared for the Forum membership by the rapporteurs and includes a collection of individually authored papers[2] and commentary. Sections of the workshop summary not specifically attributed to an individual reflect the views of the rapporteurs and not those of the Forum on Microbial Threats, its sponsors, or the IOM. The contents of the unattributed sections are based on presentations and discussions at the workshop.

The workshop summary is organized into chapters as a topic-by-topic summation of the presentations and discussions that took place at the workshop. Its purpose is to present lessons from relevant experience, to delineate a range of pivotal issues and their respective problems, and to offer potential responses as discussed and described by workshop participants.

Although this workshop summary provides an account of the individual presentations, it also reflects an important aspect of the Forum philosophy. The workshop functions as a dialogue among representatives from different sectors and allows them to present their beliefs about which areas may merit further attention. The reader should be aware, however, that the material presented here expresses the views and opinions of the individuals participating in the workshop and not the deliberations and conclusions of a formally constituted IOM study committee. These proceedings summarize only the statements of participants in the workshop and are not intended to be an exhaustive exploration of the subject matter or a representation of consensus evaluation.

THE LIFE AND LEGACIES OF JOSHUA LEDERBERG

This workshop continued the tradition established by the late Joshua Lederberg, this Forum's first chairman, of wide-ranging discussion among experts from many disciplines and sectors, honoring him by focusing on fields of inquiry to which he had made important contributions. At the same time, this gathering was unique in the history of the Forum, for it also offered participants a chance to reflect upon Lederberg's life (see Box WO-1) and his extraordinary contributions to science, academia, public health, and government. Formal remarks by David Hamburg of Cornell University's Weill Medical College, Stephen Morse of Columbia University, and Adel Mahmoud of Princeton University (collected in Chapter 1) inspired open discussion of Lederberg's life and legacy, as well as personal reminiscences about his role as mentor, advisor, advocate, and friend.

Recalling the words of Ralph Waldo Emerson, who likened institutions to the lengthened shadows of their founders (Emerson, 1841), Morse observed that Lederberg's influential shadow reaches into many places, but is most imposing in

---

[2]Some of the individually authored manuscripts may contain figures that have appeared in prior peer-reviewed publications. They are reprinted as originally published.

BOX WO-1
Joshua Lederberg: An Extraordinary Life

- Born on May 23, 1925, in Montclair, New Jersey, to Zvi Lederberg, an orthodox rabbi, and Esther Schulman, a homemaker and descendant of a long line of rabbinical scholars; Lederberg's family moved to the Washington Heights area of upper Manhattan when he was six months old.
- From 1938-1940, attended Stuyvesant High School in New York City (a public, highly competitive school of science and technology).
- In 1941, enrolled at Columbia University, majoring in zoology.
- In 1943, enrolled in the United States Navy's V-12 training program, which combined an accelerated premedical and medical curriculum to fulfill the armed services' projected need for medical officers.
- In 1944, received his bachelor's degree in zoology at Columbia and began medical training at the university's College of Physicians and Surgeons.

the area of infectious diseases, as epitomized by the Forum. Indeed, Forum member Stanley Lemon,[3] of the University of Texas Medical Branch in Galveston, observed that the Forum's mission—"tackling tough problems and addressing them with the best of science from the academic perspective and the active involvement of government"—is now borne by scores of people who can only hope to carry out what Lederberg once undertook single-handedly.

As stated previously, it was largely due to Lederberg's efforts, and particularly his co-chairmanship of the IOM Committee on Emerging Microbial Threats to Health, that the idea for a Forum became a reality. In recognition of the profound

---

[3]Vice-Chair from July 2001 to June 2004; Chair from August 2004 to July 2007.

- In 1946, during a year-long leave of absence from medical school, carried out experiments on *Escherichia coli* in the laboratory of Edward Tatum at Yale University. Lederberg's findings demonstrated that certain strains of bacteria can undergo a sexual stage, and that they mate and exchange genes.
- In 1947, having extended his collaboration with Tatum for another year in order to begin mapping the *E. coli* chromosome, received his Ph.D. degree from Yale. He then received an offer of an assistant professorship in genetics at the University of Wisconsin, which caused him to abandon his plans to return to medical school in order to pursue basic research in genetics. He was accompanied by his new wife, Esther Zimmer Lederberg, who received her doctorate in microbiology at Wisconsin and who also rose to prominence in that field.
- In 1957, founded and became chairman of the Department of Medical Genetics at Wisconsin and was elected to the National Academy of Sciences.
- In 1958, became the first chairman of the newly established Department of Genetics at Stanford University's School of Medicine, days before being awarded the Nobel Prize in Physiology or Medicine, along with Tatum and George Beadle, for "discoveries concerning genetic recombination and the organization of the genetic material of bacteria."
- In 1966, his marriage to Esther Lederberg ended in divorce; in 1968 he married Marguerite Stein Kirsch, a clinical psychologist, with whom he had two children.
- From 1966-1971, published "Science and Man," a weekly column on science, society, and public policy in *The Washington Post*.
- In 1978, accepted the presidency of Rockefeller University.
- In 1989, awarded the National Medal of Science.
- In 1990, retired from the presidency and continued at Rockefeller as Raymond and Beverly Sackler Foundation Scholar.
- In 2006, awarded the Presidential Medal of Freedom.
- On February 2, 2008, died of pneumonia at New York-Presbyterian Hospital.

SOURCE: NLM (2008); photo courtesy of The Rockefeller University.

impact of *Emerging Infections: Microbial Threats to Health in the United States* (IOM, 1992)—which provided the U.S. government with a basis for developing a national strategy on emerging infections and informed the pursuit of international negotiations to address this threat—the Centers for Disease Control and Prevention (CDC) and the National Institute of Allergy and Infectious Diseases (NIAID) asked the IOM to create a forum to serve as a follow-on activity to the national disease strategy developed by these agencies. In 1996, the IOM launched the Forum on Emerging Infections (now the Forum on Microbial Threats). Lederberg chaired the Forum for its first five years and remained an avid participant in its workshops and discussions until his failing health precluded travel.

Even in his physical absence, the Forum has continued—and undoubtedly will continue—to be inspired by Lederberg's expansive vision: a command of science that forged connections between microbiology and a broad range of disciplines, that was profoundly informed by history and literature, and that embraced the fullness of human imagination and possibility.

Scientist

"Joshua Lederberg has been the dominant force that shaped our thinking, responses, and intellectual understanding of microbes for much of the last half of the twentieth century," Mahmoud remarked. From his early, Nobel Prize–winning work on bacterial recombination, accomplished while he was barely 20, through the last years of his life, when he continued to provide much sought-after advice to global policy makers on emerging infectious diseases and biological warfare, Lederberg extended his command of microbiology to profoundly influence a host of related fields, including biotechnology, artificial intelligence, bioinformatics, and exobiology. Exobiology, the study of extraterrestrial life, was one among many widely used terms coined by Lederberg, according to Stephen Morse. He also noted along with several other participants that the hero of the classic science fiction novel *The Andromeda Strain*[4] (Crichton, 1969), Dr. Jeremy Stone, may well have been based on Lederberg. Ultimately, Lederberg viewed his wide-ranging scientific interests through the lens of evolution. According to Morse, the unifying theme of Lederberg's scientific studies was to characterize sources of genetic diversity and natural selection.

Nowhere is Lederberg's comprehensive view of microbial evolution and its consequences more evident than in his essay, "Infectious History" (Lederberg, 2000), which informed the workshop's agenda and serves as a framework for this workshop overview. Referring to that landmark publication as "the Bible of infectious diseases," Mahmoud observed that it laid out "fundamental concepts that we are still debating about [including] the evolutionary biology and the ecology of microbes."

From his earliest years, Lederberg embodied scientific curiosity and innovation, David Hamburg noted. He recalled Lederberg's knack for "turning an issue on its head, and thereby illuminating it," and added that he "took deep, deep satisfaction in discovery, his own and others," which was apparent in his relentless questioning. Lederberg "was a great challenger of the scientific community to pursue many ramifications of questions that appeared to be, at least for the time being, answered but were never answered for him," Hamburg said. "This inter-

---

[4]*The Andromeda Strain* (1969), by Michael Crichton, is a techno-thriller novel documenting the efforts of a team of scientists investigating a deadly extraterrestrial microorganism that rapidly and fatally clots human blood. The infected show Ebola-like symptoms and die within two minutes (see http://en.wikipedia.org/wiki/The_Andromeda_Strain; accessed December 15, 2008).

related set of attributes characterized Josh all his life and had much to do with his great accomplishments."

Hamburg recounted that Lederberg entered medical school at Columbia University with this intense curiosity and sense of discovery, as well as a desire to improve the lot of humanity and to relieve human suffering. Fascinated with bacterial genetics, however, Lederberg took a one-year leave from medical school to work on *Escherichia coli* with Edward Tatum, at Yale University, in 1946. "This was groundbreaking, highly imaginative work on the nature of microorganisms, especially their mechanisms of inheritance," Hamburg said. "It opened up bacterial genetics, including the momentous discovery of genetic recombination," a line of inquiry that paved the way for Lederberg's being awarded the Nobel Prize in Physiology or Medicine in 1958, along with Tatum and George Beadle for "discoveries concerning genetic recombination and the organization of the genetic material of bacteria."

Following an extremely successful first year of research in Tatum's laboratory, Lederberg decided to take another year away from medical school and continue to explore bacterial genetics. "We lost the budding physician in Joshua Lederberg by the end of the second year, because he was offered a faculty position at the University of Wisconsin," Mahmoud explained, "but that did not stop Joshua Lederberg from being at the forefront of those concerned about human health and well-being."

According to Forum member Jo Handelsman, professor of bacteriology at the University of Wisconsin, Lederberg's influence reverberates to this day. "He left behind the great legacy of his research and the spirit of a truly great mind in science," she said, as well as stories that have attained the status of "urban legends." At Wisconsin, Lederberg also established the legendary habit of appearing to sleep during seminars, after which he would ask difficult and probing questions. This habit was still in evidence in the early 1990s during his co-chairmanship of the first IOM study on emerging infections, according to Forum member Enriqueta (Queta) Bond, president of the Burroughs Wellcome Fund. "I was the executive officer at the Institute of Medicine when the first *Emerging Infections* report was done," she recalled. "I remember coming to one of the first meetings of the committee, and . . . Josh would sit there and you would think, 'Is he awake? He's supposed to be chairing this committee.' . . . Then you would get the zingers from Josh: just the perfect question to move the agenda, develop the next topic, and so forth."

Indeed, Morse said, Lederberg "was never happier than when he was absorbing knowledge and questioning it. I like to think of this, with all of us here, as being an important part of Josh's legacy," he added. Hamburg recalled Lederberg's "rare capacity to range widely with open eyes and open mind, and also dig deeply at times into specialized topics; to combine these capacities in research, education, and intellectual synthesis led to so much fruitful stimulation in a variety of fields."

"He believed that there are no limits to what the human mind can accomplish, especially when its power is hitched to a willingness to think boldly and unconventionally, and to hard work," Mahmoud said. "Until almost the day he died, Joshua could be found in his office, in his apartment, working. His mind was always thinking, always probing, always questioning." Indeed, during his last days, Lederberg offered insightful advice to his longtime friend Hamburg, who was editing the final draft of his recently published book, *Preventing Genocide: Practical Steps Toward Early Detection and Effective Action* (Hamburg, 2008). "We had a couple of very intensive hours in which he asked his usual penetrating questions and clarified key issues, and then was obviously quite exhausted," Hamburg recalled. "We were prepared to take him back home. He said, 'No. I'd like to rest for an hour or so and come back. I have one more chapter I want to discuss.'"

"We did that," Hamburg continued. "It was vintage Josh. He mobilized himself to address an important problem with a friend that he valued and made an important contribution. The final changes in the book—all improvements—were due to that conversation."

Academic

Another tribute to Lederberg's remarkable capacities was institutional innovation, Hamburg observed. When Lederberg created departments of genetics in the medical schools at the University of Wisconsin and Stanford University, Hamburg recalled, "[the field of] genetics had been marginal or nonexistent in medical schools. There was a widely shared assumption, in the middle of the twentieth century, that genetics might be intrinsically interesting, but it would never have much practical significance for medicine."

"In teaching and in institution building, Lederberg emphasized the mutually beneficial interplay of basic and clinical research," Hamburg continued. Lederberg, he said, helped clinical departments at Stanford University's School of Medicine build interdisciplinary groups and identify research opportunities and promising lines of innovation. He fostered many lines of inquiry within his own Department of Genetics at Stanford—including molecular genetics, cellular genetics, clinical genetics, population genetics, immunology, neurobiology, and exobiology (particularly in relation to the National Aeronautics and Space Admininstration's [NASA's] Mariner and Viking missions to Mars)—and hired a superb group of internationally-known researchers, including Walter Bodmer and Eric Shooter from the United Kingdom, Luca Cavalli-Sforza from Italy, and Gus Nossal from Australia, Hamburg recalled. He also recruited from within the university, including speaker Stanley Cohen, who eventually succeeded Lederberg as chairman of the genetics department at Stanford. By taking this action, Hamburg said, Lederberg "was not robbing another department, but rather opening up an opportunity that Stan [Cohen] wanted and needed, and, of course, in which he made tremendous contributions."

While at Stanford, Lederberg also made a major contribution to undergraduate education, establishing a cross-disciplinary program in human biology that remains one of the university's most sought-after majors. Hamburg—who as chairman of Stanford's psychiatry and behavioral science department, assisted in this effort along with Donald Kennedy, then the chairman of Stanford's biology department—remarked that the program might not have had such a long and illustrious history if Lederberg had not insisted that it include endowed chairs.

Following his years at Stanford, Lederberg's "rich experience, knowledge, skill, and wisdom were brought to bear on Rockefeller University under his presidency, broadening the scope of its great faculty, opening new opportunities for young people, and greatly improving the facilities," Hamburg said. Although admitting that he did not at first think university administration was the best use of his friend's talents, Hamburg recognized that Lederberg adapted well to his new responsibilities and proved adept both as a financial and a human resources manager who was deeply concerned about the personal well-being of his faculty.

While it seems that nothing was too big for Lederberg to tackle, Forum member Gerald Keusch of Boston University described how he had benefited from Lederberg's willingness to address what might have seemed a small issue. During the mid-1990s, National Institutes of Health (NIH) director Harold Varmus was thinking about the impact on NIH of shrinking the number of institutes and centers, beginning with the Fogarty International Center. "Harold is a very smart person and knew there were going to be problems in trying to change the status quo. How to proceed? You form a committee to give you the recommendation that allows you to go ahead and act," Keusch recalled. "So he asked Josh and Barry Bloom[5] to do a review of the Fogarty and all international programs at the NIH." Lederberg and Bloom proceeded to conduct an exhaustive study, which ultimately recommended that the Fogarty be strengthened, not disbanded. As a result, a new position was created—for which Keusch was hired—to direct the Fogarty International Center and serve as the NIH's associate director for international research.

After five years in this position, Keusch asked Lederberg and Bloom to return and review the Fogarty's progress. Although unwell and not traveling as he once had, Lederberg did not hesitate "to come back to do an honest, objective review and [once again] come out strongly in favor of the Fogarty's international mission," Keusch said. "You might have thought, in 1996, that Fogarty and the

---

[5]In the mid-1990s, Barry Bloom was a Howard Hughes Medical Institute investigator and served on the National Advisory Board of the Fogarty International Center at the National Institutes of Health; see http://www.hsph.harvard.edu/administrative-offices/deans-office/dean-barry-r-bloom/.

international programs at NIH would not have attracted [Lederberg's] attention. But they did, and I think the Fogarty is certainly the better for it, [as] is NIH."

Global Citizen

Achieving the Nobel Prize at the age of 33 gave Lederberg a global perspective that he fully embraced in the subsequent half-century, according to Mahmoud. In so doing, Lederberg undertook multiple roles, including advisor to governments, institutions, and industry, as well as educator of the general public.

"Every president from John F. Kennedy to the current administration sought Joshua's advice and consultation," Hamburg said. "He chaired and studied issues from space science to human and artificial intelligence, to human-microbe interplay." Lederberg advised many agencies in the United States, most notably the NIH, the Centers for Disease Control and Prevention (CDC), the National Science Foundations (NSF), NASA, the Office of Science and Technology Policy (OSTP), and the Department of the Navy. He also served as an advisor to the World Health Organization (WHO) and was particularly influential as that organization attempted to establish regional surveillance centers for emerging infectious diseases. Forum member James Hughes, of Emory University, remarked that Lederberg was "very engaged in Geneva, to the point that he took it upon himself to meet with the director-general of the WHO at the time, Dr. Hiroshi Nakajima. I am sure this is one of the reasons that WHO went on to develop its emerging infections focus."

"Josh used to go to Washington sometimes three times a week, back and forth, to give scientific advice," Morse recalled. "He was the model of the scientific adviser. His advice was honest and dispassionate and in no way self-interested. His interest was furthering the cause of science and humanity." Morse observed Lederberg had been concerned that samples obtained from space or spaceships might contain extraterrestrial life forms. NASA asked Lederberg how to decontaminate such samples and what precautions should be taken with them. "He gave very freely of his advice," Morse said. "This led, I think, to one of the most interesting job descriptions I have ever seen. NASA created a position called 'planetary quarantine officer.'[6] Those of us who talk about emerging infections on this world have to realize that Josh's purview extended far beyond that."

Emerging infections on Earth did, however, feature prominently in Lederberg's advisory efforts, as many participants readily acknowledged. According to Mahmoud, "It was Josh, and Josh alone, who articulated and brought to the forefront of the scientific agenda the subject of emerging and reemerging infections."

Concern about emerging infections has grown following the appearance of new diseases, such as HIV/AIDS, and the reemergence of others, such as dengue,

---

[6]This position was later renamed by NASA as "Planetary Protection Officer, Earth."

and from appreciation of the complex determinants of their emergence—including microbial adaptation to new hosts (HIV infection, severe acute respiratory syndrome [SARS]), population immunity pressures (influenza A), travel (acute hemorrhagic conjunctivitis), animal migration and movement (West Nile virus infection, H5N1 avian influenza), microbial escape from antibiotic pressures (multidrug-resistant and extensively drug-resistant tuberculosis), mechanical dispersal (Legionnaires' disease), and others (panel, Figure WO-1; Morens et al., 2008).[7]

Lederberg was also "a pioneer in biological warfare and bioterrorism defense, applying his farsighted vision to efforts to understand the danger and find ways to cope with it," Morse said, long before that threat was widely acknowledged. "He strongly influenced the negotiation of the biological weapons disarmament treaty."[8]

When Lederberg first voiced his concerns regarding emerging microbial threats in the late 1980s, Mahmoud recalled, "half of the scientific community was just smiling [as if to say], 'the old man is just babbling about the subject.'" Instead, the advent of "a fundamental platform," the 1992 IOM report, "really opened the way for a new way of thinking about microbes . . . [and also] forced the whole community to come back, in 2003, for the second report on the subject." Lederberg also co-chaired the committee that produced this second report, *Microbial Threats to Health* (IOM, 2003), along with current Forum co-chair, Margaret ("Peggy") A. Hamburg of the Nuclear Threat Initiative/Global Health and Security Initiative (and daughter of David Hamburg).

At an early conference on emerging viruses, in 1989, "somebody asked Josh, when should we declare that a virus is a new species or a new unknown virus?" Morse recalled, to which Lederberg gave the Solomonic answer, "When it matters." "That was very much Josh's way, to cut through all of the red tape and all of the inconsistencies and see straight to the heart of the matter," Morse concluded.

Lederberg strongly believed in educating the public about science and encouraging public discussion of complex and politically and emotionally charged topics, Peggy Hamburg said. The tangible evidence of this belief can be found in the columns on science and society that Lederberg wrote for *The Washington Post* between 1966 and 1971, and that have been collected by the National Library of Medicine at its website, "Profiles in Science" (NLM, 2008). As David Hamburg remembered, "many in the scientific community thought, why would a person of his gifts devote that kind of time to the public?" Lederberg believed, however,

---

[7]For more information, see also IOM (1992, 2003); Morens et al. (2004); Parrish et al. (2008); and Stephens et al. (1998).

[8]The Convention on the Prohibition of the Development, Production and Stockpiling of Bacteriological (Biological) and Toxin Weapons and on their Destruction; signed on April 10, 1972; effective March 26, 1975. As of July 2008, there were 162 states party to this international treaty to prohibit an entire class of weapons.

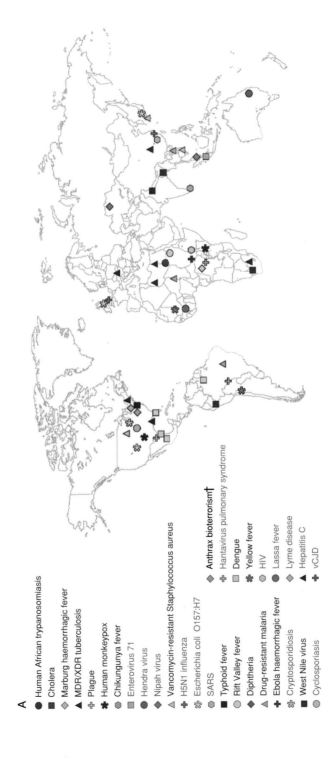

A

● Human African trypanosomiasis
■ Cholera
◆ Marburg haemorrhagic fever
▲ MDR/XDR tuberculosis
✚ Plague
✳ Human monkeypox
⬡ Chikungunya fever
▢ Enterovirus 71
● Hendra virus
◆ Nipah virus
◁ Vancomycin-resistant Staphylococcus aureus
✚ H5N1 influenza
✳ Escherichia coli O157:H7
⬡ SARS
■ Typhoid fever
◆ Rift Valley fever
◁ Drug-resistant malaria
✚ Ebola haemorrhagic fever
✳ Cryptosporidiosis
■ West Nile virus
● Cyclosporiasis

◆ Anthrax bioterrorism†
✚ Hantavirus pulmonary syndrome
▢ Dengue
✳ Yellow fever
⬡ HIV
● Lassa fever
◆ Lyme disease
▲ Hepatitis C
✚ vCJD

B

✳ The French pox (syphilis), 1494
⬡ The American plague (yellow fever), 1793
● Hueyzahuatl (smallpox), 1520†
■ Anthrax, 1770†
◆ Cholera, 1832
△ HIV/AIDS, circa 1930
✛ Spanish influenza, 1918
⬡ Measles, 1875
■ The Black Death (plague), 1347–50
◆ The Plague of Athens (unidentified disease), 430 BC

Newly emerging
Re-emerging
Deliberately emerging†

FIGURE WO-1 Newly emerging, reemerging or resurging, and deliberately emerging diseases. (A) Selected emerging diseases of public-health importance in the past 30 years (1977-2007), with representative examples of where epidemics occurred. (B) Selected emerging diseases of public-health importance in previous centuries (430 B.C. to 1981 A.D.). MDR = multidrug-resistant; SARS = severe acute respiratory syndrome; vCJD = variant Creutzfeldt-Jakob disease; XDR = extensively drug-resistant.
SOURCE: Reprinted from Morens et al. (2008) with permission from Elsevier.

that an informed public was essential not only for good science policy, but ultimately for human survival, Hamburg said. It should be noted that many of his columns in *The Washington Post* addressed the health implications of environmental conditions.

Forum member Terence Taylor, director of the International Council for Life Sciences, asked Dr. David Hamburg to speculate on what Lederberg would say to the next U.S. president if he had been asked to name priorities for enhancing scientific advice to the nation's leadership. Hamburg observed that Lederberg "had a fundamental concern about the relationship between scientific expertise and political leadership. On the one hand, he thought it was enormously important for our political leaders to have access to and appreciation of the scientific community. . . . However, he was equally concerned that we as scientists might inadvertently mislead political leaders, with even the best of intentions."

Hamburg believed that Lederberg would stress the importance of a diversity of expert advice, through processes that invite experts from different arenas to challenge each other. "He never, I think, to the end of his life, was satisfied that we had found the right formula for that," Hamburg added, "but I'm sure he would tell the new president, 'Make much more use of the scientific community than your predecessor has done and do it with much less ideology or political slant. . . . Don't just pick people you know, but reach out to get people that you don't know and have never heard of that you have some reason to believe are excellent.'" Moreover, Hamburg said, Lederberg would encourage further efforts toward improving the yet-unresolved and vital relationship between scientific expertise and political leadership.

"I simply know of no eminent scientist of such immense stature, who gave so much serious analysis of public policy and social problems," Hamburg concluded. "Our country and the world are in his debt. Those of us here today profoundly appreciate what he did, not just for science, but for humanity. His life exemplified the finest attributes of the great institution in which we meet today to honor his memory."

Mentor, Colleague, and Friend

In the course of remembering Lederberg's prodigious accomplishments, workshop participants also reflected upon the ways in which he had touched their lives and careers, further revealing his extraordinary character. Forum co-chair Peggy Hamburg—whose experiences with Lederberg evolved from those of a young daughter of a colleague (exploring tidal pools) to that of a professional peer (co-chairing the IOM Committee on Microbial Threats to Health in the 21st Century)—recalled a man who "loved to go out and walk and talk and study the life on the beach and in the tide pools and see what you could discover in there and how it changed." He was also the first person she knew who owned a computer: "I remember being brought over and sort of ushered into the room, as though it was almost a temple." As a child, Lederberg seemed to her "the epitome of a mad scientist." However, "as I got older and learned more, I realized that he was, in fact, this extraordinary presence in the field of science."

Peggy Hamburg went on to observe, "I would say that one of the things that I actually appreciated most about Josh was that even though he had gotten to know me when I was just a kid, he was able to make the transition—and I don't know exactly when it happened—to really treating me as a peer and a colleague. That is, I think, quite extraordinary, particularly in someone of his generation."

As one might expect, Lederberg's leisure interests were largely intellectual, including technology in all its forms and reading widely and voraciously (he had a particular fondness for the *Times Literary Supplement*), according to Morse. "A kind of recreation for him was to meet someone about whom he had heard good things, in a completely, even wildly different field . . . and through conversation with that person, to get some idea of what was going on in many different fields," David Hamburg recalled.

He was also a phenomenal correspondent, as attested by many workshop participants who received handwritten notes, telephone calls, and e-mails from him over the years. Lederberg had special memo pads upon which he would write notes that were challenged and challenging to many people, Hamburg included. "At first I thought he was just picking on me," he said. "He explained to me that he didn't really expect that the person receiving it would respond or necessarily act on it, but he thought from what he knew of the person's interest that this was something that he or she ought to know about. It was kind of a way of needling us all to broaden our horizons."

When he became director of what was then the Hospital Infections Program at the Centers for Disease Control and Prevention, Hughes was at first amazed to be receiving notes from Lederberg, who had been a figure of awe to Hughes as a medical student at Stanford. Then, Hughes said, "I began to get notes from him asking very interesting and challenging questions that I had never been asked before and that, of course, I never knew the answers to and had great difficulty finding anyone else who knew the answers to his questions."

Speaker Bruce Levin, of Emory University, was equally challenged by communications he received from Lederberg. "It was always a delight for me to receive those e-mails," he said. "The questions Josh asked would sometimes keep me busy for a day, making me think about things I thought I knew, but really didn't. While I don't know whether he got much out of my answers, I know I learned a great deal by thinking about his questions."

"Josh's notes have always been insightful," added Cohen. "I can't imagine how he found the time to write all of the notes he has written to all of us over so many years, and keep track of our interests, and pick out exactly the relevant things to say at particular times. . . . I really miss them."

Speakers Mark Woolhouse, of the University of Edinburgh, and Margaret

McFall-Ngai, of the University of Wisconsin, were both surprised to hear from Lederberg when their work caught his attention. In Woolhouse's case, it was a catalog of human pathogen species (see Woolhouse and Gaunt in Chapter 5), which caused him to reflect that while his group employs various forms of sophisticated mathematics and modeling in many of their studies, "Josh Lederberg liked our work because we can count. So when somebody of that eminence says he likes your work because you can count, you count some more."

McFall-Ngai was a young associate professor, in 1998, when Lederberg e-mailed her after reading a piece she wrote for the *American Zoologist* (McFall-Ngai, 1998). "At first I thought it was spam," she admitted. "Why would Joshua Lederberg write to me? I was getting ready to trash it and I thought, okay, I'll open this up. It started an e-mail volley between him and me, several back-and-forths, about the role of beneficial microbes."

In Memoriam

Recalling his own childhood in Egypt, Mahmoud observed that imposing monuments, such as the Great Sphinx, were a testament to the enormous egos of the rulers who ordered their construction. "They wanted to be sure that long after they were gone, people would be able to gaze upon their mighty works and remember that a great man once ruled here," he said. "Joshua Lederberg, of course, needs no [such] monuments to ensure that his life and work are long remembered. In a very real sense, his accomplishments are embedded in the DNA of many whose lives have been shaped because of his work. That work and those concepts will be passed on to every generation yet to come, long after the Great Sphinx has crumbled into dust."

"Joshua believed very strongly in the work of this Forum," Mahmoud continued. "He had great confidence in the ability of scientists and researchers to continue to solve some of the riddles that still confront science in the fight against infectious diseases. By remembering him with this tribute, we are also remembering the many things that his life and career can teach all of us. I hope that every time we meet at this Forum, Joshua Lederberg will be an inspiration and a reminder that our work can truly change the world, just as his life and career certainly did."

MICROBIAL ECOLOGY AND ECOSYSTEMS

Perhaps one of the most important changes we can make is to supercede the 20th-century metaphor of war for describing the relationship between people and infectious agents. A more ecologically informed metaphor, which includes the germs'-eye view of infection, might be more fruitful. Consider that microbes occupy all of our body surfaces. Besides the disease-engendering colonizers of our skin, gut, and mucous membranes, we are host to a poorly cataloged ensemble of symbionts to which we pay scant attention. Yet they are equally part of the superorganism genome with which we engage the rest of the biosphere.

Joshua Lederberg, "Infectious History" (2000)

More than a century of research, sparked by the germ theory of disease and rooted in historic notions of contagion that long precede Pasteur and Koch's nineteenth-century research and intellectual synthesis, underlies current knowledge of microbe-host interactions. This pathogen-centered understanding attributed disease entirely to the actions of "invading" microorganisms, thereby drawing the lines of battle between "them" and "us," the injured hosts (IOM, 2006a). The paradigm of the systematized search for the microbial basis of disease, followed by the development of antimicrobial and other therapies to eradicate these disease-causing "agents," is now firmly established in human and veterinary clinical practice.

The considerable impact of this approach, assisted by improvements in sanitation, diet, and living conditions in the industrialized world, once led us to believe that we humans were engaged in a war against pathogenic microbes, and that we were winning (IOM, 2006a; Lederberg, 2000). By the mid-1960s, experts opined that, since infectious disease was all but controlled, researchers should focus their attention on other chronic disease challenges, such as heart disease, cancer, and psychiatric disorders.

This optimism coupled with several decades of complacency was profoundly shaken by the appearance in the early 1980s of HIV/AIDS, and was dealt a further blow with the emergence and spread of multidrug-resistant bacteria (IOM, 2006a). As these experiences began to lead researchers to reexamine the host-microbe relationship, additional reasons to do so began to accumulate: pandemic threats from newly emergent (e.g., SARS) and reemergent (e.g., influenza) infectious diseases; lethal outbreaks of Ebola, hantavirus, and other exotic viruses of animal origin; and a new appreciation for the infectious etiology of a variety of chronic diseases, including the association of peptic ulcer with *Helicobacter pylori* infection, liver cancer with hepatitis B and C viruses, and Lyme arthritis with *Borrelia burgdorferi*, to name a few.

In certain lines of inquiry the advantages of adopting an ecological framework for understanding the dynamic equilibria of host-microbe-environment interactions have become evident. Studies of the microbiota of the human gastrointestinal tract—a complex, dynamic, and spatially diversified community comprising at least 10^{13} organisms of more than 1,000 species, most of which are anaerobic bacteria—reveal that these microbes comprise an exquisitely tuned metabolic "organ" that mediates both energy harvest and storage (Bäckhed et al., 2004). Research on the biocontrol agent *Bacillus cereus* suggests that it reduces disease in alfalfa and soybeans by modifying the composition of the microbial community associated with the plants' roots to make it resemble that of the surrounding soil (Gilbert et al., 1994). Such promising discoveries were anticipated by Lederberg, who also noted that superinfections associated with antibiotic therapy attested to the protection naturally conferred by microbial communities in dynamic equilibria. "Understanding these phenomena affords openings for our advantage, akin to the ultimate exploitation by Dubos and Selman Waksman of intermicrobial competition in the soil for seeking early antibiotics," he wrote (Lederberg, 2000). "Research into the microbial ecology of our own bodies will undoubtedly yield similar fruit."

This challenge has been taken up, and elaborated upon, by several workshop presenters, including Forum chair David Relman of Stanford University, who noted that the scientific community has known for hundreds of years—beginning with van Leeuwenhoek's observations of the morphological diversity of microbes in his own dental plaque—that a complex microbiota exists within the human body. Equally complex "host-less" microbial communities exist in the form of biofilms—complex aggregations of microorganisms that grow on solid substrates—as described by speaker Jill Banfield of the University of California, Berkeley. The diversity of mutually beneficial host-microbe interactions was reflected in a pair of presentations by Margaret McFall-Ngai and Jean-Michel Ané, both of the University of Wisconsin, Madison, who described the symbiotic relationships between bacteria and eukaryotes that either allow squids to camouflage themselves from aquatic predators, or enable plants to acquire nutrients through their roots.

Communities of Microbes and Genes

Exploring the Human Microbiome[9]

Microbes colonize the human body during its first weeks to years of life and establish themselves in relatively stable communities in its various microhabitats (Dethlefsen et al., 2007). The human microbiome is far from being fully appreciated or definitively described. Research to date suggests that while site-specific communities (such as skin, mouth, intestinal lumen, small intestine, and large bowel, to name a few) of most individual humans contain characteristic microbial families and genera, the exact mix of species and strains of microbes present in any given individual may be as unique as a fingerprint. The microbiomes of other

---

[9]The microorganisms that live inside and on humans are known as the microbiota; together, their genomes are collectively defined as the microbiome, a term coined by Lederberg (Hooper and Gordon, 2001). However, since most of the organisms that make up the microbiome have resisted cultivation in the laboratory, and thus are known only by their genomic sequences, the microbiota and the microbiome are largely one and the same.

terrestrial vertebrates are dominated by organisms related to, but distinct from, those found in humans. This suggests that host species have co-evolved with their microbial flora and fauna.

Through their explorations of the human microbiome, Relman and cowork-ers seek to understand the role of indigenous microbial communities associated with human health, disease, and the various transition states in between. By understanding essential features of symbiotic relationships between microbial communities and their human hosts, they hope eventually to be able to predict host phenotypes—such as health status—that are associated with particular fea-tures of indigenous communities, and potentially manipulate these communities to restore or preserve health. This effort is at an early stage of development, with research focused on identifying elements of microbial communities that can be monitored and measured to assess physical and metabolic interactions within and among microbial communities and between human and microbial cells.

One such important, and measurable, characteristic of microbial communi-ties is their diversity, as reflected in the number of different ribosomal RNA sequences present in a given location in the human body. These highly-conserved sequences also reveal microbial ancestry and phylogenetic relatedness, permit-ting the construction of phylogenetic trees (see Relman in Chapter 2, especially Figure 2-1). The organisms represented by these sequences remain largely uncul-tivated. Sequences derived by Relman and coworkers in 2005, from the microbial inhabitants of human colonic tissue, suggested that approximately 80 percent had not yet been cultured, and about 60 percent had not been previously described (Eckburg et al., 2005).

This analysis also revealed a striking diversity of microbes at the genus and species level, but affiliated with relatively few phyla, a pattern apparently com-mon among indigenous microbial communities of vertebrates, but not among microbial communities found in external environments. The dearth of microbial phyla on or within the human body probably results from multiple influences, including selection, environmental factors, and even early opportunistic environ-mental exposures to particular microorganisms. Further, samples collected from various locations in the gut of several subjects revealed greater variation in the diversity of microbial communities between these hosts than was present within an individual host (Eckburg et al., 2005). Similarly distinct gut communities were found by Relman and coworkers in each of 14 babies, whose feces were sampled periodically throughout the first year of life (Palmer et al., 2007). The composi-tion and temporal patterns of the microbial communities varied widely from baby to baby, especially early in the first year of life, but the patterns converged by the end of the first year towards a distinct signature for each baby, as well as towards a generic adult signature (Figure WO-2).

Clinical problems associated with the human microbiota include chronic peridontitis, Crohn's disease and other forms of inflammatory bowel disease, tropical sprue, antibiotic-associated diarrhea, bacterial vaginosis, and premature

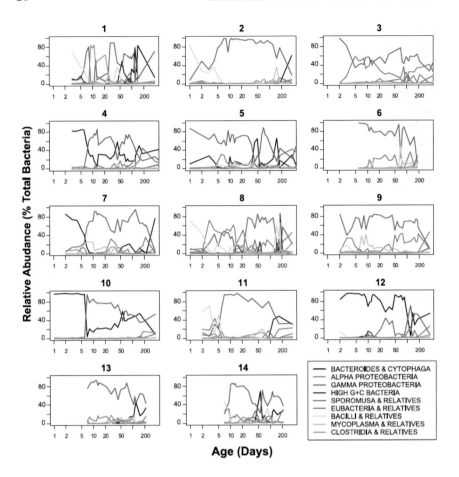

FIGURE WO-2 Temporal profiles of the most abundant level 3 taxonomic groups. Level 3 taxonomic groups were selected for display if their mean (normalized) relative abundance across all baby samples was greater than 1 percent. The x-axis indicates days since birth and is shown on a log scale, and the y-axis shows estimated (normalized) relative abundance. For some babies, no values are plotted for the first few days because the total amount of bacteria in the stool samples collected on those days was insufficient for microarray-based analysis.
SOURCE: Palmer et al. (2007).

labor and delivery (see Relman in Chapter 2). Considerable evidence suggests that the indigenous microbiota is altered during states of infectious disease, especially those diseases that involve the mucosal or skin surfaces that serve as a contact boundary. In some cases, it appears that the indigenous microbiota propagates the disease process. Treatment of such clinical problems with antibiotics, or diversion of the luminal flow away from a segment of bowel, reduces inflammation and other symptoms. These features suggest a system in which the collective microbial community acts as a pathogen, and in which disease results from community disturbance, rather than from infection by a specific organism or group of organisms. It also invites an ecological view of infectious disease control that seeks to restore community equilibrium following disturbance.

In order to study the effects of such disturbances, Relman and coworkers have examined patterns of microbial diversity in the human gut before, during, and after deliberate, periodic, exposure of healthy human subjects to the antibiotic ciprofloxacin. They identified approximately 5,800 different species or strains of bacteria from these samples, of which only 6 percent had been seen before. All three subjects studied so far showed significant reduction in the number of bacterial species present following antibiotic treatment, the result of which was a partial elimination of the differences in community structure that distinguished the three host individuals.

"What makes the human microbiome so intrinsically interesting, at least to me, is the degree to which it may reflect who we are as individuals and as a host species," Relman said, and this individuality has implications for health and disease. While the human microbiome remains largely uncharacterized, Relman held out hope that, thanks to the progress of microbiology since van Leeuwenhoek, we now possess sufficient experimental technology and clinical opportunities to explore the microscopic *terra incognita*[10] within and upon us all.

Biofilms and the Processes That Shape Them

Microbe-microbe relationships include nutritional interactions (e.g., the stepwise processing of plant polysaccharides in the human gut by members of the microbiota) and genetic exchanges that occur through transformation, phage transduction, and conjugation (IOM, 2006a). The last of these processes, bacterial conjugation, first described by Lederberg and coworkers, earned him the Nobel Prize in 1958. Indeed, horizontal gene transfer—also known as lateral gene transfer (Eisen, 2000)—among members of some microbial communities appears to be an extremely pervasive process, but perhaps not to the extent as to call into question whether the concept of speciation applies to communal microbes (Eppley et al., 2007).

Investigations of microbial biofilm communities—which grow on substrates such as rocks in freshwater streams, drains, and teeth[11]—are providing insights into the ecological and evolutionary processes that shape microbial communities. The microbial constituents of the biofilm known as dental plaque include

---

[10]Latin term for "unknown land."

[11]Biofilms are not restricted to streams, drains, and teeth. They are also found on natural and man-made objects, including catheters and other indwelling devices.

hundreds of species and strains of bacteria,[12] as well as various methanogens (Archaea) whose collective metabolic activities are associated with gum disease and tooth decay (Lepp et al., 2004). Biofilms containing iron- and sulfur-oxidizing microbes also thrive in mines and in watersheds where mine wastes drain, resulting in the release of acids and toxic metals into creeks and streams (Banfield, 2008a). This process, called acid mine drainage (AMD), impairs biodiversity and ecological productivity in aquatic ecosystems and, in some cases, precludes inhabitation by macroorganisms altogether (Klemow, 2008). Horizontal gene exchange appears to help microbes in these biofilms adapt to this extreme environment (Lo et al., 2007).

Banfield and coworkers use mine-derived biofilms as a model system to examine how relatively simple microbial communities (that is, communities dominated by a few types of organisms) organize themselves and how their members interact with each other and their physical surroundings (Banfield, 2008b). Biofilms "grow"—that is, they add or accumulate increasingly large populations of microbes—in stages. In this setting, a biofilm nucleus begins at a stream's margins and extends across the water's surface toward its center, while simultaneously increasing in thickness. In her workshop presentation, Banfield described her group's efforts to characterize this process by comparing genomic and protein profiles of biofilms at early and late stages of development.

Using metagenomic[13] methods, Banfield and coworkers have constructed near-complete collective genomes from several different mine-derived biofilm communities (see Chapter 2 Overview). These proved to be dominated by members of bacterial *Leptospirillum* groups II and III, but the biofilm communities also contained several uncultivated Archaea species, as well as some novel organisms. In comparisons of 27 early- and late-stage biofilms, the researchers found that early biofilms, which were dominated—in some cases, almost exclusively—by *Leptospirillum* group II, later developed into more complex communities with more diverse members, including greater numbers of *Leptospirillum* group III bacteria and more species of Archaea.

To examine how this changing cast of organisms functions in the community, and how their functions change as the community develops, the researchers used proteomic[14] methods to determine whether, much like the community's

---

[12]Anton van Leeuwenhoek was the first to see and describe plaque bacteria through a microscope in 1674. For more information about this inventor, see http://inventors.about.com/library/inventors/blleeuwenhoek.htm (accessed December 15, 2008).

[13]Metagenomics involves obtaining DNA from communities of microorganisms, sequencing it in a "shotgun" fashion, and characterizing genes and genomes comparisons with known gene sequences. With this information, researchers can gain insights into how members of the microbial community may interact, evolve, and perform complex functions in their habitats (Jurkowski et al., 2007; NRC, 2007).

[14]Analogous to genomic methods, proteomics permits the identification of expressed proteins from an individual or community.

taxonomic composition, the genes being expressed by its members changed over the course of development (see Chapter 2 Overview). Significant shifts in protein expression correlated with the sequential domination of the community by two different but closely-related strains from *Leptospirillum* group II. This result suggests that different suites of proteins, as well as genotypes, perform different functions at different times in these communities, Banfield concluded.

Further characterization by Banfield and coworkers of 27 biofilms of various stages of development, sampled from eight different microenvironments at the same iron mine, revealed the presence of six distinct genotypes (Lo et al., 2007); each contained blocks of sequence from the two closely-related *Leptospirillum* group II strains. Many of the biofilm samples were found to contain only one genotype; others had several (Denef et al., 2009). The researchers also examined the distribution of genotypes across the eight sampling sites (Figure WO-3). Over the course of more than two years, they consistently found the same genotype at one site—despite the fact that biofilms at this site would have had constant exposure to other genotypes. Thus, Banfield concluded, there appeared to be strong local selection for this particular genotype, which has "achieved a fine level of adaptation to environmental opportunity" (Figure WO-3).

Banfield's group has also examined the role of viruses in biofilms, and particularly the viral "predators" of the dominant bacterial species in these communities. Their investigations were inspired by recent reports (Makarova et al., 2006; Mojica et al., 2005) that the genomes of most Bacteria and Archaea contain repeat regions, known as clustered regularly interspaced short palindromic repeats (CRISPRs). Derived from coexisting viruses, CRISPRs appear to provide immunity (perhaps via RNA interference) to their possessors for the virus of its derivation. Thus, Banfield said, "a microbe has a level of immunity to a virus, so long as it has the spacers that match it or silence it. It has been shown experimentally by the Danisco Group that should a mutation occur such that the spacer is no longer effective, the virus may proliferate and the microbe will suffer" (Barrangou et al., 2007). However, she added, another component of the bacterial system, CRISPR-associated proteins, rapidly sample the local viral DNA and incorporate new spacers, conferring the population with a range of immunity levels to different mutant viruses as they arise (Tyson and Banfield, 2008).

Taking advantage of the correspondence in CRISPR sequences between viruses and their host microbes (see Chapter 2 Overview), Banfield and coworkers identified the sequences of viruses that target Bacteria and Archaea present in acid mine drainage biofilms (Andersson and Banfield, 2008; Figure WO-4). Their investigation revealed a picture of microbial interaction within the biofilm, where a "cloud of viruses" maintains high levels of sequence diversity by various means in order to defeat host microbes, while the hosts counter by rapidly acquiring viral spacers, and, thereby, immunity. Overall, Banfield said, this dynamic system is probably in stasis; nevertheless, she added, "it's clearly an example of co-evolution in a virus and host community."

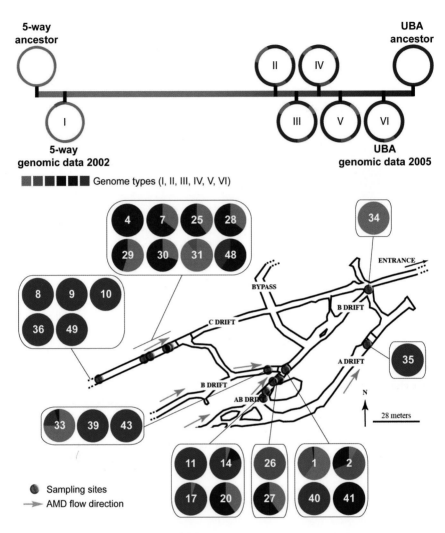

FIGURE WO-3 Six genotypes of *Leptospirillum* group II bacteria were detected in the Richmond Mine (Iron Mountain, CA) by proteomic-inferred genome typing and inferred to have arisen via homologous recombination between parental genotypes. The types (shown schematically as mixtures of red and blue genome segments) were observed in biofilms at locations shown on the map. Each pie chart displays the genome type composition in each sample (samples 34 and 35 were typed via community genomics). The selection for specific genotypes, despite system-wide dispersal of all types, indicates that recombination serves as a mechanism for fine-scale adaptation.

SOURCE: Adapted with permission from Denef et al. (2009).

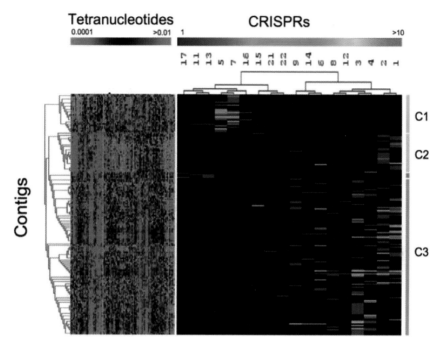

FIGURE WO-4 Virus-host associations in AMD biofilms. Putative viral (SNC) contigs were clustered based on tetranucleotide frequencies (left panel), and CRISPRs were clustered based on patterns of SNC contig matching (right panel). Columns in the left panel represent tetranucleotides (reverse complementary pairs are combined); colors indicate frequencies (gray indicates absence). Columns in the right panel represent CRISPRs; colors indicate number of distinct spacer sequences matching SNC contigs. The majority of Cluster 1 (C1) contigs belong to the AMDV1 population, Cluster 2 (C2) to AMDV2, and Cluster 3 (C3) to AMDV3, AMDV4, and AMDV5.
SOURCE: Andersson and Banfield (2008).

Models of Coexistence and Cooperation

Quoting Heinrich Anton de Bary[15] in his 1879 monograph *Die Erscheinung der Symbios*, Ané defined symbiosis as "a prolonged living-together of different

---

[15]Heinrich Anton de Bary (January 26, 1831-January 19, 1888) was a German botanist whose researches into the roles of fungi and other agents in causing plant diseases earned him distinction as a founder of modern mycology and plant pathology. De Bary determined the life cycles of many fungi, for which he developed a classification that has been retained in large part by modern mycologists. Among the first to study host-parasite interactions, he demonstrated ways in which fungi penetrate host tissues (see www.britannica.com/EBchecked/topic/54513/heinrich-anton-de-bary, accessed March 10, 2009).

organisms that is beneficial for at least one of them." He noted that this general description applies to a continuum of interactions ranging from the extreme of strict mutualism, which benefits both partners, to the opposite extreme of parasitism, which benefits one partner and is detrimental to the other (Figure WO-5). Individual symbioses evolve over time, and under the influence of a variety of environmental, physiological, and developmental factors.

In order to capture certain aspects of this complexity and gain insights into the bases of symbioses, biologists develop models of colonization. These fall into two main categories, according to McFall-Ngai: *constructed models*, based on germ-free hosts such as mice and zebrafish, which allow investigators to study colonization as microbes are introduced in a controlled manner (see, for example, IOM, 2006a); and *natural models*, which permit researchers to observe the process of colonization, typically by only a few microbial phylotypes, as it occurs naturally in a variety of hosts and sites. Existing models of the latter type include the guts of certain insects (such as the gypsy moth, described below, and by Handelsman in Chapter 4), as well as the two systems described in workshop presentations and discussed below: plant roots and the light organ of the Hawaiian squid.

Plant Root Symbionts

In relationships somewhat analogous to those that exist between mammals and their gastrointestinal microbiota, plants establish mutualistic associations

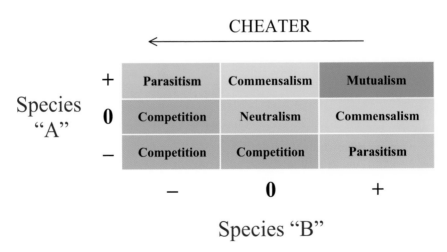

FIGURE WO-5 The symbiotic continuum.
SOURCE: Figure courtesy of John Meyer, North Carolina State University.

with several microorganisms (see Chapter 2 Overview). The roots of most higher plant species form arbuscular mycorrhiza, associations with specific fungal species that significantly improves the plant's ability to acquire phosphorus, nitrogen, and water from the soil (Brelles-Mariño and Ané, 2008). This type of interaction dates back approximately 460 million years and has played a central role in the evolution of land plants, according to Ané.

A more recent association, over the past 60 million years, involves legumes and nitrogen-fixing bacteria named rhizobia. The bacteria induce and colonize new organs on the plant's roots, called nodules; there, they receive energy in the form of carbon from the plant and convert atmospheric nitrogen to ammonia for the plant's use. This partnership furnishes much of Earth's biologically available nitrogen and boosts productivity in non-leguminous crops that are grown in rotation with legumes.

Symbiotic relationships between plants and bacteria or fungi are established through chemical and genetic "cross-talk." As shown in Figure WO-6, legume roots release compounds that trigger nitrogen-fixing rhizobia to express modified chitin oligomers called Nod factors, which in turn facilitate infection of the root by the bacteria, as well as nodule development (Brelles-Mariño and Ané, 2008; Riely et al., 2006). Plants also produce chemical signals called strigolactones that increase branching of fungal hyphae, and thereby increase their contact with arbuscular mycorrhizal fungi. The fungi release diffusible compounds known as

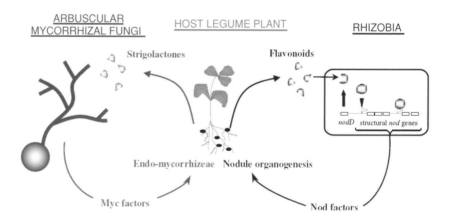

FIGURE WO-6 Symbiotic relationship between plants and bacteria. Legume roots release compounds that trigger nitrogen-fixing rhizobia to express modified chitin oligomers called Nod factors, which in turn facilitate infection of the root by the bacteria, as well as nodule development.
SOURCE: Figure courtesy of Jean-Michel Ané.

Myc factors, which, when recognized by the plant, activate symbiosis-related genes.

The discovery that a largely shared signaling pathway makes possible both arbuscular mycorrhization and legume nodulation—despite their apparent differences—has led to the conclusion that plants have a single, highly-conserved genetic program for recognizing beneficial microbes, according to Ané. Both microbial Nod and Myc factors also appear to have common features, including the ability to promote plant growth, which may benefit microbes by increasing the availability of infection sites, he said.

Plant-microbe symbioses do not exist in a vacuum, but are challenged by "cheaters" and parasites (see Chapter 2 Overview). The cheaters include individual rhizobial colonists of legume nodules that do not fix nitrogen efficiently, and thereby act as parasites, receiving carbohydrates without offering anything in return (and without expending the considerable energy involved in fixing nitrogen), Ané explained. However, their hosts appear to have ways of detecting these microbial freeloaders and "sanctioning" them. Some researchers have hypothesized that the plant decreases oxygen supplies to under-performing nodules (Kiers et al., 2003). While the actual mechanism remains unknown, Ané said, he suspects that the plant may starve the cheaters by reducing their access to carbohydrates.

Parasites on plant roots include root-knot nematodes, nearly ubiquitous pathogens that account for up to 10 percent of global crop loss, according to Ané. Evidence suggests that these nematodes infect legume roots by using genetic pathways adapted for rhizobial colonization, perhaps by producing molecular mimics of Nod factors (Weerasinghe et al., 2005). Human pathogens, including *Salmonella* and *E. coli* O157:H7, also take advantage of the symbiotic signaling pathway to colonize legume roots, such as alfalfa sprouts, that have been linked to several outbreaks of foodborne illness (Taormina et al., 1999). Characterizing the plant and microbe genes involved in these infections, and understanding how these pathogens override or constrain the plant's defenses against invading microbes, may reveal ways to prevent such outbreaks.

The Squid and the Bacterium

The Hawaiian squid *Euprymna scolopes* forms a persistent association with the gram-negative luminous bacterium *Vibrio fischeri* (Nyholm and McFall-Ngai, 2004). Incorporated in the squid's light organ, the bacterium emits luminescence that resembles moonlight and starlight filtering through ocean waters, camouflaging the squid—a nocturnal animal—from predators. In her presentation, McFall-Ngai described the process by which the bacterium colonizes the squid's light organ, which begins within an hour after hatching and appears to occur in stages, each enabling greater specificity between host and symbiont, as shown in Figure WO-7. McFall-Ngai referred to this progression as "a fairly well-orchestrated

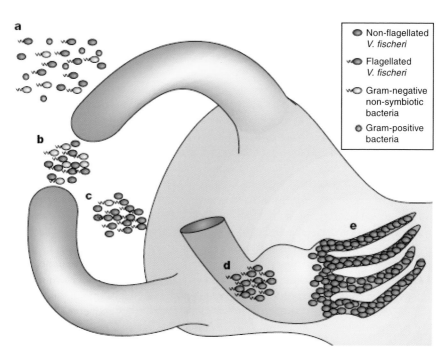

FIGURE WO-7 The "winnowing." This model depicts the progression of light-organ colonization as a series of steps, each more specific for symbiosis-competent *Vibrio fischeri*. (a) In response to gram-positive and gram-negative bacteria (alive or dead) the bacterial peptidoglycan signal causes the cells of the ciliated surface epithelium to secrete mucus. (b) Only viable gram-negative bacteria form dense aggregations. (c) Motile or non-motile *V. fischeri* out-compete other gram-negative bacteria for space and become dominant in the aggregations. (d) Viable and motile *V. fischeri* are the only bacteria that are able to migrate through the pores and into the ducts to colonize host tissue. (e) Following successful colonization, symbiotic bacterial cells become non-motile and induce host epithelial cell swelling. Only bioluminescent *V. fischeri* will sustain long-term colonization of the crypt epithelium.
SOURCE: Reprinted from Nyholm and McFall-Ngai (2004) with permission from Macmillan Publishers Ltd. Copyright 2004.

minuet between the host and the symbiont" that induces the maturation of the squid's light organ, as well as developmental changes that appear to exclude colonization of the organ by other bacterial cells. She noted that much of this process is signaled by microbe-associated molecular patterns (MAMPs).[16]

---

[16]Investigations of the innate immune system, which enables both plants and animals to detect pathogens and mount defensive responses, have identified a series of receptor proteins that recognize

Using a range of molecular approaches, McFall-Ngai and coworkers are engaged in characterizing the colonization of the squid light organ and the maintenance of its symbionts in exacting detail, "to get an hour-by-hour view of the conversation that the host has with its bacterial partner," McFall-Ngai explained. Among the many squid genes that are transcriptionally upregulated in response to colonization by *V. fisheri*, the researchers identified 18 genes that are also upregulated during the colonization of both mouse and zebrafish guts, as determined in constructed models (Chun et al., 2008; see McFall-Ngai in Chapter 2). These shared genes encode proteins that are components of cellular pathways involved in transcriptional regulation, oxidative stress, and apoptosis—responses that typically have been associated with pathogenesis, McFall-Ngai noted. Instead, she said, these pathways constitute a "language of symbiosis," by which a host "talks" to a bacterial colonist, which in most cases is not a pathogen (see the following section for an extended discussion of microbial pathogenesis and the host response).

"What I think this demands of biologists . . . is to go back and question our basic premises about how bacteria and animals work together, what virulence factors really are, and such host behaviors as inflammation, tolerance, and carriage," she concluded. "The horizon, then, is how these characters of host and symbiont are controlled to result in a mutualistic, commensal, or beneficial association."

MICROBIAL EVOLUTION, ADAPTIVE MECHANISMS, AND THE EMERGENCE OF VIRULENCE AND RESISTANCE

Most successful parasites travel a middle path. It helps for them to have aggressive means of entering the body surfaces and radiating some local toxicity to counter the hosts' defenses, but once established they also do themselves (and their hosts) well by moderating their virulence.

Joshua Lederberg, "Infectious History" (2000)

As they explored the effects of adaptation, virulence, and antimicrobial resistance on the host-microbe equilibrium, workshop participants were reminded of Lederberg's important contributions to research on these topics. Presenter Stanley Falkow, of Stanford University, described how Lederberg's discovery of bacterial conjugation and characterization of plasmids—the machinery of horizontal gene

---

conserved molecular patterns specific to bacteria, viruses, and fungi. These signaling elements, which are displayed on the surfaces of pathogenic, commensal, and mutualistic microbes, are known as microbe-associated molecular patterns, or MAMPs. The binding of MAMPs by host receptor proteins elicits a transcriptional response that in some cases triggers host defenses against pathogens, but in others—such as the squid light organ—is associated with host colonization (Didierlaurent et al., 2001; Nyholm and McFall-Ngai, 2004; Yokoyama and Colonna, 2008).

transfer, and, thereby, the means to virulence[17]—built upon prior discoveries of bacterial transformation and mutagenesis and helped to set the stage for present-day research on bacterial pathogenicity[18] (see Falkow in Chapter 3). Stanley Cohen, Lederberg's colleague in the Department of Genetics at Stanford University, pointed out that Lederberg had invented the term "plasmid" for extrachromosomal genetic elements. He noted Lederberg's long-standing concerns about the challenge posed by disease-producing microbes and discussed Lederberg's early work demonstrating the genetic basis for antimicrobial drug resistance.

Many in attendance at this workshop, and certainly the scientists whose presentations are summarized herein, would echo the following remark by Falkow: "I consider all of the scientists whose discoveries expanded Lederberg's initial work on bacterial conjugation to be giants standing on his shoulders, and they made possible my own experimental work."

The Nature of Bacterial Pathogenicity

As Lederberg's observation above suggests, and studies of indigenous microbial communities attest, coexistence between host and microbe is a dynamic equilibrium (Blaser, 1997; Lederberg, 2000). In the case of microbes that cause persistent, asymptomatic infections, physiological or genetic changes in either host or microbe may shift the relationship toward microbial invasion of host tissue, which typically results in an immune response that destroys the invading microbes, but which may also injure or kill the host (Dethlefsen et al., 2007; Merrell and Falkow, 2004).

Research in the decades following Lederberg's ground-breaking work on bacterial conjugation has revealed the following fundamental characteristics of bacterial pathogens, as noted by Falkow (see also Chapter 3):

- Bacteria manipulate the normal functions of host cells in ways that benefit the bacteria (see Falkow's Figure 3-1 in Chapter 3).
- Horizontal gene transfer via mobile genetic elements has shaped the evolution of bacterial specialization.
- Pathogenicity is generally conferred through the inheritance of blocks of genes, called pathogenicity islands.

In order to establish themselves within their hosts, reproduce, and find a new suitable host, pathogenic and commensal bacteria alike must overcome many similar challenges posed by the host's immune system and by competition with

---

[17]Virulence is the degree of pathogenicity of an organism as evidenced by the severity of resulting disease and the organism's ability to invade the host tissues.

[18]Pathogenicity reflects the ongoing evolution between a parasite and host, and disease is the product of a microbial adaptive strategy for survival.

other microbes. Pathogens have an inherent ability—largely conferred by the products of pathogenicity islands, known as virulence factors—to breach host barriers and defenses that commensals cannot penetrate, Falkow explained. When pathogenic bacteria cross the intestinal epithelium of their mammalian hosts, usually through areas known as Peyer's patches, they are engulfed by phagocytes: immune cells that destroy invaders by digesting them. Successful pathogens are able to avoid this fate and survive and sometimes replicate within phagocytes, however, and thereafter are distributed to the liver and spleen. Some pathogens establish persistent, systemic—and sometimes asymptomatic—infections in their hosts and may be shed for the remainder of the host's life. "In my view," Falkow said, "pathogens choose to live in a dangerous place [exposed to the host's immune system] to avoid competition and to get nutrients."

However, Falkow also observed, several members of the human bacterial microbiota that typically live uneventfully in the nasopharynx—including *Streptococcus pneumoniae, Neisseria meningitidis, Haemophilus influenzae* type b, and *Streptococcus pyogenes*—sometimes cause disease. These microbes have virulence factors, suggesting that they interact with the host's immune system, and they persistently infect a significant proportion of the human population, the vast majority of whom are asymptomatic carriers. The existence of such "commensal pathogens" suggests that virulence factors represent one form of a larger class of adaptive factors that allow microbes to colonize and survive in particular niches, and that these factors have been selected on this basis, rather than for their ability to produce disease in host organisms. Indeed, Falkow remarked, it may also be the case that the continual interaction of persistent, asymptomatic bacterial infections with the host immune system keeps it "primed for defensive action."

The conceptualization of virulence factors as colonization factors underlies the larger notion of a distinction between pathogenicity and disease, Falkow observed. "I submit that medicine's focus on disease really distracts us from understanding the biology of pathogenicity," he said. "Disease does not encompass the biological aspects of pathogenicity and the evolution of the host-parasite relationship." Thus, he continued, "If the nature of microbial pathogenicity is schizophrenic—characterized by inconsistent or contradictory elements—then it is important to study every aspect of its biology, and not be distracted by its role in causing disease."

Microbial Virulence and the Host Response

Just as there is more to microbial pathogenicity than disease, there is more to infectious disease than the actions of virulence factors on host cells and systems. Rather, as workshop presenter Bruce Levin, of Emory University, bluntly asserted, virulence almost always results from "screw-ups" by the host's immune system. These immunological failings include responding more vigorously than needed, as occurs in bacterial sepsis; responding incorrectly to a pathogen, as occurs in lepromatous leprosy; or responding to the wrong signals, as occurs in toxic shock syndrome (see Margolis and Levin, 2008, reprinted in Chapter 3). "Sometime in the future, we will look at antimicrobial chemotherapy as a primitive approach to treating infections; we will treat diseases such as sepsis and meningitis by controlling the host response," Levin predicted. "It's going to be a hard job, and we can't do it by episodic dosing. Effective treatment will require real-time monitoring and response to changes in the immune response. I believe it will be possible to treat infections in this way, but to do so we have to know a lot more about the immune response and its control than we do now."

Tying this "it's the host's fault" perspective into existing hypotheses for the evolution of virulence, which focus primarily on the parasite, raises some interesting issues (see Margolis and Levin in Chapter 3). According to "conventional wisdom," as described by May and Anderson (1983), virulence is an early stage in the association between a parasite and its host after which, over the course of evolution, a successful parasite "learns" not to bite the hand that feeds it. Levin suggested that the host, too, could evolve such that its immune system "learns" not to overreact to the parasite, and that eventually, on "equilibrium day," all such host-parasite relationships would achieve mutualism (Levin et al., 2000). He also considered the trade-off hypothesis, which postulates that a too-virulent parasite will kill its host too rapidly to permit efficient transmission. Natural selection in the parasite population, therefore, favors some—but not too much—virulence. Levin further wondered whether this trade-off could be achieved by the parasite evolving restraint in its production of agents that inflame the host's immune system.

By contrast, in his prepared remarks Levin presented experimental findings suggesting that the host effects of certain bacterial products (e.g., Shiga toxin produced by *Escherichia coli* O157:H7; Steinberg and Levin, 2007) appear to have evolved coincidentally as virulence determinants, having been selected for different functions and the advantages that they confer upon a microbe. Other virulent microbes (e.g., Falkow's "commensal pathogens") may have been selected within the host, under local circumstances that favor more pathogenic members of a colonizing population, even if they are at a disadvantage in the community of hosts, Levin said.

But how to account, in evolutionary terms, for the disadvantages of host "immunoperversity": the tendency to overreact to pathogens, resulting in host morbidity and mortality? This phenomenon may be an artifact of the relative slowness of human evolution, Levin explained, coupled with the low efficiency of infectious disease-mediated selection in our species. It may also result from selection pressures associated with maintaining a large microbiota, McFall-Ngai suggested. "By and large, the invertebrates (with the exception of the termites and the cockroaches, which do have large consortia) generally have very limited persistent coevolved interactions with microbes," she said. Thus, it is possible that the adaptive immune system of jawed vertebrates evolved as a mechanism by which to control the large populations of microbes—a task that may require extreme responses that occasionally result in disease.

Pathogen Evolution, as Illustrated by *Salmonella*

Setting aside the inconsistencies and contradictions inherent to pathogenicity, Falkow and fellow workshop speakers Gordon Dougan and Julian Parkhill, of the Wellcome Trust Sanger Institute in Cambridge, United Kingdom, described approaches to discovering how certain microbes have evolved to cause disease in their hosts (see Chapter 3). In particular, each presentation discussed bacterial pathogens of the genus *Salmonella*. Serovars[19] of *Salmonella enterica* include *S. typhimurium*, which infects a wide range of hosts and is a major cause of gastroenteritis in humans, and *S. typhi*, the human-specific agent of the systemic infection typhoid fever (Lawley et al., 2006; Monack et al., 2004b).

In humans, *S. typhimurium* infections are generally (but not always) contained within the intestinal epithelium, while *S. typhi* evades destruction by the immune system and is transported, via the liver and spleen, to the gall bladder and bone marrow, in which the bacteria can persist (Figure WO-8; Monack et al., 2004b). Thus, significant numbers of people infected with typhoid—including those asymptomatically infected with *S. typhi*—become chronic carriers of the pathogen and reservoirs of a disease that poses a considerable threat to public health. From the perspective of *S. typhi*, however, this "stealth" strategy is essential to its survival. Workshop presentations described how evolution—both ancient and recent—has shaped pathogenicity in *Salmonella*, from its initial acquisition of genes that confer invasiveness to the loss of gene function in some serovars, leading to a reduction in host range and increasing virulence, to the recent challenge of antibiotics, which the bacterium has quickly met with resistance.

Genes Make a Pathogen

In order to identify genes and gene products that enable *Salmonella* to establish systemic infection, Falkow and coworkers employed a mouse model of persistent, systemic infection by *S. typhimurium*, which resembles that of *S. typhi* in humans (Monack et al., 2004a). Using a microarray-based strategy, they screened the entire *Salmonella* genome for genes associated with different stages of persistent *Salmonella* infection (see Falkow in Chapter 3). Some of the genes they identified enable *Salmonella* to excrete proteins that kill macrophages during initial infection, while others allow the bacterium to replicate and persist within the vacuoles of macrophages, invisible to the host's immune system. Using existing technology, "we can now identify all such genes quite readily," Falkow said, "but we may not be able to determine their exact function."

---

[19]Strains distinguished serologically, based on the antigens displayed on their surfaces.

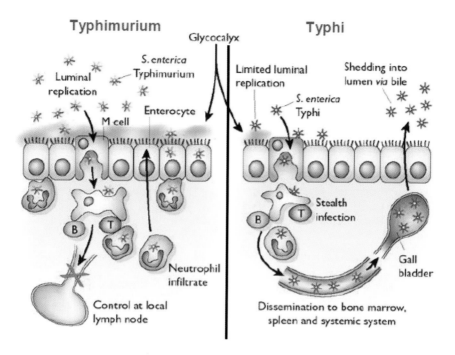

FIGURE WO-8 Comparison of pathogenesis of infection associated with *Salmonella typhimurium* versus *S. typhi* (human restricted).
SOURCE: Reprinted from Young et al. (2002) with permission from Macmillan Publishers Ltd. Copyright 2002.

Loss of Function Leads to Specialization

While bacterial pathogenicity is associated with the acquisition of novel virulence genes (typically through horizontal transfer), research by Parkhill and colleagues indicates that host-restricted, virulent pathogens such as *S. typhi* have evolved those characteristics following a loss of function in genes that control interactions with host cells (thereby limiting their host range) and that modulate the expression of virulence factors (see Box WO-2 and Chapter 3). The genomes of *S. typhi* and another systemic, host-restricted pathogen, *S. paratyphi* A—each independently descended from *S. enterica*—contain approximately 200 inactivated genes, of which about 30 are common to both serovars. Many of these encode functions involved in determining virulence or host range. Most of these shared pseudogenes, however, do not bear the same inactivating mutations, suggesting that their loss conferred a selective advantage.

BOX WO-2
Host-Restriction Versus Virulence in *Bordetella spp.*

Genomic comparisons of three *Bordetella* species by workshop presenter Julian Parkhill and colleagues provide clues to the evolution of virulence in bacterial pathogens (Parkhill et al., 2003). *B. pertussis*, the primary causative agent of whooping cough in humans, can survive only within its single host species. *B. parapertussis* also causes whooping cough; some strains are restricted to humans, others to sheep. Both of these species have apparently evolved from the less virulent *B. bronchiseptica*, which causes chronic and often asymptomatic disease in a wide range of animals. These three species are genetically identical at the 16S RNA level. By most measures (except phenotype), they constitute a single species.

There is little sequence variation among strains of either *B. pertussis* or *B. parapertussis* worldwide, indicating that these species are very recently evolved; by contrast, *B. bronchiseptica* strains vary considerably. Their genome structures also differ significantly: compared with *B. bronchiseptica,* the *B. pertussis* genome is approximately 25 percent smaller and contains a large number of genes inactivated by mutation (pseudogenes). A large majority of these pseudogenes resulted from single mutations, indicating that they were recently inactivated (since further mutations have not accumulated in the inactive gene). Approximately one-third of the pseudogenes in *B. pertussis* were inactivated by an insertion sequence (IS) element, a mobile genetic sequence that, once introduced into a bacterial genome, can reproduce and reinsert in multiple locations. The *B. pertussis* genome contains 240 copies of a single IS element, which have undoubtedly produced high levels of recombination and deletion. *B. parapertussis* contains different IS elements that have also proliferated in its genome.

In evolving from *B. bronchiseptica* toward host restriction and greater virulence, *B. pertussis* and *B. parapertussis* did not acquire novel virulence factors, but instead lost function in genes associated with host interaction (thereby narrowing their host ranges) and in regulation of the expression of virulence factors, such as the pertussis toxin. Parkhill and coworkers hypothesize that these changes occurred when humans began living in close proximity to each other, increasing opportunities for pathogen transmission, and easing selection against virulence (which formerly might have killed a host before it could transmit the pathogen). These circumstances could have created an evolutionary bottleneck, causing the increased fixation of advantageous (and potentially disadvantageous) point mutations and IS element insertions in the population, giving rise to the new, more virulent, and host-restricted species.

Similar events appear to have influenced the evolution of a variety of human pathogens, including *S. typhi* and *S. paratyphi*, independent derivatives of the ancestral *S. enterica*, and *Yersinia pestis,* the causative agent of plague, from its ancestor *Y. pseudotuberculosis*. An evolutionary bottleneck—perhaps the result of domestication—may also have enabled the descent of *Burkholderia mallei*, a pathogen restricted to horses, which causes glanders, from *B. pseudomallei*, a broad host-range pathogen. Likewise, the planting of crops as monocultures may have separated *Clavibacter michiganensis* subspecies *sepedonicus,* an endophytic pathogen of potato with a narrow host range, from *C. michiganensis* subspecies *michiganensis,* an epiphytic pathogen with a broad host range.

Emergence of Resistance

As is typical of human-restricted (and therefore recently evolved) pathogens, *S. typhi* strains exhibit scant genetic variation (see Chapter 3). A sequencing study conducted by Dougan and colleagues that compared 200 gene fragments of approximately 500 base pairs each from 105 globally representative *S. typhi* isolates identified only 88 single nucleotide polymorphisms (SNPs; Roumagnac et al., 2006). Considerable numbers of these SNPs—at least 15 independent mutations to the same crucial gene encoding a DNA gyrase subunit—arose following the introduction of fluoroquinolone antibiotics in the late 1980s (see the following section for a general discussion of antimicrobial resistance).

Another route to antibiotic resistance appears recently to have been taken by non-typhoidal serovars of *Salmonella*, including *S. typhimurium*, Dougan noted. These strains cause invasive infections—instead of the usual gastroenteritis—and have become a major cause of morbidity and mortality in African children (Gordon et al., 2008; Graham, 2002). Sequences of strains causing non-typhoidal salmonellosis (NTS) proved genetically distinct from *Salmonella* strains (of the same serovars) that cause gastroenteritis in Western populations: they bore plasmids containing two distinct genetic elements (integrons[20]) that conferred resistance to multiple antibiotics, as well as to quaternary ammonium disinfectants. Dougan warned that these resistance genes could spread rapidly through horizontal transfer to other *Salmonella* strains following the planned introduction of large-scale antibiotic prophylaxis for HIV-infected African children.

Antibiotic Resistance: Origins and Countermeasures

Reports of antibiotic-resistant bacterial infections followed within a few years of the first widespread use of penicillin at the close of World War II. By the mid-1950s, multidrug-resistant bacterial strains began to emerge (Figure WO-9) and have since become ubiquitous. Indeed, mortality rates due to bacterial infections threaten to return to the levels of the pre-antibiotic era, according to speaker Julian Davies of the University of British Columbia.

Perhaps these developments could have been anticipated based on Lederberg's work on bacterial conjugation, a key route by which plasmids carrying drug-resistance genes are horizontally transferred between bacteria. Certainly, the ongoing impact of antibiotic resistance has confirmed the importance of understanding its evolutionary, genetic, and ecological origins, as several workshop presentations attested.

---

[20]Integrons are gene elements that facilitate horizontal gene transfer by allowing bacteria to integrate and express DNA in the form of "gene cassettes": mobile genes bearing attC recombination sites. Integrons catalyze the integration of foreign genes into a DNA molecule that is already recognized by the native replication machinery of the chromosome or plasmid, and under the control of a promoter that allows gene expression in the host (Nemergut et al., 2008).

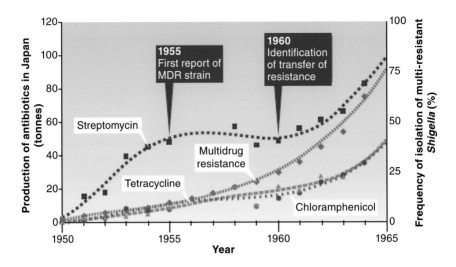

FIGURE WO-9 The relationship between antibiotic resistance development in *Shigella* dysentery isolates in Japan and the introduction of antimicrobial therapy between 1950 and 1965. In 1955, the first case of plasmid determined resistance was characterized. MDR = multidrug resistance. Transferable, multi-antibiotic, resistance was discovered five years later in 1960.
SOURCE: Reprinted from Davies (2007) with permission from Macmillan Publishers Ltd. Copyright 2007.

The biochemical mechanisms by which bacteria achieve resistance are many and varied, and the genes to accomplish each of them can be acquired by horizontal transfer, Davies said. Mechanisms conferring resistance include increased efflux of antibiotic, enzymatic inactivation, target modification, target overexpression, sequestration, and intracellular localization. Yet although we have gained considerable understanding of the biochemical and genetic bases of antibiotic resistance, we have failed dismally to control the development of antibiotic resistance, or to stop its transfer among bacterial strains, Davies observed. Novel antibiotics are unlikely to be developed without significant financial incentives for the pharmaceutical industry, which has largely abandoned infectious disease therapeutic discovery for more profitable targets, such as chronic conditions (Spellberg et al., 2008). Workshop participants considered a variety of means to address these considerable challenges, including investigating the environmental origins of antibiotic resistance, identifying sources of novel antibiotics, and developing alternatives to conventional antibiotic therapies.

Environmental Sources of Resistance Genes and Antibiotics

As Lederberg and others have shown, genes that confer resistance to clinical antibiotics exist in bacterial populations that have never encountered these compounds. Many such naturally-occurring resistant bacterial strains have been isolated (or activities recognized through metagenomic methods, as will be subsequently described) from the soil—as were the bacterial strains from which antibiotics were initially derived (Dantas et al., 2008; D'Costa et al., 2006; Riesenfeld et al., 2004). In naïve bacterial populations, "resistance" genes are likely to encode other functions (e.g., metabolism, regulation) that nevertheless offer a selective advantage, Davies explained. "Resistance genes in the environment, in general, are not resistant," he said. "They become resistant when picked up and overexpressed in a foreign cytoplasm."

Opportunities for such acquisitions are presented by the flow of water among the various environments in which bacterial resistance genes exist, Davies observed. In particular, wastewater treatment plants—which he described as "an incredible mixing pot of genes and plasmids"—provide an ideal opportunity for pathogenic bacteria to acquire new resistance genes, and new virulence genes as well (see Davies in Chapter 4). He noted recent studies by Szczepanowski and coworkers, who isolated and sequenced antibiotic-multiresistant plasmids from bacteria present in sludge in wastewater treatment plants, and found that they also contained several virulence-associated genes and integrons (Szczepanowski et al., 2004, 2005). Such plasmids, moreover, were detectable in effluents released from the treatment plant into the environment (Szczepanowski et al., 2004). Researchers from the same laboratory have also performed a metagenomic analysis of such bacteria and determined that their collective plasmid DNA encoded resistances to all major classes of antimicrobial drugs (Szczepanowski et al., 2008).

The pervasiveness of antibiotic resistance in the environment suggests that antibiotics—that is, molecules with antibiotic activity—are equally abundant in nature, produced by bacteria (and also by plants) to serve a variety of purposes, Davies said. Thus, to find novel antibiotics, his laboratory is pursuing a strategy of identifying organisms that produce bioactive compounds, then analyzing these compounds for their antibiotic properties. Similarly, Handelsman (see Chapter 4) described a process by which she and coworkers are searching the soil metagenome—DNA derived from soil, mainly of bacterial and archaeal origins, digested and ligated into a vector used to transform *Escherichia coli*—for both antibiotic and antibiotic resistance activities. One compound they have discovered is a single enzyme possessing two antibiotic resistance domains: one that disables penicillin-like compounds; the other, cephalosporin-like compounds. Although never before seen, such an enzyme may someday find its way into the human microbiome (or microbial community), Handelsman said, and if so, its potential to confer broad-spectrum antibiotic resistance might pose a serious threat to public health.

Alternatives to Antimicrobials

In his workshop presentation, Stanley Cohen noted that the public health crisis of antimicrobial drug resistance in bacteria and viruses has resulted largely from the practice of treating infectious diseases with therapeutics designed to attack pathogens, resulting in the spread of mutant microbes that are insensitive to drug therapy. He described an approach that recognizes that to be successful, many pathogens require the cooperation of host cells, which furnish the invader with genes and gene products necessary for pathogen propagation and transmission. Interfering with host functions that are recruited by pathogens provides an alternative to drugs that target pathogens (see Chapter 4). Possible toxicity from the targeting of such host genes, which Cohen termed host-oriented therapeutics, must be addressed, he acknowledged—just as they must for infectious disease therapies aimed at microbial targets. However, he reminded workshop participants that drugs targeting normal host cell functions are used routinely in the treatment of other types of disease.

To find genes in eukaryotic host organisms that enable pathogens to propagate, Cohen and coworkers developed antisense RNA-based methods for identifying mammalian cells that show altered biological properties when mutagenized with randomly-integrating retroviruses (Li and Cohen, 1996). This random homozygous knockout (RHKO) strategy and other global gene inactivation strategies, as described in Cohen's contribution to this volume (see Chapter 4), have enabled investigation of host-cell genes and genetic pathways required for viral and bacterial pathogenicity. Tsg101, the first gene isolated using RHKO, has been implicated in the egress of a broad range of viruses (including HIV and Ebola) from infected cells and is currently being pursued as a therapeutic target (Lu et al., 2003). Using an assay that identifies mammalian cells resistant to anthrax toxicity, Cohen and coworkers have also used host-gene-inactivation approaches to discover the previously unsuspected role of the host cell surface protein ARAP3 and LRP6 in anthrax toxin internalization. Cells deficient in the function of the genes encoding these proteins demonstrate increased survival in the presence of anthrax toxin (Lu et al., 2004; Wei et al., 2006). Research is underway to determine the potential efficacy of therapies directed against these proteins (see Cohen in Chapter 4).

A further departure from conventional antimicrobial therapeutics was introduced by Handelsman, who is investigating the possible manipulation of indigenous microbial communities to defend their hosts against pathogens. "To figure out how to use communities in order to protect us, we need to understand the process of invasion," she said. Her laboratory is pursuing this line of inquiry as part of ongoing efforts to characterize the composition, dynamics, and functions of model indigenous communities. Their studies of the microbial community of the gypsy moth gut have yielded intriguing evidence that commensal bacteria interact in ways that can influence host health—and in surprising ways, including collaborating with invaders to kill their hosts (see Handelsman in Chapter 4).

Handelsman and coworkers have also demonstrated various ways by which they can alter the response of their model microbial community to pathogens. "We find that perturbation by a number of means, including antibiotic addition, can increase the susceptibility of the community to invasion," she explained. Exposure to antibiotics also increased microbial diversity in their community model. Based on these results, the researchers plan to undertake experiments to measure community robustness—a composite of resistance to change, stability, and resilience—and to identify organisms and genes that create robust communities. They also hope to reveal the genetic attributes of invading microbes that permit them to overcome a robust community.

INFECTIOUS DISEASE EMERGENCE: ANALYZING THE PAST, UNDERSTANDING THE PRESENT, ANTICIPATING THE FUTURE

The future of humanity and microbes likely will unfold as episodes of a suspense thriller that could be titled Our Wits Versus Their Genes.

Joshua Lederberg, "Infectious History" (2000)

Having considered a wide spectrum of examples of microbial evolution and co-adaptation, and the diverse outcomes for microbes and hosts as individuals, species, and communities, workshop participants considered the global effects of host-microbe relationships that manifest as emerging infectious diseases. As defined by presenter Stephen S. Morse of Columbia University, such diseases have shown a rapid increase in the number of new cases (incidence) or in their geographic range; often, they are caused by novel—that is, previously undiscovered—pathogens, and their emergence is often driven by anthropogenic factors (e.g., land use, travel and trade, food handling; see, for example, IOM, 1992, 2004a,b, 2006b,c, 2007, 2008a,b). Workshop presentations and discussions examined knowledge gained from recent experiences of disease emergence, charted progress toward predicting how and where new diseases might emerge, and explored key challenges and opportunities for understanding and addressing future infectious threats.

"We have had years of complacency about infectious diseases," Morse noted, thanks to the advent of antibiotics, immunizations, and improved public health measures, all of which have led chronic diseases to displace infections as the major cause of death in wealthy countries (Figure WO-10). However, this has not been the case in much of the world, where acute infectious diseases have remained the primary cause of morbidity and mortality. Figure WO-11 depicts the global distribution of a number of recently emerged diseases, as well as "reemerg-

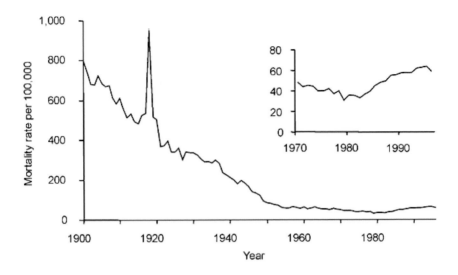

FIGURE WO-10 Deaths resulting from infectious diseases decreased markedly in the United States during most of the twentieth century. However, between 1980 and 1992, the death rate from infectious diseases increased 58 percent. The sharp increase in infectious disease deaths in 1918 and 1919 was caused by an influenza pandemic, which killed more than 20 million people.
SOURCE: Hughes (2001).

ing" diseases[21] and the "deliberate" emergence of *Bacillus anthracis*, an agent of bioterrorism (see also Morse in Chapter 5). A somewhat different perspective on disease emergence was provided by Mark Woolhouse of the University of Edinburgh, whose group has characterized 87 "novel" pathogens—organisms that have been discovered since 1980, and, thus, do not include antimicrobial-resistant strains of previously-known pathogens—among the approximately 1,400 human pathogens recognized in 2007 (see Woolhouse and Gaunt, 2007, reprinted in Chapter 5).

Are emerging diseases really on the rise? Yes, according to analyses by presenter Peter Daszak of the Consortium for Conservation Medicine. "This is a trend that has gone up, and it should continue to go up," he observed. While more than half of emergent diseases can be attributed to antibiotic resistance, Daszak said, zoonoses—infectious diseases that can be transmitted from vertebrate animals to humans—have also increased significantly, particularly those

---

[21]Familiar diseases that have recently expanded their geographic range and/or demonstrated intensified virulence due to such factors as reduced public health measures or antibiotic resistance.

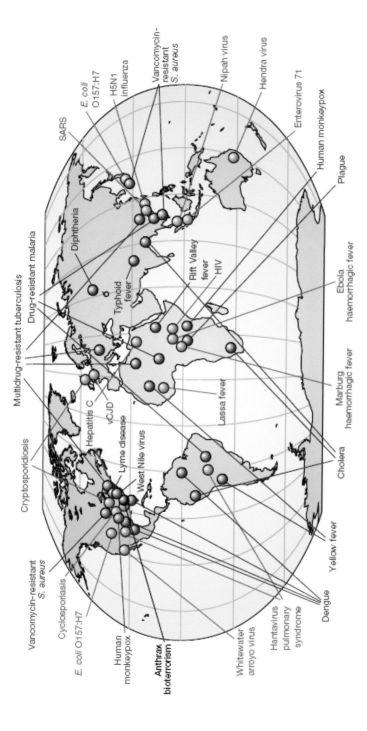

FIGURE WO-11 Global examples of emerging and reemerging infectious diseases, some of which are discussed in the main text. Red represents newly emerging diseases; blue, reemerging or resurging diseases; black, a "deliberately emerging" disease. SOURCE: Reprinted from Morens et al. (2004) with permission from Macmillan Publishers Ltd. Copyright 2004.

emerging from wildlife, such as SARS, Ebola hemorrhagic fever, and Nipah viral encephalitis (see Daszak in Chapter 5).

The IOM reports *Emerging Infections: Microbial Threats to Health in the United States* (IOM, 1992) and *Microbial Threats to Health* (IOM, 2003), produced by ad hoc committees co-chaired by Lederberg, provided a crucial framework for understanding the drivers of infectious disease emergence. The list of six "factors in emergence" in the first report was expanded to 13 in the second, as shown in Box WO-3, which also conceptualized interrelationships among these factors in the model shown in Figure WO-12. As the following summary of workshop presentations and discussions illustrates, this framework has guided, and continues to guide, research to elucidate the origins of emerging infectious threats. These concepts also inform the analysis of recent patterns of disease emergence in order to identify risks for future disease emergence events and, thereby, target surveillance to enable early detection and response in the event of an outbreak.

BOX WO-3
Factors in Emergence

1992
- Microbial adaptation and change
- Economic development and land use
- Human demographics and behavior
- International travel and commerce
- Technology and industry
- Breakdown of public health measures

2003
- Microbial adaptation and change
- Human susceptibility to infection
- Climate and weather
- Changing ecosystems
- Human demographics and behavior
- Economic development and land use
- International travel and commerce
- Technology and industry
- Breakdown of public health measures
- Poverty and social inequality
- War and famine
- Lack of political will
- Intent to harm

SOURCE: IOM (2003).

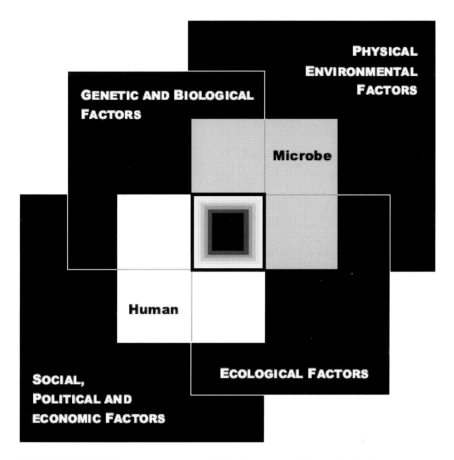

FIGURE WO-12 The convergence model. At the center of the model is a box representing the convergence of factors leading to the emergence of an infectious disease. The interior of the box is a gradient flowing from white to black; the white outer edges represent what is known about the factors in emergence, and the black center represents the unknown (similar to the theoretical construct of the "black box" with its unknown constituents and means of operation). Interlocking with the center box are the two focal players in a microbial threat to health—the human and the microbe. The microbe-host interaction is influenced by the interlocking domains of the determinants of the emergence of infection: genetic and biological factors; physical environmental factors; ecological factors; and social, political, and economic factors.
SOURCE: IOM (2003).

Patterns of Emergence

The Process

As several workshop speakers observed, infectious disease emergence occurs incrementally, and can be accelerated or hindered at various stages in the process (see Chapter 5). The introduction of a pathogen to a new host population drives many emergent events, particularly those involving zoonoses (for example, see the case studies of hantavirus pulmonary syndrome and influenza described by Morse in his contribution to Chapter 5). The "zoonotic pool" provides a rich source of extremely diverse pathogens, Morse noted. Woolhouse added that humans share nearly 60 percent of their pathogen species—and nearly 80 percent of the 87 "novel" pathogen species—with nonhuman vertebrates (see Woolhouse and Gaunt in Chapter 5). Opportunities for such introductions are provided by many of the "factors in emergence" described in *Microbial Threats to Health* (IOM, 2003) and Box WO-3. Woolhouse and Gaunt found that, among their group of "novel" pathogen species, "the most commonly cited drivers [of emergence] fall within the following IOM categories:

- economic development and land use;
- human demographics and behavior;
- international travel and commerce;
- changing ecosystems;
- human susceptibility; and
- hospitals."

Although humans are exposed to many potential novel pathogens, a relatively small number succeed in causing severe infectious disease, which also requires pathogen establishment and transmission among humans. Woolhouse described a "pathogen pyramid" (depicted in Figure 5-8 of Woolhouse and Gaunt's contribution to Chapter 5), in which about 500 out of the total 1,400 pathogens capable of infecting humans are also able to be transmitted to another human. Of these, fewer than 150 have the potential to cause epidemic or endemic disease. The potential for novel pathogens to become established in, and transmitted among, humans is also influenced by factors of emergence, particularly human migration and travel (which disseminate localized diseases), the use of hospitals (which intensify exposure to pathogens), and medical technologies such as injection equipment (which, if contaminated, can also serve to transmit disease). Environmental changes may also expand the geographic range occupied by species that serve as hosts or vectors for infectious diseases. In the case of SARS and HIV/AIDS, it is unknown how (or how many times) these diseases entered human populations, but it is clear that human migration and health care practices served to amplify their emergence (see Morse in Chapter 5).

The Pathogens

Woolhouse and coworkers used a rigorous, formal methodology to produce and refine their catalog of the nearly 1,400 recognized human pathogen species, of which the 87 "novel" species constitute a subset (see Woolhouse and Gaunt in Chapter 5). As shown in Figure WO-13, while known human pathogens are dominated by bacterial species, the vast majority of novel pathogen species are viruses. Indeed, the researchers identified four attributes of these novel pathogens that they expect will describe most future emergent microbes: a preponderance of RNA viruses; pathogens with nonhuman animal reservoirs; pathogens with a broad host range; and pathogens with some (perhaps initially limited) potential for human-to-human transmission (Woolhouse and Gaunt, 2007). "These new pathogens that we are reporting are coming from the same sorts of places, the same sorts of animal populations, we have always shared our viruses and other pathogens with," Woolhouse explained, adding that host proximity, not taxonomy, seems to be the main driver for pathogens to jump between host species.

The Origin of Novelty

While anthropogenic factors provide plenty of fuel for infectious disease emergence, pathogens clearly can and do evolve, and at a far faster rate than do humans, as Lederberg (2000) observed. Viruses evolve fastest of all, which may contribute to their lead position among novel pathogen species, Woolhouse said. He noted that the evolution of viral pathogens could occur either subsequent to

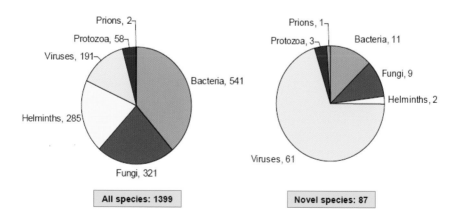

FIGURE WO-13 While known human pathogens are dominated by bacterial species, the vast majority of novel pathogen species are viruses.
SOURCE: Adapted from data in Woolhouse and Gaunt (2007).

the infection of humans (adaptation) or in reservoir hosts prior to the first infection of humans (resulting in human-infecting variants; Woolhouse and Antia, 2007).

Microbes evolve new functions through various mechanisms, according to speaker Jonathan Eisen, of the University of California, Davis. These include *de novo* invention of new genes, small sequence changes that lead to changes in function of existing genes, duplication and divergence of one or both of the duplicates, and shuffling and swapping of large domains of different genes. In addition to generating novelty within their own genomes, microbes engage, through symbiosis, in sharing functions among species.

Perhaps the best known example of this sharing is horizontal gene transfer, which allows one organism to acquire genes encoding new functions from another. In addition, microbes greatly facilitate the acquisition of functions by "macrobes" such as plants and animals by engaging in symbiotic interactions. For example, from the gut of the glassy-winged sharpshooter—the insect vector of Pierce's disease of grapes—Eisen and coworkers characterized a pair of mutually-dependent symbionts that enable their host to survive on nutrient-poor sap from the water-conducting (xylem) tissues of plants (Wu et al., 2006). Eisen also discussed how the rate of the generation of novelty can be further accelerated in many microbes through the loss of mutation-damping mismatch repair genes.

These comparisons between endosymbiotic and free-living bacterial species were undertaken using phylogenomic analysis, a method for integrating information about the evolution of an organism (expressed through its relationships with other organisms, or phylogeny) and the contents of its genome (see Eisen in Chapter 5). Eisen then described some of the phylogenomic methods he and colleagues developed and used in the analysis of various microbial genomes. For example, using one such method, Eisen and coworkers compared the genomic sequence of *Vibrio cholerae* to that of closely related bacteria, searching for gene families that had recently expanded in copy number, suggesting a recent diversification of function. They found that a family of proteins involved in chemical gradient sensing—the methyl-accepting chemotaxis proteins (MCPs)—had undergone such an expansion (Heidelberg et al., 2000).

"If a gene is not under strong selection within a bacterial genome, it usually disappears relatively rapidly over evolutionary time," Eisen said. Moreover, the presence of large numbers of closely related genes suggests that some might have evolved new functions recently. Indeed, Eisen continued, the fact that genes usually disappear rapidly if they are not under some strong selective pressure also enables researchers to make useful inferences about a class of genes that encode "conserved hypothetical proteins"—proteins that are conserved across species for which a function has yet to be identified.

The way this works, he explained, is as follows: if a pathway is not being used, the genes required for the pathway are usually lost as a unit. Furthermore, if a microbe acquires a process by lateral gene transfer, it will have to acquire all the genes required for that process. It follows that one can identify sets of genes that likely work together in a pathway by looking for the correlated gain and loss of genes over evolutionary time. In his contribution to Chapter 5, Eisen describes the use of a method based on these principles, called phylogenetic profiling, to identify genes for sporulation in *Carboxydothermus hydrogenoformans*, a hydrogen-producing bacterium (Wu et al., 2005).

Potential for Prediction

The considerable knowledge gained over recent years regarding ecological and evolutionary processes that drive disease emergence makes possible the measurement and prediction of future patterns of disease emergence, Daszak asserted. "I'm not going to try to say that we can predict in great detail where, when, and what the next new emerging disease is going to be," he said, "but I think we can use ecological approaches to make some predictions about future trends." Bringing together information from genetic sequencing, phylogenetic analyses, and ecological studies, Daszak's group has produced predictions on disease emergence on scales ranging from the local to the global (see Daszak in Chapter 5).

At the global level, Daszak and coworkers developed a predictive model identifying regions ("hotspots") where new infectious diseases are likely to emerge (Jones et al., 2008). From a database of 335 emerging infectious disease (EID) "events"[22] that occurred between 1940 and 2004, the researchers found that zoonoses were responsible for a majority (60 percent) of these events, of which 72 percent originated in wildlife—including SARS and Ebola hemorrhagic fever. Bacteria caused more than 50 percent of these disease events, and many were associated with the development of antibiotic resistance. The hotspots shown in Figure WO-14 were identified based on correlations between the disease events and five socio-economic, environmental, and ecological variables. Concentrated in lower-latitude developing countries, these areas largely lack infectious disease surveillance and control efforts, which are disproportionately focused on the world's healthiest, wealthiest citizens. Thus, Daszak concluded, "we are misallocating our global efforts to deal with emerging infections."

Our Wits Versus Their Genes

Discussion at the workshop's conclusion (and immediately following the session on emerging infectious diseases) brought together many of Lederberg's passions, in the tradition of *Infectious History* (Lederberg, 2000). As Daszak remarked of the potential for predicting the next emerging zoonosis, "I think this

---

[22]Defined as the original case or cluster of cases representing an infectious disease in human populations for the first time.

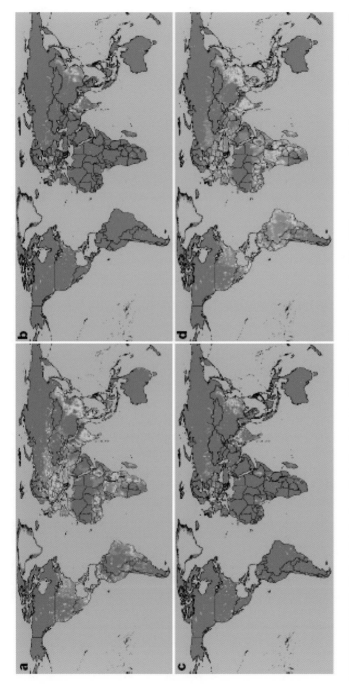

FIGURE WO-14 Global distribution of relative risk of an EID event. Maps are derived for EID events caused by (a) zoonotic pathogens from wildlife; (b) zoonotic pathogens from nonwildlife; (c) drug-resistant pathogens; and (d) vector-borne pathogens.
SOURCE: Reprinted from Jones et al. (2008) with permission from Macmillan Publishers Ltd. Copyright 2008.

is the next 'Holy Grail' for emerging disease, a way to fuse different disciplines: evolution, ecology, virology, microbiology." Workshop participants explored how best to pursue this prize, as well as the larger objective of anticipating, detecting, and responding to emerging infectious threats, and the even greater goal of developing a truly comprehensive understanding of the relationships between and among microbes and host species.

Addressing Immediate Infectious Threats

The paramount importance of surveillance,[23] recognized in *Microbial Threats to Health* as the "foundation for infectious disease prevention and control" (IOM, 2003), was evident in workshop discussions as well. Woolhouse cited a recent U.K. government study (Foresight, 2006) that deemed disease detection, identification, and monitoring essential for timely, effective, and cost-efficient response to infectious disease outbreaks, and which recommended support for a range of new technologies (e.g., genomic and post-genomic analyses, hand-held diagnostic devices, high-throughput screening) to enable the detection and diagnosis of multiple infections. "These are clearly the tools that we are going to need for effective global surveillance," he concluded.

Despite strong scientific support for strengthening infectious disease surveillance, from local to global levels, several participants noted that mustering the political will to expand such activities will be a daunting challenge. Indeed, support for public health—which Forum member Gerald Keusch of Boston University described as "the active art of making sure nothing happens"—is increasingly difficult to obtain. Additional efforts beyond those directed at improving surveillance strategies and technologies will be needed to alter a political climate in which descriptive surveillance programs have been dismissed as "stamp collecting" and a dearth of outbreaks and epidemics is interpreted as a reason to curtail "unnecessary" public health initiatives.

Participants offered various strategies for educating policy makers and the public on such matters as the accelerating emergence of infectious diseases, the true cost of infectious disease in comparison to that of surveillance, and the cost-effectiveness of broad-based surveillance for multiple infectious threats (e.g., zoonoses). Daszak is currently working with economists to develop a cost-benefit analysis for a global infectious disease surveillance program, which he views as a form of insurance—and, as he noted, begs the question: Who pays? "If you look at what causes emerging disease, quite often it is livestock production or trade," he observed. "Does the group doing the trade or the livestock production pay for [infectious disease surveillance] as part of their global insurance package?"

Similarly thorny questions are plaguing the establishment of surveillance

---

[23]Defined as "the continuing scrutiny of all aspects of occurrence and spread of a disease that are pertinent to effective control" (IOM, 2003).

systems mandated for signatories to the recently revised International Health Regulations (IHRs), the legal framework for international cooperation on infectious disease surveillance (WHO, 2008a,b). It is far from clear how many countries will muster the resources to develop such systems. Although the possibility of incentives for resource-poor countries has been raised, such a program has yet to be established. Even if the goal of global infectious disease surveillance is realized, however, another problem looms: how to interpret the ensuing flood of information. There will be a critical need for technology capable of processing enormous amounts of data, quickly enough that it can be acted upon to prevent infectious outbreaks.

Surveillance programs that target species and ecosystems at the highest risk for infectious disease outbreaks can conserve time and resources, including computing capacity. While current predictive efforts focus on identifying pathogen species and families likely to emerge as disease threats, "smart surveillance" will become even smarter when conducted in ecosystems determined to be ripe for disease emergence, Daszak noted. "With the technology we have now to discover new viruses, I think it makes sense to target [emergence-prone] ecological niches a bit better," he said.

Expanding Knowledge of Microbial Evolution and Co-Adaptation

Metagenomic studies suggest that the vast majority of microbial species have yet to be identified (DOE, 2007; see also NRC, 2007). Finding and characterizing new microbes will undoubtedly enrich and expand our understanding of microbial biology, ecology, and evolution. Parkhill noted that his group and others are attempting to use high-throughput sequencing techniques to search for unknown viruses; however, he added, "the jury is still out [on this approach]. You can find things you know about, but it's very difficult to recognize things you don't know about." As technology advances to enable the reading of ever-longer sequences, the genomes of unknown microbes should become easier to identify; however, Parkhill noted, such efforts are expected to produce so much data, so quickly, that it would overwhelm current analytical systems.

Knowledge of microbial genomes, and the functions they encode, is also severely limited. Among 40 phyla of Bacteria, for example, most of the available genomic sequences were from only three phyla as of 2002, Eisen said. He added that sequencing of Archaea and Eukaryote genomes has proceeded in a similarly sporadic manner. "This is not a very good sampling of the diversity of life in terms of genome sequence space," Eisen remarked. The difficulty of culturing most microbes presents a major obstacle to accomplishing this goal. His group has embarked on an effort to sequence representative cultured bacterial and archaeal species from the missing phyla, thereby filling out a genomic "tree of life" that can be used to study the biology of organisms (see Eisen in Chapter 5).

In addition to understanding and surveying microbial biodiversity at the taxonomic and genomic level, Eisen also aspires to the more challenging goal of sampling functional diversity across a broad range of microbes. "What we need are more experiments across the tree of life . . . for all processes that we are interested in," he said, "not just in a couple of model organisms. [Microbial] diversity is immense." He proposed developing a "field guide to the microbes"—akin to those for birds—that not only describes their taxonomy, but their behavior, ecology, and distribution patterns. With such comprehensive knowledge, he concluded, "we can really integrate all this information and maybe predict the future."

Participants also discussed the feasibility of obtaining and applying detailed knowledge of intraspecies diversity, as evidenced by phylogeny. The method used by Dougan and coworkers (see previous discussion and Chapter 3) to discern the complete phylogeny of 200 *S. typhi* strains, known as DNA-based signature typing, could potentially be used to conduct genetic studies of bacteria in the field or to type bacteria for diagnostic or surveillance purposes, Dougan said; moreover, it could be used in clinical settings to investigate associations between specific mutations in a pathogen strain and its phenotype, manifested in the characteristics (e.g., transmissibility, virulence) of disease. It also illustrates Lederberg's hopeful description of the present era of infectious history: "Together with wiser insight into the ground rules of pathogenic evolution, we are developing a versatile platform for developing new responses to infectious disease" (Lederberg, 2000).

APPENDIX WO-1
INFECTIOUS HISTORY[24]

Joshua Lederberg, Ph.D.[25]

In 1530, to express his ideas on the origin of syphilis, the Italian physician Girolamo Fracastoro penned *Syphilis, sive morbus Gallicus* (Syphilis, or the French disease) in verse. In it he taught that this sexually transmitted disease was spread by "seeds" distributed by intimate contact. In later writings, he expanded this early "contagionist" theory. Besides contagion by personal contact, he described contagion by indirect contact, such as the handling or wearing of clothes, and even contagion at a distance, that is, the spread of disease by something in the air.

Fracastoro was anticipating, by nearly 350 years, one of the most important

---

[24]This article was originally published in *Science* 288(5464):287-293. Reprinted with permission from AAAS. Copyright 2000.

[25]Joshua Lederberg (1925-2008) was a Sackler Foundation Scholar heading the Laboratory of Molecular Genetics and Informatics at The Rockefeller University in New York City, and a Nobel laureate (1958) for his research on genetic mechanisms in bacteria. He worked closely with the Institute of Medicine and the Centers for Disease Control and Prevention on analytical and policy studies on emerging infections.

turning points in biological and medical history—the consolidation of the germ theory of disease by Louis Pasteur and Robert Koch in the late 1870s. As we enter the 21st century, infectious disease is fated to remain a crucial research challenge, one of conceptual intricacy and of global consequence.

The Incubation of a Scientific Discipline

Many people laid the groundwork for the germ theory. Even the terrified masses touched by the Black Death (bubonic plague) in Europe after 1346 had some intimation of a contagion at work. But they lived within a cognitive framework in which scapegoating, say, of witches and Jews, could more "naturally" account for their woes. Breaking that mindset would take many innovations, including microscopy in the hands of Anton van Leeuwenhoek. In 1683, with one of his new microscopes in hand, he visualized bacteria among the animalcules harvested from his own teeth. That opened the way to visualize some of the dreaded microbial agents eliciting contagious diseases.

There were pre-germ-theory advances in therapy, too. Jesuit missionaries in malaria-ridden Peru had noted the native Indians' use of *Cinchona* bark. In 1627, the Jesuits imported the bark (harboring quinine, its anti-infective ingredient) to Europe for treating malaria. Quinine thereby joined the rarified pharmacopoeia—including opium, digitalis, willow (*Salix*) bark with its analgesic salicylates, and little else—that prior to the modern era afforded patients any benefit beyond placebo.

Beginning in 1796, Edward Jenner took another major therapeutic step—the development of vaccination—after observing that milkmaids exposed to cowpox didn't contract smallpox. He had no theoretical insight into the biological mechanism of resistance to the disease, but vaccination became a lasting prophylactic technique on purely empirical grounds. Jenner's discovery had precursors. "Hair of the dog" is an ancient trope for countering injury and may go back to legends of the emperor Mithridates, who habituated himself to lethal doses of poisons by gradually increasing the dose. We now understand more about a host's immunological response to a cross-reacting virus variant.

Sanitary reforms also helped. Arising out of revulsion over the squalor and stink of urban slums in England and the United States, a hygienic movement tried to scrub up dirt and put an end to sewer stenches. The effort had some health impact in the mid-19th century, but it failed to counter diseases spread by fleas and mosquitoes or by personal contact, and it often even failed to keep sewage and drinking water supplies separated. It was the germ theory—which is credited to Pasteur (a chemist by training) and Koch (ultimately a German professor of public health)—that set a new course for studying and contending with infectious disease. Over the second half of the 19th century, these scientists independently synthesized historical evidence with their own research into the germ theory of disease.

Pasteur helped reveal the vastness of the microbial world and its many practical applications. He found microbes to be behind the fermentation of sugar into alcohol and the souring of milk. He developed a heat treatment (pasteurization, that is) that killed microorganisms in milk, which then no longer transmitted tuberculosis or typhoid. And he too developed new vaccines. One was a veterinary vaccine against anthrax. Another was against rabies and was first used in humans in 1885 to treat a young boy who had been bitten by a rabid dog.

One of Koch's most important advances was procedural. He articulated a set of logical and experimental criteria, later restated as "Koch's Postulates," as a standard of proof for researchers' assertions that a particular bacterium caused a particular malady. In 1882, he identified the bacterium that causes tuberculosis; a year later he did the same for cholera. Koch also left a legacy of students (and rivals) who began the systematic search for disease-causing microbes: The golden age of microbiology had begun.

Just as the 19th century was ending, the growing world of microbes mushroomed beyond bacteria. In 1892, the Russian microbiologist Dmitri Ivanowski, and in 1898, the Dutch botanist Martinus Beijerinck, discovered exquisitely tiny infectious agents that could pass through bacteria-stopping filters. Too small to be seen with the conventional microscope, these agents were described as "filtrable [sic] viruses."

With a foundation of germ theory in place even before the 20th century, the study of infectious disease was ready to enter a new phase. Microbe hunting became institutionalized, and armies of researchers systematically applied scientific analyses to understanding disease processes and developing therapies.

During the early acme of microbe hunting, from about 1880 to 1940, however, microbes were all but ignored by mainstream biologists. Medical microbiology had a life of its own, but it was almost totally divorced from general biological studies. Pasteur and Koch were scarcely mentioned by the founders of cell biology and genetics. Instead, bacteriology was taught as a specialty in medicine, outside the schools of basic zoology and botany. Conversely, bacteriologists scarcely heard of the conceptual revolutions in genetic and evolutionary theory.

Bacteriology's slow acceptance was partly due to the minuscule dimensions of microbes. The microscopes of the 19th and early 20th centuries could not resolve internal microbial anatomy with any detail. Only with the advent of electron microscopy in the 1930s did these structures (nucleoids, ribosomes, cell walls and membranes, flagella) become discernible. Prior to that instrumental breakthrough, most biologists had little, if anything, to do with bacteria and viruses. When they did, they viewed such organisms as mysteriously precellular. It was still an audacious leap for René Dubos to entitle his famous 1945 monograph "The Bacterial Cell."

The early segregation of bacteriology and biology per se hampered the scientific community in recognizing the prospects of conducting genetic investigation with bacteria. So it is ironic that the pivotal discovery of molecular genetics—that genetic information resides in the nucleotide sequence of DNA—arose from studies on serological types of pneumococcus, studies needed to monitor the epidemic spread of pneumonia.

This key discovery was initiated in 1928 by the British physician Frederick Griffith. He found that extracts of a pathogenic strain of pneumococcus could transform a harmless strain into a pathogenic one. The hunt was then on to identify the "transforming factor" in the extracts. In 1944, Oswald Avery, Colin MacLeod, and Maclyn McCarty reported in the *Journal of Experimental Medicine* that DNA was the transforming factor. Within a few years, they and others ruled out skeptics' objections that protein coextracted with the DNA might actually be the transforming factor.

Those findings rekindled interest in what was really going on in the life cycle of bacteria. In particular, they led to my own work in 1946 on sexual conjugation in *Escherichia coli* and to the construction of chromosome maps emulating what had been going on in the study of the genetics of fruit flies, maize, and mice for the prior 45 years. Bacteria and bacterial viruses quickly supplanted fruit flies as the test-bed for many of the subsequent developments of molecular genetics and the biotechnology that followed. Ironically, during this time, we were becoming nonchalant about microbes as etiological agents of disease.

Despite its slow emergence, bacteriology was already having a large impact. Its success is most obviously evidenced by the graying of the population. That public health has been improving—due to many factors, especially our better understanding of infectious agents—is graphically shown by the vital statistics. These began to be diligently recorded in the United States after 1900 in order to guide research and apply it to improving public health. The U.S. experience stands out in charts (Figure WO-15) depicting life expectancy at birth through the century. The average life-span lengthened dramatically: from 47 years in 1900 to today's expectation of 77 years (74 years for males and 80 for females).[26] Similar trends are seen in most other industrialized countries, but the gains have been smaller in economically and socially depressed countries.

Other statistics reveal that the decline in mortality ascribable to infectious disease accounted for almost all of the improvement in longevity up to 1950, when life expectancy had reached 68. The additional decade of life expectancy for babies born today took the rest of the century to gain. Further improvements now appear to be on an asymptotic trajectory: Each new gain is ever harder to come by, at least pending unpredictable breakthroughs in the biology of aging.

The mortality statistics fluctuated considerably during the first half of the last

---

[26]This sex difference in life expectancy is partly explained by the ability of two X chromosomes to buffer against accumulated recessive mutations and is illustrated by the prevalence in males of color blindness and hemophilia. Another factor is the gender-related difference in self-destructive behaviors.

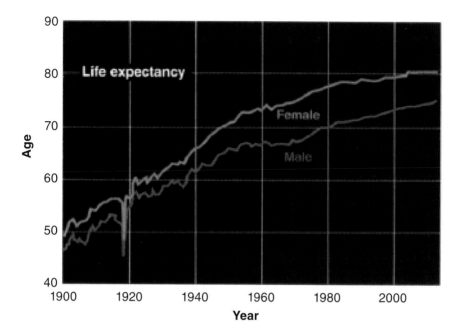

FIGURE WO-15 Longer life. Anti-infection medicine and other factors have helped dramatically lengthen the average life expectancy in the United States.
SOURCE: Social Security Office of the Chief Actuary.

century. Much of this instability was due to sporadic outbreaks of infections such as typhoid fever, tuberculosis, and scarlet fever, which no longer have much statistical impact. Most outstanding is the spike due to the great influenza pandemic of 1918-19 that killed 25 million people worldwide—comparable to the number of deaths in the Great War. Childhood immunization and other science-based medical interventions have played a significant role in the statistical trends also. So have public health measures, among them protection of food and water supplies, segregation of coughing patients, and personal hygiene. Overall economic growth has also helped by contributing to less crowded housing, improved working conditions (including sick leave), and better nutrition.

As infectious diseases have assumed lower rankings in mortality statistics, other killers—mostly diseases of old age, affluence, and civilization—have moved up the ladder. Heart disease and cancer, for example, have loomed as larger threats over the past few decades. Healthier lifestyles, including less smoking, sparer diets, more exercise, and better hygiene, have been important countermeasures. Prophylactic medications such as aspirin, as well as medical and surgical interventions, have also kept people alive longer.

The 1950s were notable for the "wonder drugs"—the new antibiotics penicillin, streptomycin, chloramphenicol, and a growing list of others that at times promised an end to bacteria-based disease. Viral pathogens have offered fewer routes to remedies, except for vaccines, such as Jonas Salk's and Albert Sabin's polio vaccines. These worked by priming immune systems for later challenges by the infectious agents. Old vaccines, including Jenner's smallpox vaccine, also were mobilized in massive public health campaigns, sometimes with fantastic results. By the end of the 1970s, smallpox became the first disease to be eradicated from the human experience.

Confidence about medicine's ability to fight infectious disease had grown so high by the mid-1960s that some optimists were portraying infectious microbes as largely conquered. They suggested that researchers shift their attention to constitutional scourges of heart disease, cancer, and psychiatric disorders. These views were reflected in the priorities for research funding and pharmaceutical development. President Nixon's 1971 launch of a national crusade against cancer, which tacitly implied that cancer could be conquered by the bicentennial celebrations of 1976, was an example. Few people now sustain the illusion that audacious medical goals like conquering cancer or infectious disease can be achieved by short-term campaigns.

TABLE WO-1 An Infectious Disease Timeline

1300s	
1346	Black Death begins spreading in Europe.
1400s	
1492	Christopher Columbus initiates European-American contact, which leads to transmission of European diseases to the Americas and vice versa.
1500s	
1530	Girolamo Fracastoro puts forward an early version of the germ theory of disease.
1600s	
1627	Cinchona bark (quinine) is brought to Europe to treat malaria.
1683	Anton van Leeuwenhoek uses his microscopes to observe tiny animalcules (later known as bacteria) in tooth plaque.
1700s	
1796	Edward Jenner develops technique of vaccination, at first against smallpox.

TABLE WO-1 Continued

1800s	
1848	Ignaz Semmelweis introduces antiseptic methods.
1854	John Snow recognizes link between the spread of cholera and drinking water supplies.
1860s	Louis Pasteur concludes that infectious diseases are caused by living organisms called "germs." An early practical consequence was Joseph Lister's development of antisepsis by using carbolic acid to disinfect wounds.
1876	Robert Koch validates germ theory of disease and helps initiate the science of bacteriology with a paper pinpointing a bacterium as the cause of anthrax.
1880	Louis Pasteur develops method of attenuating a virulent pathogen (for chicken cholera) so that it immunizes but does not infect; in 1881 he devises an anthrax vaccine and in 1885, a rabies vaccine. Charles Laveran finds malarial parasites in erythrocytes of infected people and shows that the parasite replicates in the host.
1890	Emil von Behring and Shibasaburo Kitasato discover diphtheria antitoxin serum, the first rational approach to therapy for infectious disease.
1891	Paul Ehrlich proposes that antibodies are responsible for immunity.
1892	The field of virology begins when Dmitri Ivanowski discovers exquisitely small pathogenic agents, later known as viruses, while searching for the cause of tobacco mosaic disease.
1899	Organizing meeting of the Society of American Bacteriologists—later to be known as the American Society for Microbiology—is held at Yale University.
1900s	
1900	Based on work by Walter Reed, a commission of researchers shows that yellow fever is caused by a virus from mosquitoes; mosquito-eradication programs are begun.
1905	Fritz Schaudinn and Erich Hoffmann discover bacterial cause of syphilis—*Treponema pallidum.*
1911	Francis Rous reports on a viral etiology of a cancer (Rous sarcoma virus).
1918-1919	Epidemic of "Spanish" flu causes at least 25 million deaths.
1928	Frederick Griffith discovers genetic transformation phenomenon in pneumococci, thereby establishing a foundation of molecular genetics.

Continued

TABLE WO-1 Continued

1929	Alexander Fleming reports discovering penicillin in mold.
1935	Gerhard Domagk synthesizes the antimetabolite Prontosil, which kills *Streptococcus* in mice.
1937	Ernst Ruska uses an electron microscope to obtain first pictures of a virus.
1941	Selman Waksman suggests the word "antibiotic" for compounds and preparations that have antimicrobial properties; 2 years later, he and colleagues discover streptomycin, the first antibiotic effective against tuberculosis, in a soil fungus.
1944	Oswald Avery, Colin MacLeod, and Maclyn McCarty identify DNA as the genetically active material in the pneumococcus transformation.
1946	Edward Tatum and Joshua Lederberg discover "sexual" conjugation in bacteria.
1948	The World Health Organization (WHO) is formed within the United Nations.
1952	Renato Dulbecco shows that a single virus particle can produce plaques.
1953	James Watson and Francis Crick reveal the double helical structure of DNA.
Late 1950s	Frank Burnet enunciates clonal selection theory of the immune response.
1960	Arthur Kornberg demonstrates DNA synthesis in cell-free bacterial extract. François Jacob and Jacques Monod report work on genetic control of enzyme and virus synthesis.
1970	Howard Temin and David Baltimore independently discover that certain RNA viruses use reverse transcription (RNA to reconstitute DNA) as part of their replication cycle.
1975	Asilomar conference sets standards for the containment of possible biohazards from recombinant DNA experiments with microbes.
1979	Smallpox eradication program of WHO is completed; the world is declared free of smallpox.
1981	AIDS first identified as a new infectious disease by U.S. Centers for Disease Control and Prevention.
1982	Stanley Prusiner finds evidence that a class of infectious proteins, which he calls prions, cause scrapie in sheep.
1983	Luc Montagnier and Robert Gallo announce their discovery of the human immunodeficiency virus that is believed to cause AIDS.

TABLE WO-1 Continued

1984	Barry Marshall shows that isolates from ulcer patients contain the bacterium later known as *Helicobacter pylori*. The discovery ultimately leads to a new pathogen-based etiology of ulcers.
1985	Robert Gallo, Dani Bolognesi, Sam Broder, and others show that AZT inhibits HIV action in vitro.
1988	Kary Mullis reports basis of polymerase chain reaction (PCR) for detection of even single DNA molecules.
1995	J. Craig Venter, Hamilton Smith, Claire Fraser, and colleagues at The Institute for Genomic Research elucidate the first complete genome sequence of a microorganism: *Haemophilus influenzae*.
1996	Implied link between bovine spongiform encephalopathy ("mad cow disease") and human disease syndrome leads to large-scale controls on British cattle.
1999	New York experiences outbreak of West Nile encephalopathy transmitted by birds and mosquitoes.
2000s	
c. 2000	Antibiotic-resistant pathogens are spreading in many environments.

NOTE: For more extensive chronological listings, see "Microbiology's fifty most significant events during the past 125 years," poster supplement to *ASM News* 65(5), 1999.

Wake-Up Calls

The overoptimism and complacency of the 1960s and 1970s was shattered in 1981 with the recognition of AIDS. Since then, the spreading pandemic has overtaken one continent after another with terrible costs. Its spread has been coincident with another wake-up call—the looming problem of antibiotic-resistant microbes. This was a predictable consequence of the evolutionary process operating on microbes challenged by the new selection pressure of antibiotics, arising in part from medical prescriptions and in part from unregulated sales and use in feed for crop animals.

AIDS's causative agent, the human immunodeficiency virus (HIV), is a member of the retrovirus family. These viruses had been laboratory curiosities since 1911, when Francis Peyton Rous discovered the Rous sarcoma virus (RSV) in chickens. Early basic research on retroviruses later helped speed advances in HIV research. By the time AIDS began to spread, RSV had been studied for years as a model for cancer biology, because it could serve as a vector for transferring oncogenes into cells. That work accelerated the characterization of HIV as a retrovirus, and it also helped guide our first steps toward medications that slow HIV infection.

AIDS and HIV have spurred the most concentrated program of biomedical research in history, yet they still defy our counterattacks. And our focus on extirpating the virus may have deflected less ambitious, though more pragmatic, aims, including learning to live with the virus by nurturing in equal measure the immune system that HIV erodes. After all, natural history points to analogous infections in simians that have long since achieved a mutually tolerable state of equilibrium.

Costly experiences with AIDS and other infectious agents have led to widespread reexamination of our cohabitation with microbes. Increased monitoring and surveillance by organizations such as the U.S. Centers for Disease Control and Prevention (CDC) and the World Health Organization (WHO) have revealed a stream of outbreaks of exotic diseases. Some have been due to the new importation of microbes (such as cholera in the Southern Hemisphere); some to older parasites (such as *Legionella*) that have been newly recognized as pathogenic; and some to newly evolved antibiotic-resistant pneumonia strains.

Even maladies that had never before been associated with infectious agents recently have been revealed as having microbial bases. Prominent among these are gastric ulcers, which previously had been attributed almost entirely to stress and other psychosomatic causes. Closer study, however, has shown a *Helicobacter* to be the major culprit. Researchers are now directing their speculations away from stress and toward *Chlamydia* infection as a cause of atherosclerosis and coronary disease.

The litany of wake-up calls goes on. Four million Americans are estimated to be infected with hepatitis C, mainly by transfusion of contaminated blood products. This population now is at significant risk for developing liver cancer. Those harboring hepatitis C must be warned to avoid alcohol and other hepatotoxins, and they must not donate blood.

Smaller but lethal outbreaks of dramatic, hypervirulent viruses have been raising public fear. Among these are the Ebola virus outbreak in Africa in 1976 and again in 1995 and the hantavirus outbreak in the U.S. Southwest in 1993. In hindsight, these posed less of a public health risk than the publicity they received might have suggested. Still, studying them and uncovering ecological factors that favor or thwart their proliferation is imperative because of their potential to mutate into more diffusible forms.

Our vigilance is mandated also by the facts of life: The processes of gene reassortment in flu viruses, which are poorly confined to their canonical hosts (birds, swine, and people), goes on relentlessly and is sure to regenerate human-lethal variants. Those thoughts were central in 1997 when the avian flu H5N1 transferred into a score of Hong Kong citizens, a third of whom died. It is likely that the resolute actions of the Hong Kong health authorities, which destroyed 2 million chickens, stemmed that outbreak and averted the possibility of a world-wide spread of H5N1.

Complacency is not an option in these cases, as other vectors, including wildfowl, could become carriers. In Malaysia, a new infectious entity, the Nipah virus, killed up to 100 people last year; authorities there killed a million livestock to help contain the outbreak. New York had a smaller scale scare last summer with the unprecedented appearance of bird- and mosquito-borne West Nile encephalitis, although the mortality rate was only a few percent of those infected. We need not wonder whether we will see outbreaks like these again. The only questions are when and where?

These multiple wake-up calls to the infectious disease problem have left marks in vital statistics. From mid-century to 1982, the U.S. mortality index (annual deaths per 100,000) attributable to infection had been steady at about 30. But from 1982 to 1994, the rate doubled to 60. (Keep in mind that the index was 500 in 1900 and up to 850 in 1918-19 due to the Spanish flu epidemic.) About half of the recent rise in deaths is attributable to AIDS; much of the rest is due to respiratory disease, antibiotic resistance, and hospital-acquired infection.

Our Wits Versus Their Genes

As our awareness of the microbial environment has intensified, important questions have emerged. What puts us at risk? What precautions can and should we be taking? Are we more or less vulnerable to infectious agents today than in the past? What are the origins of pathogenesis? And how can we use deeper knowledge to develop better medical and public health strategies? Conversely, how much more can the natural history of disease teach us about fundamental biological and evolutionary mechanisms?

An axiomatic starting point for further progress is the simple recognition that humans, animals, plants, and microbes are cohabitants of the planet. That leads to refined questions that focus on the origin and dynamics of instabilities within this context of cohabitation. These instabilities arise from two main sources loosely definable as ecological and evolutionary.

Ecological instabilities arise from the ways we alter the physical and biological environment, the microbial and animal tenants (humans included) of these environments, and our interactions (including hygienic and therapeutic interventions) with the parasites. The future of humanity and microbes likely will unfold as episodes of a suspense thriller that could be titled *Our Wits Versus Their Genes*.

We already have used our wits to increase longevity and lessen mortality. That simultaneously has introduced irrevocable changes in our demographics and our own human ecology. Increased longevity, economic productivity, and other factors have abetted a global population explosion from about 1.6 billion in 1900 to its present level above 6 billion. That same population increase has fostered new vulnerabilities: crowding of humans, with slums cheek by jowl with jet setters' villas; the destruction of forests for agriculture and suburbanization, which has led to closer human contact with disease-carrying rodents and ticks; and routine long-distance travel.

Travel around the world can be completed in less than 80 hours (compared to the 80 days of Jules Verne's 19th-century fantasy), constituting a historic new experience. This long-distance travel has become quotidian: Well over a million passengers, each one a potential carrier of pathogens, travel daily by aircraft to international destinations. International commerce, especially in foodstuffs, only adds to the global traffic of potential pathogens and vectors. Because the transit times of people and goods now are so short compared to the incubation times of disease, carriers of disease can arrive at their destination before the danger they harbor is detectable, reducing health quarantine to a near absurdity.

Our systems for monitoring and diagnosing exotic diseases have hardly kept pace with this qualitative transformation of global human and material exchange. This new era of global travel will redistribute and mix people, their cultures, their prior immunities, and their inherited predispositions, along with pathogens that may have been quiescent at other locales for centuries.

This is not completely novel, of course. The most evident precedent unfolded during the European conquest of America, which was tragically abetted by pandemics of smallpox and measles imported into native populations by the invading armies. In exchange, Europeans picked up syphilis's *Treponema*, in which Fracastoro discerned contagion at work.

Medical defense against the interchange of infectious disease did not exist in the 16th century. In the 21st century, however, new medical technologies will be key parts of an armamentarium that reinforces our own immunological defenses. This dependence on technology is beginning to be recognized at high levels of national and international policy-making. With the portent of nearly instant global transmission of pathogenic agents, it is ever more important to work with international organizations like WHO for global health improvement. After all, the spread of AIDS in America and Europe in the 1980s and 1990s was due, in part, to an earlier phase of near obliviousness to the frightful health conditions in Africa. One harbinger of the kind of high-tech wit we will need for defending against outbreaks of infectious disease is the use of cutting-edge communications technology and the Internet, which already have been harnessed to post prompt global alerts of emerging diseases (see osi.oracle.com:8080/promed/promed. home).

Moving Targets

"Germs" have long been recognized as living entities, but the realization that they must inexorably be evolving and changing has been slow to sink in to the ideology and practice of the public health sector. This lag has early roots. In the 19th century, Koch was convinced that rigorous experiments would support the doctrine of monomorphism: that each disease was caused by a single invariant microbial species rather than by the many that often showed up in culture. He argued that most purported "variants" were probably alien bacteria that had floated into the petri dishes from the atmosphere.

Koch's rigor was an essential riposte to careless claims of interconvertibility—for example, that yeasts could be converted into bacteria. It also helped untangle confusing claims of complex morphogenesis and life cycles among common bacteria. But strict monomorphism was too rigid, and even Koch eventually relented, admitting the possibility of some intrinsic variation rather than contamination. Still, for him and his contemporaries, variation remained a phenomenological and experimental nuisance rather than the essence of microbes' competence as pathogens. The multitude of isolable species was confusing enough to the epidemic tracker; it would have been almost too much to bear to have to cope with constantly emerging variants with altered serological specificity, host affinity, or virulence.

Even today it would be near heresy to balk at the identification of the great plague of the 14th century with today's *Yersinia pestis*; but we cannot readily account for its pneumonic transmission without guessing at some intrinsic adaptation at the time to aerosol conveyance. Exhumations of ancient remains might still furnish DNA evidence to test such ideas.

We now know and accept that evolutionary processes elicit changes in the genotypes of germs and of their hosts. The idea that infection might play an important role in natural selection sank in after 1949 when John B. S. Haldane conjectured that the prevalence of hemoglobin disorders in Mediterranean peoples might be a defense against malaria. That idea developed into the first concrete example of a hereditary adaptation to infectious disease.

Haldane's theory preceded Anthony C. Allison's report of the protective effect of heterozygous hemoglobinopathy against falciparum malaria in Africa. The side effects of this bit of natural genetic engineering are well known: When this beneficial polymorphism is driven to higher gene frequencies, the homozygous variant becomes more prevalent and with it the heavy human and societal burden of sickle cell disease.

We now have a handful of illustrations of the connection between infection and evolution. Most are connected to malaria and tuberculosis, which are so prevalent that genetic adaptations capable of checking them have been strongly selected. The same prevalence also makes their associated adaptations more obvious to researchers. A newly reported link between infection and evolution is the effect of a *ccr5* (chemokine receptor) deletion, a genetic alteration that affords some protection against AIDS. It would be interesting to know what factors—another pathogen perhaps—may have driven that polymorphism in earlier human history.

One lesson to be gleaned from this coevolutionary dynamic is how fitful and sporadic human evolution is when our slow and plodding genetic change is pitted against the far more rapidly changing genomes of microbial pathogens.

We have inherited a robust immune system, but little has changed since its early vertebrate origins 200 million years ago. In its inner workings, immunity is a Darwinian struggle: a randomly generated diversification of leukocytes that collectively are prepared to duel with a lifetime of unpredictable invaders. But these duels take place in the host soma; successful immunological encounters do not become genetically inscribed and passed on to future generations of the host. By contrast, the germs that win the battles quickly proliferate their successful genes, and they can use those enhancements to go on to new hosts, at least in the short run.

The human race evidently has withstood the pathogenic challenges encountered so far, albeit with episodes of incalculable tragedy. But the rules of encounter and engagement have been changing; the same record of survival may not necessarily hold for the future. If our collective immune systems fail to keep pace with microbial innovations in the altered contexts we have created, we will have to rely still more on our wits.

Evolving Metaphors of Infection: Teach War No More

New strategies and tactics for countering pathogens will be uncovered by finding and exploiting innovations that evolved within other species in defense against infection. But our most sophisticated leap would be to drop the manichaean view of microbes—"We good; they evil." Microbes indeed have a knack for making us ill, killing us, and even recycling our remains to the geosphere. But in the long run microbes have a shared interest in their hosts' survival: A dead host is a dead end for most invaders too. Domesticating the host is the better long-term strategy for pathogens.

We should think of each host and its parasites as a superorganism with the respective genomes yoked into a chimera of sorts. The power of this sociological development could not be more persuasively illustrated than by the case of mitochondria, the most successful of all microbes. They reside inside every eukaryote cell (from yeast to protozoa to multicellular organisms), in which they provide the machinery of oxidative metabolism. Other bacteria have taken similar routes into plant cells and evolved there into chloroplasts—the primary harvesters of solar energy, which drive the production of oxygen and the fixed carbon that nourishes the rest of the biosphere.

These cases reveal how far collaboration between hosts and infecting microbes can go. In the short run, however, the infected host is in fact at metastable equilibrium: The balance could tip toward favorable or catastrophic outcomes.

On the bad side, the host's immune response may be excessive, with autoimmune injuries as side effects. Microbial zeal also can be self-defeating. As with rogue cancer cells, deviant microbial cells (such as aggressive variants from a gentler parent population) may overtake and kill the host, thereby fomenting their own demise and that of the parent population.

Most successful parasites travel a middle path. It helps for them to have aggressive means of entering the body surfaces and radiating some local toxicity to counter the hosts' defenses, but once established they also do themselves (and their hosts) well by moderating their virulence.

Better understanding of this balancing act awaits further research. And that may take a shift in priorities. For one, research has focused on hypervirulence. Studies into the physiology of homeostatic balance in the infected host qua superorganism have lagged. Yet the latter studies may be even more revealing, as the burden of mutualistic adaptation falls largely on the shoulders of the parasite, not the host. This lopsided responsibility follows from the vastly different evolutionary paces of the two. But then we have our wits, it is to be hoped, for drafting the last word.

To that end, we also need more sophisticated experimental models of infection, which today are largely based on contrived zoonoses (the migration of a parasite from its traditional host into another species). The test organism is usually a mouse, and the procedure is intended to mimic the human disease process. Instead, it is often a caricature.

Injected with a few bugs, the mouse goes belly up the next day. This is superb for in vivo testing of an antibiotic, but it bears little relation to the dynamics of everyday human disease.

Natural zoonoses also can have many different outcomes. In most cases, there will be no infection at all or only mild ones such as the gut ache caused by many *Salmonella enteritidis* species. Those relatively few infectious agents that cause serious sickness or death are actually maladapted to their hosts, to which they may have only recently gained access through some genetic, environmental, or sociological change. These devastatingly virulent zoonoses include psittacosis, Q fever, rickettsiosis, and hantavirus. Partly through lack of prior coevolutionary development with the new host, normal restraints fail.

I suggest that a successful parasite (one that will be able to remain infectious for a long time) tends to display just those epitopes (antigen fragments that stimulate the immune system) as will provoke host responses that (a) moderate but do not extinguish the primary infection, and (b) inhibit other infections by competing strains of the same species or of other species. According to this speculative framework, the symptoms of influenza evolved as they have in part to ward off other viral infections.

Research into infectious diseases, including tuberculosis, schistosomiasis, and even AIDS, is providing evidence for this view. So are studies of *Helicobacter*, which has been found to secrete antibacterial peptides that inhibit other enteric infections. We need also to look more closely at earlier stages of chronic infection and search for cross-protective factors by which microbes engage one another. HIV, for one, ultimately fails from the microbial perspective when opportunistic infections supervene to kill its host. That result, which is tragic from the human point of view, is a byproduct of the virus's protracted duel with the host's cellular immune system. The HIV envelope and those of related viruses also produce antimicrobials, although their significance for the natural history of disease remains unknown.

Now genomics is entering the picture. Within the past decade, the genomes of many microbes have been completely sequenced. New evidence for the web of genetic interchange is permeating the evolutionary charts. The functional analyses of innumerable genes now emerging are an unexplored mine of new therapeutic targets. It has already shown many intricate intertwinings of hosts' and parasites' physiological pathways. Together with wiser insight into the ground rules of pathogenic evolution, we are developing a versatile platform for developing new responses to infectious disease. Many new vaccines, antibiotics, and immune modulators will emerge from the growing wealth of genomic data.

The lessons of HIV and other emerging infections also have begun taking hold in government and in commercial circles, where the market opportunities these threats offer have invigorated the biotechnology industry. If we do the hard work and never take success for granted (as we did for a while during the last century), we may be able to preempt infectious disasters such as the influenza outbreak of 1918-1919 and the more recent and ongoing HIV pandemic.

Perhaps one of the most important changes we can make is to supercede the 20th-century metaphor of war for describing the relationship between people and infectious agents. A more ecologically informed metaphor, which includes the germs'-eye view of infection, might be more fruitful. Consider that microbes occupy all of our body surfaces. Besides the disease-engendering colonizers of our skin, gut, and mucous membranes, we are host to a poorly cataloged ensemble of symbionts to which we pay scant attention. Yet they are equally part of the superorganism genome with which we engage the rest of the biosphere.

The protective role of our own microbial flora is attested to by the superinfections that often attend specific antibiotic therapy: The temporary decimation of our home-team microbes provides entrée for competitors. Understanding these phenomena affords openings for our advantage, akin to the ultimate exploitation by Dubos and Selman Waksman of intermicrobial competition in the soil for seeking early antibiotics. Research into the microbial ecology of our own bodies will undoubtedly yield similar fruit.

Replacing the war metaphor with an ecological one may bear on other important issues, including debates about eradicating pathogens such as smallpox and polio. Without a clear strategy for sustaining some level of immunity, it makes sense to maintain lab stocks of these and related agents to guard against possible recrudescence. An ecological perspective also suggests other ways of achieving lasting security. For example, domestication of commensal microbes that bear relevant cross-reacting epitopes could afford the same protection as vaccines based on the virulent forms. There might even be a nutraceutical angle: These commensal epitopes could be offered as optional genetically engineered food additives, clearly labeled and meticulously studied.

Another relevant issue that can be recast in an ecological model is the rise in popularity of antibacterial products. This is driven by the popular idea that a superhygienic environment is better than one with germs—the "enemy" in the war metaphor. But too much antibacterial zeal could wipe out the very immunogenic stimulation that has enabled us to cohabit with microbes in the first place.

BOX WO-4
The Microbial World Wide Web

The field of molecular genetics, which began in 1944 when DNA was proven to be the molecule of heredity in bacteria-based experiments, ushered microbes into the center of many biological investigations. Microbial systems now provide our most convenient models for experimental evolution. Diverse mechanisms for genetic variation and recombination uncovered in such systems are spelled out in ponderous monographs. Assays for chemical mutagenesis (e.g., the Ames test using *Salmonella*) are now routinely carried out on bacteria, because microbial DNA is so accessible to environmental insult. Mutators (genes that enhance variability) abound and may be switched on and off by different environmental factors. The germs' ability to transfer their own genetic scripts, via processes such as plasmid transfer, means they can exchange biological innovations including resistance to antibiotics.

Indeed, the microbial biosphere can be thought of as a World Wide Web of informational exchange, with DNA serving as the packets of data going every which way. The analogy isn't entirely superficial. Many viruses can integrate (download) own DNA into host genomes, which subsequently can be copied and passed on: Hundreds of segments of human DNA originated from historical encounters with retroviruses whose genetic information became integrated into our own genomes.

What makes microbial evolution particularly intriguing, and worrisome, is a combination of vast populations and intense fluctuations in those populations. It's a formula for top-speed evolution. Microbial populations may fluctuate by factors of 10 billion on a daily cycle as they move between hosts, or as they encounter antibiotics, antibodies, or other natural hazards. A simple comparison of the pace of evolution between microbes and their multicellular hosts suggests a million fold or billion fold advantage to the microbe. A year in the life of bacteria would easily match the span of mammalian evolution!

By that metric, we would seem to be playing out of our evolutionary league. Indeed, there's evidence of sporadic species extinctions in natural history, and our own human history has been punctuated by catastrophic plagues. Yet we are still here! Maintaining that status within new contexts in which germs and hosts interact in new ways almost certainly will require us to bring ever more sophisticated technical wit and social intelligence to the contest. -J. L.

Ironically, even as I advocate this shift from a war metaphor to an ecology metaphor, war in its historic sense is making that more difficult. The darker corner of microbiological research is the abyss of maliciously designed biological warfare (BW) agents and systems to deliver them. What a nightmare for the next millennium! What's worse, for the near future, technology is likely to favor offensive BW weaponry, because defenses will have to cope with a broad range of microbial threats that can be collected today or designed tomorrow.

As a measure of social intelligence and policy, we should push for enforcement of the 1975 BW disarmament convention. The treaty forbids the development, production, stockpiling, and use of biological weapons under any circumstances. One of its articles also provides for the international sharing of biotechnology for peaceful purposes. The scientific and humanistic rationale is self-evident: to enhance and apply scientific knowledge to manage infectious disease, naturally occurring or otherwise.

Further Readings

W. Bulloch, *History of Bacteriology* (Oxford University Press, London, 1938).

R. Dubos, *Mirage of Health: Utopias, Progress, and Biological Change* (Rutgers University Press, New Brunswick, New Jersey, 1987).

J. Lederberg, R. E. Shope, S. C. Oaks Jr., Eds., *Emerging Infections: Microbial Threats to Health in the United States* (National Academy Press, Washington, D.C., 1992) (see www.nap.edu/books/0309047412/html/index.html).

G. Rosen, *A History of Public Health* (Johns Hopkins University Press, Baltimore, 1993). T. D. Brock, *The Emergence of Bacterial Genetics* (Cold Spring Harbor Laboratory Press, Cold Spring Harbor, New York, 1990).

S. S. Morse, Ed., *Emerging Viruses* (Oxford University Press, New York, 1993). *Journal of the American Medical Association* theme issue on emerging infections, August 1996.

W. K. Joklik *et al.*, Eds., *Microbiology, a Centenary Perspective* (ASM Press, Herndon, Virginia, 1999).

S. C. Stearns, Ed., *Evolution in Health and Disease* (Oxford University Press, Oxford, New York, 1999).

P. Ewald, "Evolution of Infectious Disease," in *Encyclopedia of Microbiology*, J. Lederberg, Ed. (Academic Press, Orlando, Florida, 2000).

G. L. Mandell, J. E. Bennett, R. Dolin, *Principles and Practice of Infectious Diseases* (Churchill Livingstone, 5th ed., Philadelphia, 2000).

Notable Web Sites

www.lib.uiowa.edu/hardin/md/micro.html
www.idsociety.org
www.asmusa.org
osi.oracle.com:8080/promed
www.cdc.gov/ncidod/EID

WORKSHOP OVERVIEW REFERENCES

Andersson, A. F., and J. F. Banfield. 2008. Virus population dynamics and acquired virus resistance in natural microbial communities. *Science* 320(5879):1047-1050.

Bäckhed, F., H. Ding, T. Wang, L. V. Hooper, G. Y. Koh, A. Nagy, C. F. Semenkovich, and J. I. Gordon. 2004. The gut microbiota as an environmental factor that regulates fat storage. *Proceedings of the National Academy of Sciences* 101(44):15718-15723.

Banfield, J. F. 2008a. *Geomicrobiology program UC Berkeley*, http://eps.berkeley.edu/~jill/banres.html (accessed July 31, 2008).

———. 2008b. *Subproject 1: integrating genomics, proteomics, and functional analyses into ecosystem-level studies of natural microbial communities*, http://eps.berkeley.edu/~jill/gtl/subproject_1.htm (accessed July 31, 2008).

Barrangou, R., C. Fremaux, H. Deveau, M. Richards, P. Boyaval, S. Moineau, D. A. Romero, and P. Horvath. 2007. CRISPR provides acquired resistance against viruses in prokaryotes. *Science* 315(5819):1709-1712.

Blaser, M. J. 1997. Ecology of *Helicobacter pylori* in the human stomach. *Journal of Clinical Investigation* 100(4):759-762.

Brelles-Mariño, G., and J. M. Ané. 2008. Nod factors and the molecular dialogue in the rhizobia-legume interaction. In *Nitrogen fixation research progress*, edited by G. N. Couto. Hauppauge, NY: Nova Science Publishers, Inc.

Chun, C. K., J. V. Troll, I. Koroleva, B. Brown, L. Manzella, E. Snir, H. Almabrazi, T. E. Sheetz, M. de Fatima Bonaldo, T. L. Casavant, M. B. Soares, E. G. Ruby, and M. J. McFall-Ngai. 2008. Effects of colonization, luminescence, and autoinducer on host transcription during development of the squid-vibrio association. *Proceedings of the National Academy of Sciences* 105(32):11323-11328.

Crichton, M. 1969. *The andromeda strain*. 1st ed. New York: Knopf.

Dantas, G., M. O. Sommer, R. D. Oluwasegun, and G. M. Church. 2008. Bacteria subsisting on antibiotics. *Science* 320(5872):100-103.

Davies, J. 2007. Microbes have the last word: A drastic re-evaluation of antimicrobial treatment is needed to overcome the threat of antibiotic-resistant bacteria. *EMBO Reports* 8(7):616-621.

D'Costa, V. M., K. M. McGrann, D. W. Hughes, and G. D. Wright. 2006. Sampling the antibiotic resistome. *Science* 311(5759):374-377.

Denef, V. J., N. C. VerBerkmoes, M. B. Shah, P. Abraham, M. Lefsrud, R. L. Hettich, and J. F. Banfield. 2009. Proteomics-inferred genome typing (PIGT) demonstrates inter-population recombination as a strategy for environmental adaptation. *Environmental Microbiology* 11(2):313-325.

Dethlefsen, L., M. McFall-Ngai, and D. A. Relman. 2007. An ecological and evolutionary perspective on human-microbe mutualism and disease. *Nature* 449(7164):811-818.

Didierlaurent, A., J.-C. Sirard, J.-P. Kraehenbuhl, and M. R. Neutra. 2001. How the gut senses its content. *Cellular Microbiology* 4(2):61-72.

DOE (Department of Energy). 2007. *Understanding how a cell works: background*, http://microbial genomics.energy.gov/MicrobialCellProject/background.shtml (accessed December 10, 2008).

Eckburg, P. B., E. M. Bik, C. N. Bernstein, E. Purdom, L. Dethlefsen, M. Sargent, S. R. Gill, K. E. Nelson, and D. A. Relman. 2005. Diversity of the human intestinal microbial flora. *Science* 308(5728):1635-1638.

Eisen, J. A. 2000. Horizontal gene transfer among microbial genomes: new insights from complete genome analysis. *Current Opinion in Genetics and Development* 10(6):606-611.

Emerson, R. W. 1841. Self-reliance. In *Essays, first series,* http://www.emersoncentral.com/selfreliance.htm (accessed August 22, 2008).

Eppley, J. M., G. W. Tyson, W. M. Getz, and J. F. Banfield. 2007. Genetic exchange across a species boundary in the Archaeal genus *Ferroplasma*. *Genetics* 177(1):407-416.

Foresight. 2006. *Infectious diseases: preparing for future threats*. London, UK: Office of Science and Innovation.

Gilbert, G. S., J. Handelsman, and J. L. Parke. 1994. Root camouflage and disease control. *Phytopathology* 8(3):222-225.

Gordon, M. A., S. M. Graham, A. L. Walsh, L. Wilson, A. Phiri, E. Molyneux, E. E. Zijlstra, R. S. Heyderman, C. A. Hart, and M. E. Molyneux. 2008. Epidemics of invasive *Salmonella enterica* serovar enteritidis and *S. enterica* serovar typhimurium infection associated with multidrug resistance among adults and children in Malawi. *Clinical Infectious Diseases* 46(7):963-969.

Graham, S. M. 2002. Salmonellosis in children in developing and developed countries and populations. *Current Opinion in Infectious Diseases* 15(5):507-512.

Hamburg, D. A. 2008. *Preventing genocide: practical steps toward early detection and effective action*. Boulder, CO: Paradigm Publishers.

Heidelberg, J. F., J. A. Eisen, W. C. Nelson, R. A. Clayton, M. L. Gwinn, R. J. Dodson, D. H. Haft, E. K. Hickey, J. D. Peterson, L. Umayam, S. R. Gill, K. E. Nelson, T. D. Read, H. Tettelin, D. Richardson, M. D. Ermolaeva, J. Vamathevan, S. Bass, H. Qin, I. Dragoi, P. Sellers, L. McDonald, T. Utterback, R. D. Fleishmann, W. C. Nierman, O. White, S. L. Salzberg, H. O. Smith, R. R. Colwell, J. J. Mekalanos, J. C. Venter, and C. M. Fraser. 2000. DNA sequence of both chromosomes of the cholera pathogen *Vibrio cholerae*. *Nature* 406(6795):477-483.

Hooper, L. V., and J. I. Gordon. 2001. Commensal host–bacterial relationships in the gut. *Science* 292(5519):1115-1118.

Hughes, J. M. 2001. Emerging infectious diseases: a CDC perspective. *Emerging Infectious Diseases* 7(3):494-496.

IOM (Institute of Medicine). 1992. *Emerging infections: microbial threats to health in the United States*. Washington, DC: National Academy Press.

———. 2003. *Microbial threats to health: emergence, detection, and response*. Washington, DC: The National Academies Press.

———. 2004a. *Learning from SARS: preparing for the next disease outbreak*. Washington, DC: The National Academies Press.

———. 2004b. *The threat of pandemic influenza: are we ready?* Washington, DC: The National Academies Press.

———. 2006a. *Ending the war metaphor: the changing agenda for unraveling the host-microbe relationship*. Washington, DC: The National Academies Press.

———. 2006b. *The impact of globalization on infectious disease emergence and control: exploring the consequences and opportunities*. Washington, DC: The National Academies Press.

———. 2006c. *Addressing foodborne threats to health: policies, practices, and global coordination*. Washington, DC: The National Academies Press.

———. 2007. *Global infectious disease surveillance and detection: assessing the challenges*. Washington, DC: The National Academies Press.

————. 2008a. *Vector-borne diseases: understanding the environmental, human health, and ecological connections.* Washington, DC: The National Academies Press.

————. 2008b. *Global climate change and extreme weather events: understanding the contributions to infectious disease emergence.* Washington, DC: The National Academies Press.

Jones, K. E., N. G. Patel, M. A. Levy, A. Storeygard, D. Balk, J. L. Gittleman, and P. Daszak. 2008. Global trends in emerging infectious diseases. *Nature* 451(7181):990-993.

Jurkowski, A., A. H. Reid, and J. B. Labov. 2007. Metagenomics: a call for bringing a new science into the classroom (while it's still new). *CBE Life Sciences Education* 6(4):260-265.

Kiers, E. T., R. A. Rousseau, S. A. West, and R. F. Denison. 2003. Host sanctions and the legume-rhizobium mutualism. *Nature* 425(6953):78-81.

Klemow, K. M. 2008. *Environmental effects of mining in the anthracite region: problems and possible solutions,* http://www.wilkes.edu/pages/2280.asp (accessed July 31, 2008).

Lawley, T. D., K. Chan, L. J. Thompson, C. C. Kim, G. R. Govoni, and D. M. Monack. 2006. Genome-wide screen for salmonella genes required for long-term systemic infection of the mouse. *PLoS Pathogens* 2(2):e11.

Lederberg, J. 2000. Infectious history. *Science* 288(5464):287-293.

Lepp, P. W., M. M. Brinig, C. C. Ouverney, K. Palm, G. C. Armitage, and D. A. Relman. 2004. Methanogenic archaea and human periodontal disease. *Proceedings of the National Academy of Sciences* 101(16):6176-6181.

Levin, B. R., V. Perrot, and N. Walker. 2000. Compensatory mutations, antibiotic resistance and the population genetics of adaptive evolution in bacteria. *Genetics* 154(3):985-997.

Li, L., and S. N. Cohen. 1996. Tsg101: a novel tumor susceptibility gene isolated by controlled homozygous functional knockout of allelic loci in mammalian cells. *Cell* 85(3):319-329.

Lo, I., V. J. Denef, N. C. Verberkmoes, M. B. Shah, D. Goltsman, G. DiBartolo, G. W. Tyson, E. E. Allen, R. J. Ram, J. C. Detter, P. Richardson, M. P. Thelen, R. L. Hettich, and J. F. Banfield. 2007. Strain-resolved community proteomics reveals recombining genomes of acidophilic bacteria. *Nature* 446(7135):537-541.

Lu, Q., L. W. Hope, M. Brasch, C. Reinhard, and S. N. Cohen. 2003. TSG101 interaction with HRS mediates endosomal trafficking and receptor down-regulation. *Proceedings of the National Academy of Sciences* 100(13):7626-7631.

Lu, Q., W. Wei, P. E. Kowalski, A. C. Chang, and S. N. Cohen. 2004. EST-based genome-wide gene inactivation identifies ARAP3 as a host protein affecting cellular susceptibility to anthrax toxin. *Proceedings of the National Academy of Sciences* 101(49):17246-17251.

Makarova, K. S., N. V. Grishin, S. A. Shabalina, Y. I. Wolf, and E. V. Koonin. 2006. A putative RNA-interference-based immune system in prokaryotes: Computational analysis of the predicted enzymatic machinery, functional analogies with eukaryotic RNAi, and hypothetical mechanisms of action. *Biology Direct* 1(7): http://www.biology-direct.com/content/1/1/7 (accessed August 28, 2008).

Margolis, E., and B. R. Levin. 2008. Evolution of bacterial-host interactions: virulence and the immune overresponse. In *Evolutionary biology of bacterial and fungal pathogens,* edited by F. Bequero, C. Nombela, G. H. Cassel, and J. A. Gutierrez. Washington, DC: ASM Press. Pp. 3-12.

May, R. M., and R. M. Anderson. 1983. Parasite-host coevolution. In *Coevolution,* edited by D. J. Futuyama and M. Slatkin. Sunderland, MA: Sinauer Associates.

McFall-Ngai, M. J. 1998. The development of cooperative associations between animals and bacteria: establishing détente among domains. *American Zoologist* 38(4):593-608.

Merrell, D. S., and S. Falkow. 2004. Frontal and stealth attack strategies in microbial pathogenesis. *Nature* 430(6996):250-256.

Mojica, F. J., C. Diez-Villasenor, J. Garcia-Martinez, and E. Soria. 2005. Intervening sequences of regularly spaced prokaryotic repeats derive from foreign genetic elements. *Journal of Molecular Evolution* 60(2):174-182.

Monack, D. M., D. M. Bouley, and S. Falkow. 2004a. *Salmonella typhimurium* persists within mac-rophages in the mesenteric lymph nodes of chronically infected *Nramp1*[+/+] mice and can be reactivated by IFNγ neutralization. *Journal of Experimental Medicine* 199(2):231-241.

Monack, D. M., A. Mueller, and S. Falkow. 2004b. Persistent bacterial infections: the interface of the pathogen and the host immune system. *Nature Reviews Microbiology* 2(9):747-765.

Morens, D., G. K. Folkers, and A. S. Fauci. 2004. The challenge of emerging and re-emerging infectious diseases. *Nature* 430(6996):242-249.

Morens, D., G. K. Folkers, and A. S. Fauci. 2008. Emerging infections: a perpetual challenge. *Lancet Infectious Diseases* 8(11):710-719.

Morse, S. S. 2008. Retrospective: Joshua Lederberg (1925-2008). *Science* 319(5868):1351.

Nemergut, D. R., M. S. Robeson, R. F. Kysela, A. P. Martin, S. K. Schmidt, and R. Knight. 2008. Insights and inferences about integron evolution from genomic data. *BMC Genomics* 9(261): http://www.biomedcentral.com/1471-2164/9/261 (accessed August 26, 2008).

NLM (National Library of Medicine). 2008. *The Joshua Lederberg papers: biographical information*, http://profiles.nlm.nih.gov/BB/ (accessed August 4, 2008).

NRC (National Research Council). 2007. *The new science of metagenomics: revealing the secrets of our microbial planet*. Washington, DC: The National Academies Press.

Nyholm, S. V., and M. J. McFall-Ngai. 2004. The winnowing: establishing the squid-vibrio symbiosis. *Nature Reviews Microbiology* 2(8):632-642.

Palmer, C., E. M. Bik, D. B. Digiulio, D. A. Relman, and P. O. Brown. 2007. Development of the human infant intestinal microbiota. *PLoS Biology* 5(7):e177.

Parkhill, J., M. Sebaihia, A. Preston, L. D. Murphy, N. Thomson, D. E. Harris, M. T. Holden, C. M. Churcher, S. D. Bentley, K. L. Mungall, A. M. Cerdeno-Tarraga, L. Temple, K. James, B. Harris, M. A. Quail, M. Achtman, R. Atkin, S. Baker, D. Basham, N. Bason, I. Cherevach, T. Chillingworth, M. Collins, A. Cronin, P. Davis, J. Doggett, T. Feltwell, A. Goble, N. Hamlin, H. Hauser, S. Holroyd, K. Jagels, S. Leather, S. Moule, H. Norberczak, S. O'Neil, D. Ormond, C. Price, E. Rabbinowitsch, S. Rutter, M. Sanders, D. Saunders, K. Seeger, S. Sharp, M. Simmonds, J. Skelton, R. Squares, S. Squares, K. Stevens, L. Unwin, S. Whitehead, B. G. Barrell, and D. J. Maskell. 2003. Comparative analysis of the genome sequences of *Bordetella pertussis, Bordetella parapertussis* and *Bordetella bronchiseptica. Nature Genetics* 35(1):32-40.

Parrish, C. R., E. C. Holmes, D. M. Morens, E. C. Park, D. S. Burke, C. H. Calisher, C. A. Laughlin, L. J. Saif, and P. Daszak. 2008. Cross-species virus transmission and the emergence of new epidemic diseases. *Microbiology and Molecular Biology Reviews* 72(3):457-470.

Relman, D. A., and S. Falkow. 2001. The meaning and impact of the human genome sequence for microbiology. *Trends in Microbiology* 9(5):206-208.

Riely, B. K., J.-H. Mun, and J.-M. Ané. 2006. Unravelling the molecular basis for symbiotic signal transduction in legumes. *Molecular Plant Pathology* 7(3):197-207.

Riesenfeld, C. S., R. M. Goodman, and J. Handelsman. 2004. Uncultured soil bacteria are a reservoir of new antibiotic resistance genes. *Environmental Microbiology* 6(9):981-989.

Roumagnac, P., F. X. Weill, C. Dolecek, S. Baker, S. Brisse, N. T. Chinh, T. A. Le, C. J. Acosta, J. Farrar, G. Dougan, and M. Achtman. 2006. Evolutionary history of *Salmonella typhi. Science* 314(5803):1301-1304.

Spellberg, B., R. Guidos, D. Gilbert, J. Bradley, H. W. Boucher, W. M. Scheld, J. G. Bartlett, J. Edwards Jr., and Infectious Diseases Society of America. 2008. The epidemic of antibiotic-resistant infections: a call to action for the medical community from the Infectious Diseases Society of America. *Clinical Infectious Diseases* 46(2):155-164.

Steinberg, L. M., and B. R. Levin. 2007. Grazing protozoa and the evolution of the *Escherichia coli* O157:H7 Shiga toxin-encoding prophage. *Proceedings of the Royal Society: Biological Sciences* 274(1621):1921-1929.

Stephens, D. S., E. R. Moxon, J. Adams, S. Altizer, J. Antonovics, S. Aral, R. Berkelman, E. Bond, J. Bull, G. Cauthen, M. M. Farley, A. Glasgow, J. W. Glasser, H. P. Katner, S. Kelley, J. Mittler, A. J. Nahmias, S. Nichol, V. Perrot, R. W. Pinner, S. Schrag, P. Small, and P. H. Thrall. 1998. Emerging and reemerging infectious diseases: a multidisciplinary perspective. *American Journal of the Medical Sciences* 315(2):64-75.

Szczepanowski, R., I. Krahn, B. Linke, A. Goesmann, A. Puhler, and A. Schluter. 2004. Antibiotic multiresistance plasmid pRSB101 isolated from a wastewater treatment plant is related to plasmids residing in phytopathogenic bacteria and carries eight different resistance determinants including a multidrug transport system. *Microbiology* 150(Pt 11):3613-3630.

Szczepanowski, R., S. Braun, V. Riedel, S. Schneiker, I. Krahn, A. Puhler, and A. Schluter. 2005. The 120 592 bp IncF plasmid pRSB107 isolated from a sewage-treatment plant encodes nine different antibiotic-resistance determinants, two iron-acquisition systems and other putative virulence-associated functions. *Microbiology* 151(Pt 4):1095-1111.

Szczepanowski, R., T. Bekel, A. Goesmann, L. Krause, H. Kromeke, O. Kaiser, W. Eichler, A. Puhler, and A. Schluter. 2008. Insight into the plasmid metagenome of wastewater treatment plant bacteria showing reduced susceptibility to antimicrobial drugs analysed by the 454-pyrosequencing technology. *Journal of Biotechnology* 136(1-2):54-64.

Taormina, P. J., L. R. Beuchat, and L. Slutsker. 1999. Infections associated with eating seed sprouts: an international concern. *Emerging Infectious Diseases* 5(5):626-634.

Tyson, G. W., and J. F. Banfield. 2008. Rapidly evolving CRISPRs implicated in acquired resistance of microorganisms to viruses. *Environmental Microbiology* 10(1):200-207.

Weerasinghe, R. R., D. M. Bird, and N. S. Allen. 2005. Root-knot nematodes and bacterial nod factors elicit common signal transduction events in *Lotus japonicus*. *Proceedings of the National Academy of Sciences* 102(8):3147-3152.

Wei, W., Q. Lu, G. J. Chaudry, S. H. Leppla, and S. N. Cohen. 2006. The LDL receptor-related protein LRP6 mediates internalization and lethality of anthrax toxin. *Cell* 124(6):1141-1154.

WHO (World Health Organization). 2008a. *About the IHR*, http://www.who.int/csr/ihr/prepare/en/index.html (accessed July 27, 2008).

———. 2008b. *Core capacity requirements for surveillance and response*, http://www.who.int/csr/ihr/capacity/en/index.html (accessed July 27, 2008).

Woolhouse, M., and R. Antia. 2007. Emergence of new infectious diseases. In *Evolution in health and disease*, 2nd ed., edited by S. C. Stearns and J. K. Koella. Oxford, UK: Oxford University Press. Pp. 215-228.

Woolhouse, M., and E. Gaunt. 2007. Ecological origins of novel human pathogens. *Critical Reviews in Microbiology* 33(4):231-242.

Wu, D., S. C. Daugherty, S. E. Van Aken, G. H. Pai, K. L. Watkins, H. Khouri, L. J. Tallon, J. M. Zaborsky, H. E. Dunbar, P. L. Tran, N. A. Moran, and J. A. Eisen. 2006. Metabolic complementarity and genomics of the dual bacterial symbiosis of sharpshooters. *PLoS Biology* 4(6):e188.

Wu, M., Q. Ren, A. S. Durkin, S. C. Daugherty, L. M. Brinkac, R. J. Dodson, R. Madupu, S. A. Sullivan, J. F. Kolonay, D. H. Haft, W. C. Nelson, L. J. Tallon, K. M. Jones, L. E. Ulrich, J. M. Gonzalez, I. B. Zhulin, F. T. Robb, and J. A. Eisen. 2005. Life in hot carbon monoxide: the complete genome sequence of *Carboxydothermus hydrogenoformans* z-2901. *PLoS Genetics* 1(5):e65.

Yokoyama, W. M., and M. Colonna. 2008. Innate immunity to pathogens. *Current Opinion in Immunology* 20(1):1-2.

Young, D., T. Hussell, and G. Dougan. 2002. Chronic bacterial infections: living with unwanted guests. *Nature Immunology* 3(11):1026-1032.

1

The Life and Legacies of Joshua Lederberg

OVERVIEW

The essays in this chapter offer three personal perspectives on Joshua Lederberg's many contributions to science, society, scholarship, and to the lives and careers of his colleagues, students, and friends. The first contributor, David A. Hamburg of Cornell University's Weill Medical College, recounts Lederberg's legacies as scientist and humanist through the lens of nearly 50 years of friendship. In the second essay, Stephen S. Morse, of Columbia University, recalls meeting Lederberg, who was then president of Rockefeller University, when Morse was "the most junior of junior faculty members." Thus began a friendship, rooted in a shared interest in emerging infectious diseases, that lasted for more than 20 years—a collaboration that embraced the ideas upon which the Forum on Microbial Threats was founded.

Former Forum chair Adel Mahmoud, of Princeton University, notes in the chapter's final essay that "this Forum is the brainchild of Joshua as he was exploring how to respond to the multifaceted challenges of microbes." After reviewing and celebrating the breadth of his accomplishments, Mahmoud concludes that Lederberg "needs no monument to ensure that his life and work are long remembered." Rather, his ideas and example will continue to be "an inspiration and a reminder that our work can truly change the world just as the life and career of Joshua Lederberg certainly did."

REFLECTIONS ON THE CAREER OF JOSHUA LEDERBERG

David A. Hamburg, M.D.[1]
Carnegie Corporation

I am honored to speak about Josh Lederberg on the occasion of this important meeting. It was my great privilege to have nearly half a century of joint efforts and deep friendship with him. Let me start with a citation for his achievements written three decades after he received his Nobel Prize in Medicine. In 1989, our nation's highest honor in science and technology, the National Medal of Science, was awarded to him with a concise and illuminating citation:

> For [Joshua Lederberg's] work in bacterial genetics and immune cell single type antibody production; for his seminal research in artificial intelligence in biochemistry and medicine; and for his extensive advisory role in government, industry, and international organizations that address themselves to the societal role of science.

I could add more—and will, to some extent. All of us here respect his truly great scientific achievements and creative leadership in science and public policy.

How did all of this happen? In childhood, he had prodigious intellectual gifts, along with a reverence for learning and scholarship—powerfully reinforced by his family. From then on, his life was characterized by boundless curiosity—a fresh look at everything.

He took deep satisfaction in discovery—and then raising the next question, and the next, and challenging the scientific community to pursue many ramifications. This interrelated set of attributes characterized him all his life and had much to do with his great accomplishments.

One dramatic feature of his career: he was a school dropout—medical school, that is. He entered medical school with his typical intense curiosity and sense of discovery. This was a learning moment: the emergence of the new biology. He shifted to graduate school in biology to pursue the frontiers of knowledge. There began a line of inquiry that led before long to the Nobel Prize.

This was groundbreaking, highly imaginative work on the nature of microorganisms, especially their mechanisms of inheritance. He opened up bacterial genetics, including the momentous discovery of genetic recombination. This work was one of the crucial foundations for subsequent discoveries in cellular and molecular biology. Many of us stood on his shoulders. He won the Nobel Prize in 1958 at the age of 33—one of the youngest winners in any field from any nation.

Another attribute was his remarkable capacity for institutional innovation.

---

[1]President emeritus.

He created a department of genetics in the medical school at Stanford University. Until then, genetics had been marginal—or nonexistent—in medical schools. There was a widely shared assumption in the middle of the twentieth century that genetics might be intrinsically interesting but that it would never have much practical significance for medicine. How wrong that assumption was!

While actively stimulating and fostering basic research, Josh also sought applications, and he helped to create the biotechnology industry. In teaching and in institution building, he emphasized the mutually beneficial interplay of basic and clinical research.

In this context, he was very generous in helping to establish new clinical departments and new kinds of clinical departments. At Stanford, he helped with psychiatry, pediatrics, medicine, and neurology. He inspired us with the classic experiments of Oswald Avery, Colin MacLeod, and Maclyn McCarty at Rockefeller University in the 1940s. Their clinical inquiry into pneumonia led to a great discovery on the most basic level: DNA is the genetic material. He helped us to build on basic components and to create interdisciplinary groups. He also helped us to identify research opportunities and promising lines of innovation.

He was a wide-ranging mentor. The world is full of people grateful to Josh for his powerful insights, creative suggestions, and generosity of spirit.

Within his own remarkable department at Stanford, he fostered many lines of inquiry: molecular genetics, cellular genetics, clinical genetics, population genetics, exobiology (the National Aeronautics and Space Administration's [NASA's] Mariner and Viking missions to Mars), immunology, and neurobiology.

He always had a worldwide view and brought in superb people, not only from the mysterious east of the United States, like New York City, but also, for example, Walter Bodmer (United Kingdom), Luca Cavalli-Sforza (Italy), Gus Nossal (Australia), Eric Shooter (United Kingdom), and others from afar—all of whom were major contributors. His global outlook, long-term vision, intense curiosity, and unfailing kindness inspired all of us seeking to create new kinds of clinical departments. Moreover, he did much to strengthen the scientific capability of the World Health Organization (WHO).

Thus, his rare capacity to range widely with open eyes and an open mind—and also to dig deeply into a specialized topic, and to combine these capacities in research, education, and intellectual synthesis—led to fruitful stimulation in a variety of fields and nations.

His knowledge, curiosity, and imagination have been expressed in many ways. For example, he was instrumental in the creation of a highly innovative undergraduate major at Stanford, now past its thirty-fifth year as one of the most sought after majors at Stanford, drawing in faculty from across the university. It is broadly integrative across the life sciences, linking basic science, hands-on experience (including field research), biological aspects of behavioral science, and in the senior year, applications of the life sciences to policy (e.g., in health and environmental problems). He even found a way to make this a permanent program by insisting that we find a way to get endowed chairs.

Early in the computer era his interest in computer science grew and he became a pioneer in artificial intelligence, especially in relation to biochemistry, genetics, and medicine.

He believed deeply in education of the broad public, opening complex and emotionally charged topics for informed public discussion. One major vehicle was a column in *The Washington Post* during the 1970s, in which he interpreted science for the public and for several years produced fascinating, highly informative columns.

He was a pioneer in the scientific assessment of the human impact on the environment—and especially on the health implications of environmental conditions.

All of this rich experience, knowledge, skill, and wisdom were brought to bear on Rockefeller University under his presidency, where he broadened the scope of its great faculty, opened new opportunities for young people, and greatly improved the facilities. His deep respect and concern for the well-being of faculty—young and not so young—was remarkable. This was a crucial aspect of his leadership.

Josh was a pioneer in biological warfare and bioterrorism, applying his farsighted vision in efforts to understand the danger and find ways to cope with it. He strongly influenced the negotiation of the biological weapons disarmament treaty.

He advised the U.S. government in many agencies, including: the National Institutes of Health (NIH), the National Science Foundation (NSF), NASA, the Navy Office of Science and Technology Policy (OSTP), the Department of Energy, the Defense Science Board, and others. So too on the world stage. In addition, his deep sense of science's contributions to the well-being of humanity was expressed in his role as co-chair of the Carnegie Commission on Science, Technology, and Government, producing multiple publications on most branches of government, strengthening their science and technology capacities and their decision-making processes. He served with distinction on the National Academies' Committee on International Security and Arms Control (CISAC), heading its efforts in biology.

Altogether, I know of no eminent scientist who produced so much serious analysis of public policy and social problems, giving wise advice and stimulating new lines of inquiry. Our country and the world are in his debt.

Those of us here today profoundly appreciate what he did for humanity. His life exemplified the finest attributes of the great institution in which we meet today, and we honor his magnificent legacy.

JOSH LEDERBERG REMEMBERED

Stephen S. Morse, Ph.D.[2]
Columbia University

Josh would have loved this meeting. He loved this institution. He loved the Forum on Microbial Threats and the efforts that preceded it.

I keep looking around the room thinking that Josh has got to be here somewhere. He is, in a very real sense. It was Ralph Waldo Emerson, I think, who said that an institution is the lengthened shadow of a man—well, in those days he would have said "man," but today we would say "person." In many ways, Josh's shadow was a very long one, indeed. I think we are all very much in his debt.

It's especially humbling to follow David Hamburg and to be in a room where many of the people—and I see several here—knew Josh far longer and far better than I did. During the discussion period, I hope they'll add their own thoughts, which are sure to be very valuable and instructive.

I said it's humbling to follow David Hamburg on the podium. Let me give a small anecdote to illustrate what I mean. After Josh retired as president, the Rockefeller University gave him an office suite and lab, of course. His outer office was basically a library—this was very much in Josh's character—with rows and rows of files, books, and journals on just about every subject you can imagine. That was the outer office. His ever-loyal administrative assistant, Mary Jane Zimmermann—some people referred to her as Josh's gatekeeper during his days at the Rockefeller University, but personally I always found her benevolent, and very considerate—had a desk there as well, in this library-like outer office, which was not quite the size of this meeting room.

He had several of his many awards displayed next to the door in this outer office, but when you went into his private inner office, he had only three things on the wall, as I recall. He had a certificate as a ham radio operator (apparently he was very proud of that) and his certificate as a fellow of the American Academy of Microbiology—and a picture of David Hamburg! All the other things were in the outer office, but this showed what Josh kept close to his heart.

Everybody has spoken, of course, of Josh's unique and indisputable greatness as a scientist and his interests in many areas that I think we can only touch on. He began or pioneered in many fields. Those of us who worry about emerging infections in this world, and feel that's really challenging, came to realize how far beyond even that Josh's purview extended. David Hamburg mentioned Josh's starting the field of exobiology, a term that he himself coined. There are even many people who think (although I haven't gone back to verify this) that the hero in *The Andromeda Strain* (Crichton, 1969) was based on Josh Lederberg.

---

[2]Professor of epidemiology and founding director of the Center for Public Health Preparedness at the Mailman School of Public Health.

It wouldn't surprise me. In any case, some years ago, the National Aeronautics and Space Administration (NASA) had asked him for advice on how to properly decontaminate returning spaceships and samples sent back from space, on what precautions should be taken. As you know, he always gave very generously of his time and advice. This led to one of the most interesting job descriptions I have ever seen. After receiving Josh's advice, NASA created a position called "planetary quarantine officer." I always thought that was quite impressive, rather like the film *Men in Black*. Apparently, however, unlike in the film, they were fortunately never called upon to exercise their functions.

Josh's interest in evolution, of course, has been mentioned many times. On one occasion, Josh mentioned to me that he saw the unifying theme of his science: the sources of genetic diversity (and natural selection, I'd add). I think this was apparent in many ways. It was apparent in his work in microbiology, but it was also apparent in his interest in immunology, as David mentioned in passing. Josh went down to Australia, where he met Mac Burnet, Sir Frank Macfarlane Burnet, later to win the Nobel Prize for his work on "clonal selection," which we now know is how the immune system is able to recognize and respond to the great variety of molecules that it does. The developing immune system generates a great number of cells with different, essentially random specificities and then selects from among them and maintains these populations of cells, the "immunologic repertoire." When a new antigen is presented, immune cells can bind to the antigen attach and are stimulated to replicate, hence, "clonal selection." It is basically a Darwinian system that selects from among a large number of variant cells. That idea of clonal selection, Josh told me, was actually a direct application of the evolutionary ideas that Josh brought with him and worked on when he was in Australia.

So his shadow—indeed, his presence—can be found in many places, and no place, of course, more than in the area of infectious diseases. That's why I think this particular meeting would have made him very happy, to see so many of his old friends, and particularly to see so many of the fruits of his hard work. I think all of us—and this certainly applied to Josh—do the things we do in the hope of leaving the world a better place and leaving something that will inspire future generations to keep improving the world. So this meeting, with scientists of several generations describing their work that was started by some of Josh's interests, is very much a testament to Josh's legacy.

Unlike David, I had the pleasure of knowing Josh for only a little more than 20 years. When I came to Rockefeller, Josh was the *minence grise* (a role he carried as well as he had his earlier one of child prodigy, becoming a Nobel Laureate at age 33), the president of Rockefeller, and the distinguished Nobel laureate; and I was among the most junior of junior faculty members. (I eventually worked myself up to being a more senior junior faculty member.)

It was actually just by a happy coincidence that Josh and I got involved in this issue of emerging infections. I went to a faculty party that was given periodically at the president's house. Just as I was leaving, Josh's wife, Marguerite, who is also a psychiatrist—maybe it's just a coincidence, but now that I know from David Hamburg's background, I have a feeling Josh had a special affinity for psychiatrists—reminded Josh about something he had wanted to do. She said, "Sweetheart, didn't you have some questions about virology? Steve's a virologist, you know."

Josh said, "Oh, yes." It turns out that he had had dinner with Carleton Gajdusek. Many of you may remember Gajdusek (and, sadly, his later legal problems), but he was also a very innovative and brilliant scientist himself, with many interesting ideas. He was very interested in the hemorrhagic fever viruses, such as the hantaviruses, and discovered Prospect Hill virus, the first American hantavirus. At that dinner, he was talking with Josh and suggested he should think about the researchers and workers in the university's animal facilities, who might be exposed to a hantavirus such as Seoul or Hantaan (once known as Korean hemorrhagic fever), which had been a known problem. There were schoolchildren in Russia who had contracted a hantavirus from laboratory rats while touring the animal facilities on a school trip. Obviously, Carleton, in his usual forceful manner, had succeeded in getting Josh concerned about it.

So Josh asked me that evening if this was something we should worry about. I replied, "I'll look into it."

So, of course, I went and looked into it. It turned out that it wasn't a problem for us, I was relieved to find. Not only did we not have any cases of disease, but all our rodents were routinely tested. I wrote my reply to Josh's question in a letter dated February 17, 1988, saying "I enjoyed our conversation about Korean hemorrhagic fever and other emerging viruses," thinking of those viruses and mechanisms of pathogenesis not yet identified in humans but known to exist in other species.

Josh wrote back quickly, on his personal notepaper—and here I must digress for a moment. Everybody who has received a note from Josh knows these are not to be compared with Donald Rumsfeld's now famous "snowflakes": Josh's were much more substantial. I know it's a digression, but Josh's wonderful notes deserve a digression. All of Josh's colleagues and friends know that Josh had a personal notepad with his name in light blue at the top, and that the notes were always date-stamped. There were also some markings, like hieroglyphics at the top or bottom—a check mark with two dots, or an "x" with three dots; Mary Jane once sent me a chart that explained these meant things like, "Keep copy in files" or "Send a copy and retain original." I don't know if he had that habit at Stanford.

To return to the narrative: Josh wrote me a note in his usual magisterial style, date-stamped February 22, which said "Thank you for the information, which I read with great interest. I am of course reassured. . . . We need some high-level policy attention to what needs to be done globally to deal with the threat of emerging viruses, and I would welcome your thoughts on that."

Naturally, not knowing any better, and only knowing Josh slightly at that time, I took that as a call to action. Shortly thereafter, at a Federation of American Societies for Experimental Biology (FASEB) meeting, I ran into Gaylen Bradley, a former postdoc of Josh's from his Wisconsin days, who had also been my department head when I was a postdoc (he has recently written his own biographical memoir of Josh). I asked Gaylen for advice on how to respond to this oracular statement. The obvious conclusion was to have some sort of conference to deal with this question of emerging viruses.

Some colleagues (I recall particularly Sheldon Cohen, now retired from the NIH) sent me to John LaMontagne at the National Institute of Allergy and Infectious Diseases (NIAID), who was very sympathetic and said he had similar interests. We organized a conference under NIAID's auspices, held on May 1, 1989, at the Hotel Washington (Washington, DC). We could afford it then because it was undergoing major renovations, as everybody who ever stayed there during that period for the conference knew, since they could hear the renovations going on. We got a very good rate. I know that because I never could afford to stay there afterwards.

In this big ballroom, we had perhaps 150 people, with a number of distinguished speakers on various subjects (and an equally distinguished audience). Of course, Josh was very much the star of the meeting. He opened it with a keynote address and participated in discussions at the end of the meeting. There was a summary of that meeting, for those who are interested, in the *Journal of Infectious Diseases* in 1990, and then in my book, *Emerging Viruses*, which was sort of a by-product of that meeting.

Josh gave very thoughtful and philosophical opening remarks, of course. One thing about Josh that never failed to surprise me was that he would say things that were truly gems, often profound, and they wouldn't strike you until days later, when suddenly you realized what he meant by that. It was an "aha" experience, in many ways like the joy of a scientific discovery.

I always enjoyed seeing the reactions of people having this experience for the first time. One year, I was fortunate enough to have him address my Columbia graduate class in emerging infectious diseases, as the grand finale for the semester. He talked about the toll of the 1918 influenza pandemic, its effect on life expectancy curves, and many other things. By then, I was familiar with Josh's often very philosophical and discursive style. The students listened to Josh and mostly looked very pensive. I suspect that most of the students were probably mystified by parts of his talk, but many were stimulated days, weeks, or even months later, when one of his comments hit them, and they were inspired to take some of those thoughts and pursue them.

Josh was very good at inspiring people. He had a special gift for that. In terms of mentorship, he cared deeply about the people he worked with. He was passionate about the many issues with which he was concerned, none more than the threat of microbes, perhaps, or as he summarized the situation in an article,

"Our Wits Versus Their Genes" (Lederberg, 2000). Their genes have been evolving a lot longer than our wits, I needn't tell you. In another paper he made the analogy to bacteriophage infecting a dense culture of bacteria in a tube of broth and how suddenly—and this was a classic observation—the tube became clear. That was in a *Journal of the American Medical Association* (*JAMA*) article that he wrote, in which he used the term "humankind" in the title. (Josh was not a sexist.)

Ruth Bulger, who was then director of the Board on Health Policy, and Polly Harrison, who was director of the Board on Global Health at the Institute of Medicine (IOM), came along to the 1989 meeting, and we had several discussions together. This helped to galvanize the IOM into doing a study that Josh had been advocating for some time. The study committee, which was originally the Committee on Microbial Threats to Health but which was rather quickly renamed the Committee on Emerging Microbial Threats to Health in the United States, eventually authored the famous report, *Emerging Infections: Microbial Threats to Health in the United States* that came out in October 1992. Several of you who are here today were on that Committee. As you know, the report has become a classic and, I am told, one of the IOM's best-sellers of all time. By the way, Richard Preston had an article in the *New Yorker* that was timed to coincide with the release of the report. The article was later expanded into the book *The Hot Zone* (Preston, 1995). More recently, Peggy Hamburg and Josh co-chaired a 10-year reappraisal, the report of which I think is destined to become another classic.

The report called for better infectious disease surveillance, a better understanding of pathogenesis, and a better understanding, in fact, of many, many things, including the political will to deal with emerging infections.

Science was one of Josh's real passions. As David Hamburg pointed out, no matter how sick Josh may have been in his later days, whenever the talk turned to science, he was all ears. His eyes would light up and he would be eager to take in all of that knowledge—and, of course, ask probing and often very informative questions. Josh had a knack for putting together words in wonderful ways and a knack for asking the right questions—often very profound questions. I think it was absolutely remarkable the way he combined those two talents. I'll give an example or two later.

He also had a passion for scientific advice and scientific policy, for which he gave of himself selflessly. I would always bump into him on the Delta shuttle or at a meeting such as this one, or many others, and he was always shuttling back and forth between New York and Washington. I knew he went many times to Washington. However, it wasn't until an eightieth birthday party for Josh that Richard Danzig and other friends organized at the Academy building that I realized—in fact, Marguerite told us—that Josh used to go to Washington sometimes three times a week, back and forth, to give scientific advice. He was the very model of the perfect scientific adviser. His advice was honest, dispassionate, and never self-interested. His interest was furthering the cause of science and humanity. He was always the soul of discretion as well. I think that policy and technical advice were things that those of us of a certain period—Josh's period, certainly—felt was a civic obligation. More and more, this has become a highly politicized process, but Josh could always be depended upon to give honest advice and ask good questions.

That belated eightieth birthday party was, I believe, the penultimate time he went to Washington. The last trip to Washington was when he went to pick up the Presidential Medal of Freedom (which, I recently discovered, David had also received earlier). Josh was deservedly very proud of that recognition. He had earned it.

I mentioned Josh's unique way with words. As I said earlier, I used to see him often at various meetings. Once we were invited to a World Health Organization (WHO) meeting, and bumped into each other before the meeting at a hotel that the WHO then used quite regularly, the Cornavin—some of you may know it—right by the train station in Geneva. I had just registered and Josh walked in, shook my hand, and said, "My, my, we always meet in the most expected places"—just one small example.

Back at that 1989 conference on emerging viruses, Josh was also a star of the show. There were several other Nobel laureates there, including my old friend and former professor Howard Temin. Josh and Howard had a very interesting debate, which unfortunately was not officially recorded, but as I recall, it certainly induced a lot of adrenaline. Later, somebody asked Josh, "When should we declare that a newly recognized virus is a new species?" He said, "When it matters." I quoted this to my wife, who was duly impressed, and said, "What a Solomonic answer!"

That was very much Josh's way; to cut through all the red tape and all the inconsistencies and see straight to the heart of the matter, to distinguish what was really important and what was not.

Before I close (and I fear I've already taken up more than my space quota), I think I should say a few words about the early history of the Forum on Microbial Threats, or, as it was then known, the Forum on Emerging Infections. A number of you in the room probably already know this history.

Of course, it started just the way that David showed in that slide of the Sistine Chapel ceiling. However, the Institute of Medicine did not at that time have quite so palatial surroundings. After the Committee on Microbial Threats and publication of its final report, many of us thought about what a possible follow-on could be. It's often said that American lives have no second act. Certainly, the report was a very hard act to follow, but there was recognition of a need to continue the momentum and advance the dialogue. After much deliberation, which included Josh and the then-president of the Institute of Medicine, Sam Thier, who was a great supporter of this effort, Polly Harrison, Ruth Bulger, as well as myself, Polly and Ruth suggested that it would be appropriate to start a forum that could bring together people from—I won't say all walks of life, but from academe, industry, and from the government, to talk about these issues. As you know, Josh was delighted to chair this.

The very first issue that the Forum on Emerging Infections discussed was something very close to Josh's heart: vaccine capacity for microbial threats. It led to our first report, *Orphans and Incentives*, which set out the problem and suggested some alternatives.

The rest of what happened after that, of course, is history. It was very much Josh's energy that made it possible and remains a key part of Josh's legacy.

A second thing that happened subsequent to publication of the report was that several of us who were concerned about the international ramifications of emerging infections decided to start the Program for Monitoring Emerging Diseases (ProMED) to plan and promote global surveillance of infectious diseases, especially emerging pathogens. In fact, it was the late Bob Shope, who was co-chair of that original IOM committee with Josh, who came up with that name just off the top of his head. Jim Hughes, Ruth Berkelman, and D. A. Henderson, along with a number of others, were charter members of the steering committee.

One of the most successful spin-offs of the ProMED initiative is well known to those of you who get the e-mails from ProMED-mail or read its website.[3]

Josh was never officially a member, because I thought it might be a little too political, and I didn't want to put him in an awkward position. I always kept him in the loop unofficially, and he was a great supporter of the effort, later publicly as well as privately. But I can't help thinking that part of the reason that Josh was such a great fan of ProMED-mail may have been that it was an e-mail system, and Josh was glad to see e-mail used to bring people together for a crucial and worthy purpose.

In fact, Josh was one of the earliest adopters of e-mail I know of. In those days, e-mail was almost impossible to use. You had to do all the formatting and editing of the message by hand, line by line, and send it using a 1,200 baud modem by dial-up. We had nothing more than that. I remember how technologically advanced I felt when I finally got a 2,400 baud model.

So I had not learned to use e-mail because it required such a lot of effort and there was a steep learning curve. Josh once looked at me and said, "You really should use e-mail, you know." I replied it was just too much trouble, adding "I don't even have a modem." He shamed me into it. He said, in his very typical fashion, "That's no problem. I'll *buy* you a modem."

I did have enough grant money at the time to buy myself a modem. It was from that inspiration that, in fact, ProMED-mail was later born. So Josh really can take the credit for starting many things, including that initiative.

I will add, in closing, that Josh served very happily as president of the Rockefeller University. The trustees loved him. He was one of their real favorites. I know this because I did a trustees' dinner with him on emerging infections. Of

---

[3]See www.promedmail.org.

course, he was the star of the show, and I was sort of the appendage. What a star to be the appendage for! I owe a great deal to Josh in many other ways as well. My time at the Defense Advanced Research Projects Agency (DARPA) from 1996 to 2000 was because Josh convinced the people at DARPA (especially its director at the time, Larry Lynn) that it was necessary to get into biology and seriously consider biological threats. He asked me if I would be interested in his nominating me for a job there. It was one of the most interesting chapters of my own career and, I have to say, an exceptional place to work that was committed to finding creative new ideas. I hope we managed to fund and stimulate a few. (David Relman was one of the grantees, for work on gene expression profiling in infections.)

At the time one of our concepts was to look at common pathways of patho-genesis (Stan Falkow will remember his invaluable advice on this), as well as the host response and possible host markers of infection. The rationale was fairly simple: there was a tremendous number of pathogens, in addition to whatever was lurking unrecognized out there in nature, plus the possibility of genetically engi-neered threats in the future. Approaching the threats individually (what some of my colleagues called "one bug—one drug") would become impossible. Later, this idea would be embodied in Homeland Security Presidential Directive (HSPD)-18 and other current biodefense initiatives.

We really owe all these ideas to Josh's vision in making us all think much more globally.

At the Rockefeller University, he was, as I mentioned, very influential as president, although after he stepped down I saw him on campus looking very relaxed and wearing a Rockefeller baseball cap. I need not tell you that his office as president had floor-to-ceiling bookshelves crammed with books, but there was another room down the hall in the same building. Those of you who know the Rockefeller will know it as the Cohn Library. It was in a public area and was sometimes used as a conference room. It, too, was filled with books.

One day I was waiting there for a meeting to start and idly began browsing some of the books on the shelves. I discovered that many of them were stamped with Josh's name. He had donated them to the library.

After he stepped down as president and had his own office, in deference to his many interests and skills—he could not include them all—he named his laboratory the Laboratory of Molecular Genetics and Bioinformatics, emphasiz-ing the relationship of the two. I think that was the first time those terms had been coupled, or at least the first time I had seen them together. He had always been a great supporter of both, as well as a great innovator in both of these fields.

I hope this brief account gives some indication not only of how much a poly-math he was but also how deeply he cared about people and science. I remember talking later with Torsten Wiesel, another Nobel laureate, after he became presi-dent of Rockefeller. He said, "You know, Josh was lucky. He got his Nobel Prize early so he could spend the rest of his life doing what he wanted."

What Josh wanted to do was to search for truth and inspire others in that search, for the benefit of humankind. He was never happier than when he was absorbing knowledge and questioning it. I like to think of this inspiration, with all of us here because of Josh, as his greatest legacy.

THE LIFE AND IMPACT OF A LEGEND—JOSHUA LEDERBERG

Adel Mahmoud, M.D., Ph.D.[4]
Princeton University

It is fitting to dedicate this workshop of the Forum on Microbial Threats (hereinafter, the Forum) to Joshua Lederberg and to a subject that was central to his thinking of the past decade.

This Forum is the brainchild of Joshua as he was exploring how to respond to the multifaceted challenges of microbes. Those of us old enough will remember that the Forum was Joshua's response to the first *Emerging Infections* report of the Institute of Medicine (IOM) in 1992. Joshua opened the preface to that volume as follows:

> As the human immunodeficiency virus (HIV) disease pandemic surely should have taught us, in the context of infectious diseases, there is nowhere in the world from which we are remote and no one from whom we are disconnected. Consequently, some infectious diseases that now affect people in other parts of the world represent potential threats to the United States because of global interdependence, modern transportation, trade, and changing social and cultural patterns.

He felt as if he always guided us; that the issues of infectious disease at hand will not go away and that a platform for academia, government, industry, and others to study, debate, and chart a path forward was necessary. It was my privilege and honor to have succeeded Josh as chair of the Forum.

I am not here today to talk about the Forum, and I will only touch briefly on the subject matter of our meeting. Rather, I am participating in a celebration of the life and achievements of Dr. Lederberg. One can simply state that Joshua Lederberg has been the dominant force that shaped our thinking, response, and intellectual understanding of microbes for much of the second half of the twentieth century.

Infectious Disease Research

From his earliest work when, at the age of just 20, he discovered mating and genetic recombination in *Escherichia coli*, to the discovery of viral transduction

[4]Woodrow Wilson School of Public and International Affairs and the Department of Molecular Biology, Lewis Thomas Laboratory, Room 228, Princeton, NJ 08544.

in bacteria, Joshua Lederberg helped to establish the new science of genetic engineering and its fundamental contribution to the study of infectious disease. There is a lot to share with you about these early days of Josh's career. Most of the fundamental breakthrough research on bacterial mating was performed while he was on leave from Columbia University's College of Physicians and Surgeons. When the great results of his early studies were clearly opening new horizons, he decided to extend his leave of absence from medical school for another year. That was where we lost the budding physician in Joshua; during the subsequent year, he was offered a faculty position at the University of Wisconsin, where he conducted his seminal work on viral transduction in bacteria. But that did not stop Joshua Lederberg from being at the forefront of those concerned about human health and well-being as we witnessed his leadership over the past several decades.

The "Stanford years" witnessed the maturation of the field of bacterial genetics and the expansion of Joshua's scientific horizon to areas that touched upon human health and human biology. Equally important was his role in under-graduate and graduate education. The human biology curriculum was one distinct product of that era that Joshua championed.

Joshua, the Leader

Being awarded the Nobel Prize at the age of 33 gave Joshua a global per-spective that he fully utilized in the subsequent half century. His platform became the nation and the world, and his reach and impact touched every corner. A few examples illustrate this point.

1. *Joshua the science educator:* From presidents to the global public, Joshua consulted and advised every president since John F. Kennedy. He explored and studied issues ranging from space sciences to human and artificial intelligence to the human-microbial interplay.
2. *Joshua the communicator:* Few may still remember that Joshua contrib-uted a weekly science column to *The Washington Post* in an attempt to reach the general public and make science accessible.
3. *Joshua the visionary:* Equally important were his vision and his ability to conceptualize for the nation and the scientific community. I was fortunate to have worked closely with Josh on the subject that it was he and he alone who articulated and brought to the forefront of scientific agendas (i.e., emerging infections). In the late 1980s, the concept of emerging and reemerging infections was born by Joshua's insistence that it get attention. The result is 25 years of focus that gave us two IOM reports (IOM, 1992, 2003), multiple plans from the Centers for Disease Control and Preven-tion (CDC), NIH, WHO, and our Forum. More importantly, global events proved how right and perceptive Joshua was then and now. Infections such

as hantavirus pulmonary syndrome, severe acute respiratory syndrome (SARS), H5N1 avian influenza, and multidrug-resistant *Staphylococcus aureus* (MRSA) are all global events that shape the world today and will shape it for years to come.

Perhaps, the most dominant feature of Joshua's scientific contributions over the last three decades is articulating a vision for understanding human-microbe interplay with the dimensions that for many years remained fragmentary. An ecological and evolutionary understanding of that relationship emerged in multiple publications and produced several revolutionary concepts. The introductory statement of an article written by Joshua a decade ago sums it up (Lederberg, 1998):

> Our relationship to infectious pathogens is part of an evolutionary drama. Here we are; here are the bugs. They are looking for food; we are their meat. How do we compete? They reproduce so quickly, and there are so many of them. They tolerate vast fluctuations of population size as part of their natural history; a fluctuation of 1 percent in our population size is a major catastrophe. Microbes have enormous potential mechanisms of genetic diversity. We are different from them in every respect. Their numbers, rapid fluctuations, and amenability to genetic change give them tools for adaptation that far outpace what we can generate on any short-term basis.

> So why are we still here? With very rare exceptions, our microbial adversaries have a shared interest in our survival. With very few exceptions (none among the viruses, a few among the bacteria, perhaps the clostridial spore-forming toxin producers), almost any pathogen reaches a dead end when its host is dead. Truly severe host-pathogen interactions historically have resulted in elimination of both species. We are the contingent survivors of such encounters because of this shared interest.

In a subsequent masterpiece published in the journal *Science* in 2000, Joshua added the elements of a rational understanding of what we are as superorganisms and that the microbiota constitute the total interface between humans and microbes (Lederberg, 2000). That article also points to the futility of the "war metaphor" and proposes a more fundamental, nuanced, approach:

> As our awareness of the microbial environment has intensified, important questions have emerged. What puts us at risk? What precautions can and should we be taking? Are we more or less vulnerable to infectious agents today than in the past? What are the origins of pathogenesis? And how can we use deeper knowledge to develop better medical and public health strategies? Conversely, how much more can the natural history of disease teach us about fundamental biological and evolutionary mechanisms?

> An axiomatic starting point for further progress is the simple recognition that humans, animals, plants, and microbes are cohabitants of the planet. That leads

to refined questions that focus on the origin and dynamics of instabilities within this context of cohabitation. These instabilities arise from two main sources loosely definable as ecological and evolutionary.

Ecological instabilities arise from the ways we alter the physical and biological environment, the microbial and animal tenants (humans included) of these environments, and our interactions (including hygienic and therapeutic interventions) with the parasites. The future of humanity and microbes likely will unfold as episodes of a suspense thriller that could be titled *Our Wits Versus Their Genes.*

His mind never stopped analyzing and deeply exploring what comes next. An example of this comes from his article "Metaphysical Games: An Imaginary Lecture on Crafting Earth's Biological Future" (Lederberg, 2005):

A vector of traits can be plucked from natural sources, or constructed with present or proximate future bioengineering tools [see Box 1-1]. Taken without question is the ultimate capacity to craft altered and hybrid genomes—"knockouts," "knockins," "knockdowns and knockups," and "shuffles." Augmenting these constructions is the deconstruction of what is disappointing, what does not work as predicted. Mock or denounce, but get the arguments into the open for the feasibility and utility of constructing or domesticating a target candidate. At some point, there will be exhilaration or unease about the policy fallout, and colleagues will be consulted about the best avenues for technology assessment.

A Perspective Summation

Joshua was a discoverer, a leader, and a champion in the truest and best sense of the words. His perceptiveness and impact had a fully positive effect on mankind and on history. Joshua was truly a force of nature, a force of nature that was able to unlock some of nature's most enduring secrets. Joshua believed that there are no limits to what the human mind can accomplish—especially when its power is hitched to a willingness to think boldly and unconventionally and to hard work.

Until almost the day he died, Joshua could be found in his office, in his apartment, working. His mind was always thinking, always probing, always questioning. He has always been and will always be an inspiration to generations among us and generations to come.

Thinking back to my own upbringing in Egypt, the constructions of monuments such as the Great Sphinx and the pyramids at Giza likely grew out of the enormous egos of the rulers who had them constructed. These rulers wanted to be sure that long after they were gone, people would be able to gaze upon their mighty works and remember that a great man once ruled here.

Joshua Lederberg, of course, needs no such monuments to ensure that his life and work are long remembered, because, in a very real sense, his accomplish-

---

BOX 1-1
Traits That Could Be the Foundation of
Selection in Unfamiliar Genomic Settings

Fecundity: This is the fundamental measure of Malthusian fitness. Is obviously the most complex of traits, and rarely given fully unhampered play except in natural or near-natural populations. It may also be entangled with efficiencies of diet, infection, and social arrangements. The race is not always to the swift, paradoxical selection sometimes favors a slower growing contender, with antibiotics that are more effective on bugs that have lowered their guard during the most rapid growth. What is the fastest-growing microbe? Perhaps a cousin of *Vibrio natriegens*, with a doubling time of less than 15 minutes.

Life span: Research in this part of the matrix is well filled in yeast, roundworms, fruit flies, mice, birds—for these species constitute a large part of the research agenda of the National Institute of Aging.

Chemical secretion, defense, detection, virulence, or disease susceptibility: What substance? What toxin? What function? Recall that secondary metabolism comprises a large part of contemporary applied microbiology.

Desiccation: Survival and growth capacity of cells at reduced chemical activity of water. This is closely connected with anabiosis and dormancy with explicit or tacit spore formation. Large investments by the National Aeronautics and Space Administration are wrapped up with this issue because it is generally argued that the surface environment of Mars is incompatible with proliferation of terrestrial microbes. Depending on process details, the process of lyophilization may be toxic or preservative. Additives like trehalose mitigate the toxicity of the lyophilization process and may be emulated by other genetically engineered antifreeze under extensive investigation. Persistence of viable bacteria in aerosols is a critical point for a biological weapons attack.

Motor: Power and discriminatory skills; motility some specialized cells. Some flagella are imputed to rotate at 50,000 rpm!

---

ments are embedded in the DNA of many whose lives have been shaped because of his work. That work and those concepts will be passed on to every generation yet to come, long after the Great Sphinx has crumbled into dust.

Joshua believed very strongly in the work of this Forum. He had a great confidence in the ability of scientists and researchers to continue to solve some of the riddles that still confront science in the fight against infectious diseases.

Flight, marine, navigation kinetic fragility/tenacity: One aspect testable with acoustic energy in environment. Converse, more limited range in animals but what is loudest emission, doubtless related to animal size. For smaller targets this is tested by aerosolization.

Thermal-susceptibility: Easily assayed, and heat/cold shock responses are under study.

Thermal emission and absorption: Should be secondary to metabolic rates.

Tropisms: To any part of *EM [electromagnetic] spectrum*: what are ultimate limits to sensitivity of detection?

Vision-spatial acuity: Image formation at any wavelength.

Energy emissions: May carry acoustic signals. Luminescence may be acoustically modulated, detected.

Integration of signal inputs—IQ: It involves coordinated social action. As an outrageous question, "What could be measured, say in a bacterium, that could map to what might be called 'intelligence'?"

Geno-stability and pheno-stability (or instability); promiscuity: This trait may be related to survivability of laboratory cultures, to maintenance of strain characteristics, to production of variants as objects of selection, and to ease of research protocols.

Genome size: An interesting trait in itself; related to "junk DNA."

Currency of lateral gene transfer: Bears on management of genetic change.

By remembering him with this tribute we also are remembering the many things that his life and career can teach all of us.

I hope, every time we meet at this Forum, Joshua Lederberg will be an inspiration and a reminder that our work can truly change the world, just as the life and career of Joshua Lederberg certainly did.

REFERENCES

Morse References

Crichton, M. 1969. *The andromeda strain*. New York: Knopf.
Lederberg, J. 2000. Infectious history. *Science* 288(5464):287-293.
Preston, R. 1995. *The hot zone*. First Anchor Books Ed. New York: Random House, Inc.

Mahmoud References

IOM (Institute of Medicine). 1992. *Emerging infections: microbial threats to health in the United States*. Washington, DC: National Academy Press.
———. 2003. *Microbial threats to health: emergence, detection and response*. Washington, DC: The National Academies Press.
Lederberg, J. 1998. Emerging infections: an evolutionary perspective. *Emerging Infectious Diseases* 4(3):366-371.
———. 2000. Infectious history. *Science* 288(5464):287-293.
———. 2005. Metaphysical games: an imaginary lecture on crafting earth's biological future. *Journal of the American Medical Association* 294(11):1415-1417.

2

Microbial Ecology and Ecosystems

OVERVIEW

In the spirit of Joshua Lederberg's advocacy for examining host-microbe relationships from an ecological perspective, this chapter depicts a variety of host-microbe-environment interactions as dynamic equilibria. These include the range of microbial communities that comprise the human microbiota, the taxonomically simple but genetically complex microbial communities known as biofilms, and examples of symbiotic relationships between bacteria and eukaryotes that enable plants to acquire nutrients through their roots and allow the Hawaiian bobtail squid to elude aquatic predators.

In the chapter's first paper, speaker David Relman of Stanford University describes efforts under way in his laboratory to understand the role of indigenous microbial communities associated with human health, disease, and the transition between these states. This research is currently focused on identifying elements of microbial communities that can be monitored and measured to assess physical and metabolic interactions within and among microbial communities, and between human and microbial cells.

From the RNA sequences of the microbiome (a term coined by Lederberg to encompass the collective genome of organisms living in and on the human body), Relman and colleagues infer the diversity of organisms present in various indigenous microbial communities and compare patterns of ancestry and relatedness among these communities, between individual humans, and with those of microbial communities inhabiting external environments. They also consider these patterns in light of the role of indigenous microbial communities in disorders such as inflammatory bowel disease, antibiotic-associated diarrhea, bacterial vaginosis, and premature labor and delivery.

"In many ways, the human microbiome remains *terra incognita*," Relman concludes; however, he adds, knowledge of the patterns of microbial diversity within the human body are leading to a better understanding of the spectrum of relationships between and among ourselves and the microbes that live on and in us.

Focusing on interactions within communities of microbes, speaker Jill Banfield, of the University of California, Berkeley, described research conducted in her laboratory to elucidate the structure, function, and development of biofilms, and on the role that viruses play in these communities. These inquiries, she said, pose the following questions: "How do activities of organisms change as microbial communities establish? Do different developmental stages select for different genotypes?"

Biofilms, microbial communities of relatively few types of organisms that establish themselves on nonliving substrates, serve as model systems for understanding how microbial communities organize themselves and how their members interact with each other and their physical surroundings (Banfield, 2008a). Banfield's group studies biofilms comprised of iron- and sulfur-oxidizing microbes that grow in the extremely low pH environments of mines and in the watersheds where mine wastes drain (Banfield, 2008b). These acid mine drainage (AMD) biofilms are sustained by the oxidation of ferrous to ferric iron, Banfield explained: the ferrous iron derives from a distillation of pyrite (iron disulfide, or fool's gold), and it is oxidized to iron(III) aerobically by various members of the microbial community; the iron(III) then back-reacts with pyrite, driving further dissolution and regenerating the iron(II), which is the substrate for microbial growth.

Banfield and coworkers extracted DNA from several such biofilm communities and then fragmented, cloned, and sequenced it. Based on sequence homology, they assembled genome fragments into population genomic data sets that capture the natural, inherent heterogeneity present within these microbial communities, she said. Using this approach, the researchers have been able to reconstruct several nearly complete genomes from natural biofilm communities. This genomic information provided sufficient insights about the metabolism of some members of the community to permit them to be cultured for the first time.

Biofilms "grow"—that is, they add or accumulate increasingly large populations of microbes—in stages, beginning at a stream's margins and extending across the water's surface toward its center, while simultaneously increasing in thickness. Banfield's group compared the membership of early- and late-stage biofilms by using fluorescence *in situ* hybridization (FISH) to document the changing membership of the community as these biofilms establish. They observed a transition from a very early biofilm dominated by *Leptospirillum* group II to more complex communities with more members, including higher numbers of *Leptospirillum* group III and more members of the domain Archaea.

To explore how these organisms function within the community, and how community function changes over the course of its development, Banfield's group compared the protein profiles derived from 27 biofilms in various stages of development, and sampled from eight different microenvironments in the same iron mine (Denef et al., 2009). The researchers extracted the proteins from these communities, obtained their genome sequences, and conducted mass spectrometry experiments to identify the proteins as they were predicted to occur, based on sequences present in the genome. Abundant ribosomal proteins and stress-defense proteins were found in early-stage biofilms, suggesting that "building a biofilm is a good way to reduce the metabolic cost of stress associated with dealing with this environment," Banfield said. In late developmental stages, proteins associated with mobility, movement, and sensing and responding to gradients predominated; this is to be expected, as environmental gradients are created in the thickening biofilm. Thus, Banfield surmised, "the functionality of these organisms is changing as the communities develop."

The researchers further observed a correlation between developmental shifts in protein expression and the sequential domination of the community by two different closely related strains of *Leptospirillum* group II. This led them to recognize that "in order to understand the function of a biofilm community, we need to be able to distinguish proteins at the strain level," Banfield said, because "closely related proteins are presumably doing different things at different times in these communities." Moreover, she added, this result suggests that functional distinctions exist between closely related genotypes.

Further characterization revealed the presence of six genotypes among the 27 biofilm samples, each of which was created by mixing and matching blocks of sequence from the two closely related *Leptospirillum* group II strains (Lo et al., 2007). Many of the biofilm samples were found to contain only one genotype; others had several (Denef et al., 2009). The researchers also examined the distribution of genotypes across the eight sampling sites (Figure WO-3). At one site over the course of more than two years, they consistently found the same genotype—despite the fact that biofilms at this site would have had constant exposure to other genotypes. Thus, Banfield concluded, there appeared to be strong local selection for this particular genotype, which has "achieved a fine level of adaptation to environmental opportunity."

Turning to the role of viruses in biofilms, Banfield noted that the host specificity of viruses, coupled with their generally negative host effects, suggests that viruses profoundly and dynamically shape the membership of microbial communities. Focusing on the viral "predators" of the dominant bacterial species in biofilms, her laboratory was inspired by recent reports (Makarova et al., 2006; Mojica et al., 2005) that the genomes of most bacteria and Archaea contain repeat regions, known as clustered regularly interspaced short palindromic repeats (CRISPRs). Derived from coexisting viruses, CRISPRs appear to provide immunity (perhaps via RNA interference) to their possessors from the virus of its derivation.

"A microbe is immune to a virus, so long as it has the spacers that match it or silence it," Banfield explained. "But should a mutation occur such that the spacer is no longer effective, the virus may proliferate and the microbe will suffer." Another component of the bacterial system, CRISPR-associated proteins, rapidly sample the local viral DNA and incorporate new spacers, conferring the population with a range of immunity levels to different mutant viruses as they arise (Tyson and Banfield, 2008).

Studies in Banfield's laboratory took advantage of two components of this model. First, they could be certain that a host possessing spacers from a given viral genome must be targeted by that virus. Second, they used spacer sequences to identify viral fragments in community genomic or metagenomic data sets. To do this, the researchers extracted the spacers from the CRISPR locus and used them as hooks to retrieve viral fragments from genomic DNA, which they then assembled into viral genomes. This approach has proved to be very effective, allowing Banfield's group to reconstruct many genomes of viruses that target the bacteria and Archaea present in AMD biofilms (Andersson and Banfield, 2008; Figure WO-4).

Based on the viral sequence information obtained this way, the researchers discovered that microbes within the AMD biofilm are targeted by viruses that maintain high levels of heterogeneity. "The virus maintains a high population diversity so the host immune system cannot silence it," Banfield explained. "This would, in itself, make the task of the host, the CRISPR loci, rather difficult." However, she added, there is also considerable diversity in the host genome, where they found "incredible heterogeneity, to the point that we would deduce that almost no two cells within a microbial community are the same." Thus, while a "cloud of viruses" maintains high levels of sequence diversity by various means in order to defeat host microbes within the biofilm, the viruses are countered by the rapid acquisition of new viral spacers by the microbes. Overall, Banfield said, this dynamic system is probably in stasis; nevertheless, she added, "it's clearly an example of coevolution in a virus and host community."

In relationships somewhat analogous to those that exist between mammals and their gastrointestinal microbiota, plants establish mutualistic associations with several microorganisms. These plant-microbe partnerships were the subject of a workshop presentation by Jean-Michel Ané of the University of Wisconsin, who described the signaling pathways that make possible mutually beneficial plant-root symbioses, as well as how microbial "cheaters" and parasites take advantage of these pathways (and how, sometimes, the plants fight back).

The roots of most higher plant species form arbuscular mycorrhizae, an association with fungi of the order Glomeromycota. These associations are not very host specific: many fungal species can infect one plant, and one type of fungus can affect many plants. Arbuscular mycorrhization significantly improves the plant's ability to acquire phosphorus, nitrogen, and water from the soil (Brelles-Mariño and Ané, 2008). This ancient relationship is generally considered to have permitted plants to colonize land, Ané noted; many genes known to play a role in mycorrhizal associations also influence plant development and especially affect root development.

For the past 60 million years, leguminous plants and nitrogen-fixing bacteria named rhizobia have engaged in a highly specific symbiosis. The bacteria induce and colonize new organs, called nodules, on the roots of legumes; there, the microbes receive energy in the form of carbon from the plant and convert atmospheric nitrogen to ammonia for the plant's use. This partnership furnishes much of Earth's biologically available nitrogen and boosts the productivity of nonleguminous crops that are grown in rotation with legumes.

Symbiotic relationships between plants and bacteria or fungi are established through the exchange of chemical and genetic signals among the partners. As shown in Figure WO-6, legume roots release compounds that trigger nitrogen-fixing rhizobia to express modified chitin oligomers called Nod factors, which in turn facilitate infection of the root by the bacteria, as well as nodule development (Brelles-Mariño and Ané, 2008; Riely et al., 2006). This dialogue between legumes and rhizobia was first characterized more than 15 years ago.

It has been known for about five years that plants produce chemical signals called strigolactones that increase the branching of fungal hyphae, and thereby increase their contact with arbuscular mycorrhizal fungi. These fungi also release diffusible compounds known as Myc factors, which, when recognized by the plant, activate symbiosis-related genes. Ané noted that it remains to be seen whether strigolactones stimulate Myc factor production and the structure of Myc factors has yet to be determined. He added that it will also be important to characterize the genetics of mycorrhizal fungi, because their ability to form associations with nearly all land plants makes them an important model for engineering novel plant-microbe interactions.

In the meantime, the discovery that a largely shared signaling pathway makes possible both arbuscular mycorrhization and legume nodulation—despite their apparent differences—has led to the conclusion that plants have a single, highly conserved genetic program for recognizing beneficial microbes, according to Ané. Both microbial Nod and Myc factors also appear to have common features, including the ability to promote plant growth. This may benefit microbes by increasing the availability of infection sites, a notion supported by evidence that Nod factors have been shown to increase the yields of legumes such as soybeans in the field.

Symbiotic relationships—including plant-microbe associations—run the gamut from mutualism to parasitism, as depicted in Figure WO-5. The "cheaters" in this spectrum include rhizobial colonists of legume nodules that do not fix nitrogen efficiently, Ané observed. These microbes act as parasites, receiving carbohydrates without offering anything in return (and without expending the considerable energy involved in fixing nitrogen). However, their hosts appear to have ways of detecting these microbial freeloaders and preventing them from fixing nitrogen, perhaps by decreasing oxygen supplies to underperforming nodules (Kiers et al., 2003). Although the actual mechanism by which plants "sanction" these microbial cheaters remains unknown, Ané suspects that the plant may starve the cheaters by reducing their access to carbohydrates.

Parasites on plant roots include root-knot nematodes, nearly ubiquitous pathogens that account for up to 10 percent of global crop losses, according to Ané. Evidence suggests that these nematodes infect legume roots by using genetic pathways adapted for rhizobial colonization, perhaps by producing molecular mimics of Nod factors (Weerasinghe et al., 2005). Human pathogens, including *Salmonella* and *Escherichia coli* O157:H7, also take advantage of the symbiotic signaling pathway to colonize legume roots, such as alfalfa sprouts, that have been linked to several outbreaks of foodborne illness (Taormina et al., 1999). Characterizing the plant and microbe genes involved in these infections, and understanding how these pathogens override or constrain the plant's defenses against invading microbes, may reveal ways to prevent such outbreaks.

In the chapter's final paper, speaker Margaret McFall-Ngai of the University of Wisconsin explores the question, "What are the shared characteristics of pathogenic and mutualistic interactions between microbes and their animal hosts?" She pursues this by studying a model system: the association formed between the Hawaiian bobtail squid, *Euprymna scolopes*, and the gram-negative luminous bacterium *Vibrio fischeri*, which populates the squid's light organ. Incorporated in the squid's light organ, the bacterium emits a luminescence that resembles moonlight and starlight filtering through ocean waters, camouflaging the squid—a nocturnal animal—from predators.

This coevolved partnership commences when, within minutes of hatching, the squid begins to harvest the bacterial symbiont from the environment. McFall-Ngai describes the intricate mechanism by which the host recruits and selects its symbiont, a process that depends upon the exchange of molecular signals between the partners. This system provides evidence that such interspecies signaling, although typically associated with mediating bacterial pathogenesis of animal tissues, engages a common "language" of host-microbe interactions.

WAR AND PEACE: HUMANS AND THEIR MICROBIOME

David A. Relman, M.D.[1]
Stanford University

We have known for hundreds of years—beginning with Antonie van Leeuwenhoek's observations in the late seventeenth century of the morphological diversity of microbes in human plaque—that a complex microbiota exists within the human body. Subsequently, tools such as cultivation technology enabled scientists to understand better the diverse nature of these indigenous microbes. Over the last century, a number of important observations and inferences have been made about the human microbiota and the possible benefits that it confers upon us, which include the following:

- Vitamin acquisition (Cummings and Macfarlane, 1997)
- Food degradation (Cummings and Macfarlane, 1997)
- Colonization resistance (Cummings and Macfarlane, 1997)
- Terminal differentiation of mucosa (Stappenbeck et al., 2002)
- "Education" of innate immune defenses (Mazmanian et al., 2005)
- Promotion of epithelial "homeostasis" in the gut (Rakoff-Nahoum et al., 2004)
- Regulation of energy extraction from food (Bäckhed et al., 2004)

What makes this an interesting story is its continuing evolution: we do not fully understand what other benefits might belong on this list, because these features have yet to be described, or at least confirmed. However, the current list is sufficiently compelling to suggest that the human microbiome deserves much closer examination, despite the difficulties posed by its complexity.

A major purpose of exploring the human microbiome is to understand the role of indigenous microbial communities in human health and disease—and the various transition states between them. By understanding essential features of symbiotic relationships between microbial communities and their human host—a difficult task from a practical point of view—it may eventually be possible to predict host phenotypes, such as health status, from the particular features of indigenous communities.

This work raises a number of issues. First, how do components of microbial communities determine the behavior of the whole? Second, what is the relative importance of the environment and genetics in determining the structure and behavior of these communities? Third, can these communities be manipulated to restore or preserve human health?

---

[1] Professor, Stanford University School of Medicine and chief of the Infectious Disease Section at the Veterans Affairs, Palo Alto Health Care System.

Many needs must be met in order to answer these questions. We are presently at an early stage in this effort, and so are trying to identify which elements of these communities to monitor and measure in order to determine how they interact on a physical and metabolic level. We are not yet able to perform the kinds of precise manipulations of the human microbiota that can be undertaken with simpler microbial communities and model systems, so—as I will subsequently describe—we have chosen to look at natural and clinically relevant forms of perturbation to understand the performance features of these communities.

Patterns of Diversity

The diversity of the human microbiome can be inferred from its collection of microbial ribosomal RNA sequences. From an overall perspective, these sequences are highly conserved and thereby reveal the ancestry and phylogenetic relatedness among all forms of cellular life, allowing us to place disparate organisms into phylogenetic trees such as the one shown in Figure 2-1, which represents the Bacterial domain, one of the three primary branches of the tree of life (along with the Archaea and Eukarya). The Bacteria include approximately 80 to 100 phylum-level taxa, some of which may be potentially as diverse as the plants. Figure 2-2 reveals the degree to which the membership of each bacterial phylum has so far been recovered in the laboratory or cultivated; this is usually a small fraction of the number of organisms that have been detected by the sequencing of ribosomal RNA from environmental samples. The organisms represented by these sequences remain largely uncultivated; for example, the sequences that we derived from the human colonic mucosa and human feces suggested that approximately 80 percent of the microbial inhabitants had not yet been cultivated and that about 60 percent at the time of that analysis had not been previously described (Eckburg et al., 2005).

Some important lessons have emerged from these studies of the human microbiome. One of them is the striking diversity of bacteria at the genus and species levels, and yet the limited diversity at higher taxonomic levels, such as phylum, as shown in Figure 2-3. This pattern appears to be a general feature of the indigenous microbial communities of vertebrates, and it contrasts with the taxonomic structure of microbial communities inhabiting external environments, which tend to contain a larger number of representatives from different high-order taxonomic groups (e.g., phylum, class, order, and family). The contrast between the phylogenetic trees of indigenous animal-associated microbial communities, those of environmental communities, and those of external microbial communities can be compared with the morphological differences between actual palm trees—in which branching occurs atop a long, slender trunk—and bushy hardwood trees, which branch at all levels, as shown in Figure 2-4.

There are a number of possible explanations for these differences, one of which is that habitats within vertebrates somehow encourage, allow, or tolerate

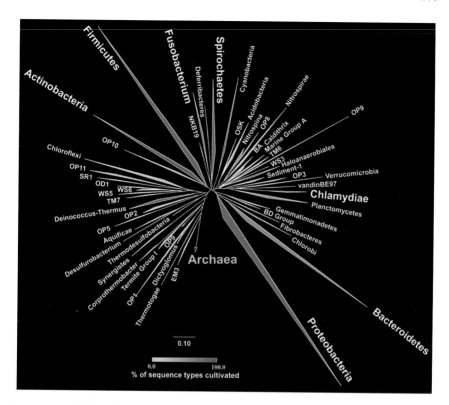

FIGURE 2-1 Domain Bacteria. The phyla that contain some of the most prominent human microbial pathogens are indicated with labels in a larger font size. Phyla without any known cultivated members are given alpha-numeric designations, for example, "TM7." SOURCE: Adapted from Handelsman et al. (2004) with permission from the American Society for Microbiology.

microbial diversification at the levels of species and strains, but only within a restricted set of high-order taxa. It is at the genus, species, and strain levels that one can distinguish the microbiota of one host from the microbiota of another in the animal world, as have researchers who recently identified distinctive features—most of which occur at the species and strain levels—of the microbiota of each of 60 different mammalian host species (Ley et al., 2008).

Why are so few phyla found on or within the human body? It may be due, in part, to selection, as has been suggested by experiments tracking the colonization of germ-free animals of various species (Rawls et al., 2006). The likely explanation probably involves multiple factors, including opportunistic environmental

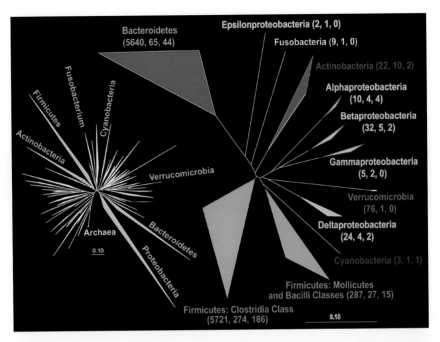

FIGURE 2-2 Bacterial diversity in human colonic tissue and stool: 11,831 16S rRNA sequences, 395 species-level operational taxonomic units, 62 percent (244) of which were novel, 80 percent uncultivated, comprising 7 phyla.
SOURCE: Eckburg et al. (2005).

exposures that may determine the composition of the microbial communities that form during the early postnatal life of the host.

A small but increasing number of studies in humans indicate individual (host)-specific features of human-associated microbial communities. When we evaluated the relatedness among the patterns of microbial diversity found in samples collected from various locations along the colonic mucosa and within the feces of several subjects, we found greater variation between than within these hosts (Eckburg et al., 2005). In a study of microbial diversity in stool samples from 14 babies, taken periodically throughout the first year of their lives, we found evidence for the emergence of individuality by the end of the first postnatal year (Palmer et al., 2007). The composition and temporal patterns of the microbial communities varied widely from baby to baby; during the first days to weeks, and there was evidence of acquisition from the mother. By the end of the first year of life, the distinct microbiota of each baby had converged toward a profile characteristic of a generic adult gastrointestinal tract.

The studies that I have discussed so far, along with others, reveal features of human indigenous microbial diversity that deserve further study: the same few

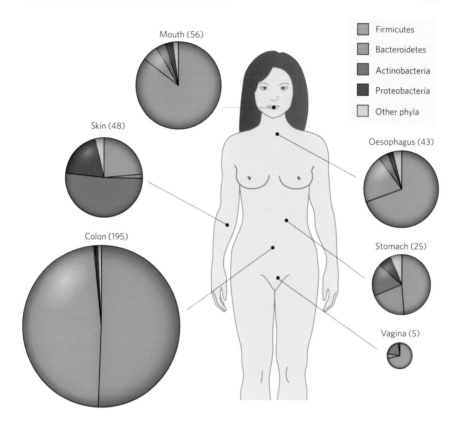

FIGURE 2-3 Site-specific distributions of bacterial phyla in healthy humans. Size of circles is proportionate to average number of species-level phylotypes per individual (in parentheses, based on data circa 2006).
SOURCE: Reprinted from Dethlefsen et al. (2007) with permission from Macmillan Publishers Ltd. Copyright 2007.

deep lineages; the excess abundance of shallow lineages (strains and species); limited archaeal diversity; individuality and host-specificity; and the importance of early events and exposures in establishing that individuality. There are multiple possible sources of variation and variability in the indigenous microbiota, including host genetics, local anatomy, pH, oxygen concentration, age, diet, place of birth and residence, occupation, contacts with other humans and animals, and perturbations such as antibiotic treatment or mechanical disruption. As metagenomic and metabolomic data become more widely available from many different individuals, we will have an opportunity to re-examine the degree to which individuals share common features of their microbiota, such as functional capa-

FIGURE 2-4 Representations of bacterial diversity. (A) Typical habitats outside of vertebrates harbor many more deep lineages than do vertebrate-associated habitats. (B) Vertebrate-associated habitats have few deep lineages but each displays a broad radiation of shallow lineages.
SOURCE: Reprinted from Dethlefsen et al. (2007) with permission from Macmillan Publishers Ltd. Copyright 2007.

bilities, in a manner that avoids some of the limitations of rRNA-based analyses (Turnbaugh et al., 2009).

Community as Pathogen

Pathogens and commensals tend to be closely related. Although they are often both found within the same genus, and sometimes within the same species, commensals and pathogens can be distinguished by their different lifestyles. Lifestyle can be defined by the manner in which the organism deals with the host response and with microbial competition. The pathogen lifestyle usually involves the deliberate induction of a host response and adaptation to, countering, or exploitation of host responses and defenses. Pathogens appear to be less well-suited for cooperation with their hosts. The commensal lifestyle, on the other hand, involves tolerance of the host response, but fewer means to counter or subvert it. On the other hand, commensals compete and cooperate with each other as part of a community and, by so doing, enhance each other's survival (Dethlefsen et al., 2007).

Clinical problems associated with disturbances in the human indigenous microbiota include chronic periodontitis, Crohn's disease and other forms of inflammatory bowel disease, tropical sprue, antibiotic-associated diarrhea, bacterial vaginosis, and premature labor and delivery. Whether disturbances in the microbiota are present at the early stages of disease, and if so, whether they play a role in causation, is not at all clear. Studies indicate that in some instances, regardless of a role in disease initiation, the indigenous microbiota plays a role in propagating the disease process. For example, treatment of some patients with Crohn's disease with antibiotics, or with surgery that diverts the microbiota away from a segment of bowel, produces partial and temporary clinical improvement by reducing inflammation and other symptoms. In these diseases, it may be appropriate to view the community itself as a pathogen, and community disturbance as the pathogenic factor, rather than a specific organism or group of organisms.

Clinicians might be well-served to consider the fitness of indigenous microbial communities, much as ecologists do, and to seek ways to restore community resistance, resilience, and a preferred (health-associated) fitness regime. With degradation of the fitness "landscape," communities become more likely to shift toward a disease-associated fitness state. Deliberate perturbation of our microbiota, such as exposure to antibiotics, provides an opportunity to assess community robustness, and, indirectly, the topology of the fitness landscape. Studies of community resilience in ecosystems such as lakes, grasslands, and coral reefs, highlight the detrimental impact of human activities, and document loss of the ability of such systems to absorb disturbance and still retain structure and function (Folke et al., 2004).

Even though microbes have evolved in the presence of antibiotics since the beginning, the concentrations of antibiotics have been tiny, compared to the massive amounts produced by humans during the last half century and applied to livestock, humans, and the environment. The detrimental consequences have yet to be fully appreciated, but include acute disease, chronic disease, and emerging resistance. In my laboratory, Les Dethlefsen has been examining the effects of antibiotics on the distal gut microbiota in healthy subjects, in order to determine how patterns of diversity vary before, during, and after these exposures, and to assess community resilience. In one clinical protocol, seven healthy subjects were each given two 5-day courses of the antibiotic ciprofloxacin separated by a 6-month interval, and their stools were sampled periodically over 10 months, with increased frequency of sampling centered around the time of treatment. 16S rDNA hypervariable region tag sequences were analyzed from a subset of the specimens from three subjects before and after the first course of ciprofloxacin using pyrosequencing technology (Dethlefsen et al., 2008; Huse et al., 2008). We identified approximately 5,700 different species or strains of bacteria from these samples, of which only 7 percent had been described previously. Examination of the patterns of abundance for the 1,450 most abundant taxa revealed that they were members of typical human-associated bacterial phyla; that the same genera (both abundant and rare) tended to be present in all three subjects; and that patterns of bacterial abundance in the subjects differed at the species level.

Approximately 30 percent of these common taxa showed a significant response (change in abundance) following antibiotic treatment, an effect that lasted in the case of most taxa, for less than four weeks. However, the abundance levels of some taxa had not recovered completely even after six months. Ciprofloxacin treatment decreased the degree of bacterial diversity by reducing the number of species present in these communities. It also decreased the evenness—the degree to which abundance levels of various taxa are shared within each specimen and subject. It appears that the features that most distinguish individuals were partially "erased" during and immediately after antibiotic use. Focusing on a particular genus, the *Bacteroides*, we found that during and just after the ciprofloxacin exposure, there was a rise and fall in species abundance in all three individuals, as shown in Figure 2-5. However, although there was not

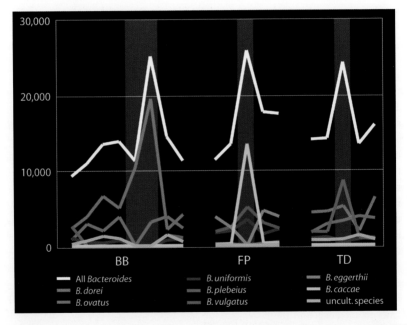

FIGURE 2-5 Individualized responses to the same antibiotic: Abundance of 16S rRNA V3 sequence tags corresponding to *Bacteroides* species—before and after ciprofloxacin administration (gray shaded zone corresponds to sampling times during or immediately after administration) 19 of the 100 most abundant V3 refOTUs belonged to the genus *Bacteroides*, and 18 of them differed significantly between individuals. However, the abundance of the *Bacteroides* genus as a whole did not differ significantly between individuals. uncult. = uncultivated.
SOURCE: Dethlefsen et al. (2008).

much individuality in the response of the *Bacteroides* as a genus, there was considerable individuality in the response of *Bacteroides* as species, with a different *Bacteroides* species in each individual host responsible for the antibiotic-induced peak in abundance. Host-specific differences prior to antibiotic exposure may explain why two healthy individuals may respond quite differently following identical antibiotic exposures.

Challenges

What makes the human microbiome so intrinsically interesting is the degree to which it may reflect who we are as individuals and as human beings, and the degree to which it may contribute to human biology. It has implications for our understanding of pathogenesis, and it suggests ways by which we might recognize early signs of disease and restore health. But there are many unanswered questions with regard to the human microbiome. We need to understand better the importance of strain diversity within the indigenous microbiota. This will require overcoming a number of technical challenges associated with sequencing and computation, and will require understanding the link between taxonomy and function. It may be that, for example, different *Bacteroides* species perform the same functions in different individuals. We know that the *Bacteroides spp.* possess a very flexible means of regulating gene expression and a considerable armamentarium of genes encoding enzymes for food degradation, and that they are particularly well tuned for responding to diverse kinds of food sources.

Second, we also need to understand better how and where to sample microbial communities within the human body. In particular, we need to determine the spatial scale within each human habitat at which clinically and ecologically relevant distinctions in microbial diversity found. And the ecological boundaries between habitats within the human body need to be better characterized. At the same time, as clinicians, we need to collect environmental, physiological, and demographic data that will enable us to interpret our molecular data more effectively.

In many ways, the human microbiome remains *terra incognita*. It is truly humbling that after hundreds of years of study we know as little, or less, about this system than we do about plant diversity in the tropical rain forest. The distinct patterns of microbial diversity that we have found within the human body appear to reflect selection, specialization, cooperation, and co-adaptation. Today, we have experimental, and even clinical settings and opportunities in which to begin to examine some of these potential features.

DECIPHERING THE COMPLEX MOLECULAR DIALOGUE OF SYMBIOSIS: ESPERANTO OR POLYGLOT?

Margaret McFall-Ngai, Ph.D.[2]
University of Wisconsin, Madison

Introduction

This contribution explores the question: What are the shared characteristics of pathogenic and mutualistic interactions between microbes and their animal hosts? Specifically, is the language of these different types of associations controlled principally by different cellular and molecular characters wherein divergent genes and pathways are used to mediate bacterial activity and host responses (i.e., are they "polyglot"); or by shared mechanisms in which the outcome is determined by when and where identical molecules and pathways are brought into play (i.e., are they Esperanto[3]-like)? The answers to these questions remain largely unknown, in part because of the complexity of many associations. However, the study of simpler model systems is shedding some light on these questions and, as such, they offer a complement to the analyses of the more complex systems (Dale and Moran, 2006; Ruby, 2008). The following remarks will explore what has been learned in one such model, the squid-vibrio association.

Because the terminology in the field of animal-microbe or plant-microbe interactions has been used in various ways, it is important to begin by providing a lexicon for the discussion of the associated biology. Most biologists in this discipline consider *symbiosis* an umbrella term that refers to the phenomenon of organisms living together, as defined by Anton deBary in *Die Erscheinung der Symbiose* (1879). The catchall set of associations (i.e., symbioses) has often been divided into three classes based on the effects of the relationship on partner fitness (i.e., the effect on the number of progeny in the next generation): mutualism, commensalism, and parasitism (McFall-Ngai and Gordon, 2005). In mutualisms, both partners benefit (i.e., the fitness increases for both); in commensalisms, one partner benefits and the other is unaffected; and in parasitisms, one partner benefits and the fitness of the other is compromised.

Historically, the field of symbiosis has been dominated by two disciplines that developed relatively independently. One has focused on the biomedically important parasitic microorganisms, such as bacterial pathogens, that have had

---

[2]Professor, Department of Medical Microbiology and Immunology, Microbial Sciences Building, 1550 Linden Drive, Madison, WI 52706; Phone: (608) 262-2393; e-mail: mjmcfallngai@wisc.edu.

[3]Esperanto is a language designed to facilitate communication between people of different lands and cultures. First published in 1887 by Dr. L. L. Zamenhof (1859-1917) under the pseudonym "Dr. Esperanto," meaning "one who hopes" (see http://www.esperanto.net/veb/faq-1.html).

profound effects on human history. These associations have been viewed largely as a binary war between the pathogen and its host. The other has concentrated on highly conspicuous mutualistic symbioses. The latter includes such phenomena as the endosymbiotic origin of the eukaryotic cell and symbioses that enable hosts to exploit specific environments, such as in the rumen symbioses of ungulates, the hydrothermal vent chemoautotrophic associations, and the coral-zooxanthellae alliances.

A sea change in our view of symbiosis began to take hold in the 1990s, largely due to the work of groups studying the human microbiota (for reviews see, Institute of Medicine, 2006). While it was long appreciated that many microbes are present in and on the human body, it was thought that these communities are principally commensal, with the microorganisms benefiting from the nutrients provided by the host habitat but having no fitness effect on the host. Strides in molecular biology in the 1990s dramatically increased the feasibility of identifying the range of microbes, determining their site-specific community composition, and characterizing their behaviors. The findings resulting from the application of these advances in molecular biology have demonstrated that the microbial communities that live with humans are anything but commensal. They appear to be an integral, coevolved part of human biology, and *health* is a term that should be applied to the collective.

The realization that humans and other vertebrates are truly composed of mutualistic communities of microbes promises to have ripple effects throughout biology (McFall-Ngai, 2008). We now are recognizing that many pathogens assault the entire assemblage, compromising the homeostasis established by complex normal alliances. How will this newfound knowledge affect our view of the biology of pathogenic associations and how pathogenesis should be treated in clinical settings? In a broader view, it is now becoming clear not only that the microbial communities associated with host animals affect the tissues with which they interact but also that their metabolic products interact with most cells of the body (Nicholson et al., 2004). These findings indicate that host physiology evolved as a result of selection pressures on the host-microbiome axis.

The Squid-Vibrio Association as an Experimental Model

Biologists now face a series of challenges imposed by the recognition of the prevalence and complexity of symbioses. At a practical level, how do we integrate this new knowledge into the theory and practice of our science? When faced with similar problems, researchers in other fields, such as developmental biology, have turned to simplified models to provide insight into the basic principles underlying particular processes. A variety of experimental models have been developed in the last several years for the study of symbioses (Dale and Moran, 2006; Ruby, 2008). They include symbioses in germ-free and gnotobiotic vertebrates, simple consortial associations[4] offered by invertebrate species, and binary associations, in which a population of a single microbial species associates with a host animal.

The light organ symbiosis between the Hawaiian bobtail squid *Euprymna scolopes* and its luminous bacterial partner, *Vibrio fischeri*, is a binary association that has been studied for the last 20 years (for reviews, see McFall-Ngai, 2007; Nyholm and McFall-Ngai, 2004; Visick and Ruby, 2006). In this relationship, populations of the bacterial symbionts associate with the apical surfaces of the host's epithelia. Its experimental tractability has made it an attractive subject for the study of many processes critical to symbiosis, including transmission between generations, ensuring specificity, inducing development, and achieving stability of the partnership. Each of these processes requires interplay between host and symbiont features (i.e., the dialogue is reciprocal and involves emergent properties) such that many, if not most, of the responses of the host to the symbiont, and vice versa, could not be predicted by studying the individual partners in isolation. Furthermore, at least in this one symbiosis, the events of this association rely on characters of both partners that erstwhile have been described as features involved in mediating pathogenesis. The following discussion describes what is known of this phenomenon.

Critical Communication with Host Epithelia

Animals acquired epithelia in the evolutionary transition from the cellular grade (i.e., the sponge body plan) to the tissue and organ grades of organization (the body plans of all Eumetazoans from anemones to mammals). From that milestone in evolution, the association of microbes with the apical surfaces of host epithelia, whether they are mutualistic, commensal, or pathogenic, has been the dominant type of animal-microbe relationship. Most often, where coevolved beneficial alliances have formed along epithelial tissues, they are horizontally transmitted between generations (i.e., the juvenile host acquires the symbionts each generation from the surrounding environment), such as the consortial symbioses of vertebrates. The squid-vibrio symbiosis is a coevolved partnership in which the host engages the symbiont within minutes of hatching into the environment. Only a few hundred *V. fischeri* are present per milliliter of seawater in the habitats where host populations occur. Furthermore, the juvenile squids are small, about 2 millimeters in total length, and the body (or mantle) cavity through which symbiont-laden water circulates has a volume of approximately 1 microliter. Thus, when the juvenile animal brings this water into the mantle cavity during ventilation, few, if any, *V. fischeri* cells are present to interact with host tissues. Thus, we theorized that the animal must have active mechanisms by which to harvest the symbiont.

---

[4]An association in which there are populations of more than one phylotype of microbe living in association with a host animal.

These mechanisms of symbiont acquisition rely on features unique to the juvenile organ. During embryogenesis, the light organ develops in association with the hindgut-ink sac complex. At hatching, the nascent organ has a series of epithelial tissues (Figure 2-6) that are critical for successful colonization: (1) a juvenile-specific superficial ciliated epithelium that facilitates the symbiont harvesting process; (2) epithelial layers that line long ciliated ducts, forming the conduit through which symbionts migrate between the surface and crypt spaces; and (3) the internal epithelium-lined crypts themselves, where the symbionts colonize throughout the life of the host.

The dynamics of this epithelial landscape mediate the successful establishment of a symbiosis within hours of the host's hatching from the egg (Figures 2-6 and 2-7). The animal begins to ventilate within seconds of hatching. Water is drawn into the mantle cavity by relaxation of the mantle muscles and then, with their contraction, is forcibly expelled through the medial funnel. The funnel circumscribes the light organ so that much of the ventilated water passes over the organ's surface. Each ciliated epithelial field, one on each side of the organ, is composed of two opposing appendages that form a ring. Within a few seconds of the first ventilation, the cilia begin to beat and their activity entrains water through the ring of epithelial tissue into the vicinity of three pores, which occur at the base of each set of appendages; these pores are the eventual sites of entry of *V. fischeri* cells into the host tissues.

How does the animal enrich for the symbiont? Also within minutes of hatching, the surface epithelium responds to the presence of the peptidoglycan from the bacterioplankton in the water by shedding copious amounts of mucus from stores in the ciliated cells of the surface epithelia. In the absence of the symbiont, the cells of any gram-negative bacterium are capable of adhering to and aggregating in the mucus. However, when *V. fischeri* cells are present, they exert competitive dominance in the mucus, such that by 2-3 hours following exposure to natural seawater, which contains ~10^6 nonspecific bacteria per milliliter with only a few hundred symbionts, most of the cells in the mucus aggregates are symbiont cells. The reasons for this exclusivity are unknown. Successful aggregation of the symbionts in the mucus requires *V. fischeri* genes encoding proteins involved in exopolysaccharide production.

After a couple of hours of residence time in the mucus, the symbionts migrate to the pores on the surface of the organ (Figure 2-6B), down ciliated ducts, and into the crypt spaces (Figure 2-6C). The population of symbiont cells then grows to fill the crypt space within hours. Partner signaling characterizes this series of events (Figure 2-7). Central to this interplay are components of the bacterial cell envelope, most notably, derivatives of lipopolysaccharide (LPS) and peptidoglycan (PGN; Koropatnick et al., 2004). While these microbe-associated molecular patterns (MAMPs) are relatively conserved among bacteria, only *V. fischeri* is able to interact with internal epithelial regions of the host; thus, only the symbiont can present these conserved molecules in locations and at

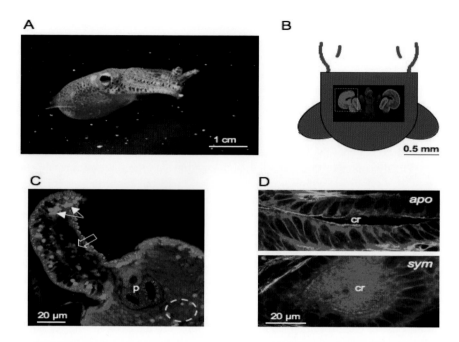

FIGURE 2-6 The squid-vibrio association. (A) An adult animal swimming in the water column. (B) A ventral view of the juvenile. The window in the center illustrates the position of the light organ (the green confocal image) in the middle of the mantle (body) cavity. The juvenile light organ has ciliated fields on each lateral face that potentiate colonization by the symbiont. The dashed box around one side defines the approximate region magnified in panel C. (C) A confocal image of one half of a juvenile light organ. This organ was colonized hours earlier with *V. fischeri*. As such, hemocytes (open arrow) can be seen in the blood sinus of the ciliated field and the condensed chromatin of apoptotic epithelial cells (solid arrows) is apparent. The dashed circle defines the approximate region of the image in panel D. P = pores. (D) Confocal images of the crypts (cr). Top, an aposymbiotic animal (apo; i.e., not colonized); bottom, a symbiotic animal (sym), with a dense population of *V. fischeri* in the crypt space.

concentrations where they result in host responses. These responses are numerous, affecting epithelia throughout the system. Most notably, at about 12 hours following initial exposure to the symbiont, the superficial epithelium receives an irreversible signal that triggers its regression; much of the signal is mediated by responses to a synergism between components of *V. fischeri* LPS and PGN. Some evidence exists that bioluminescence is also critical to triggering full morphogenesis. As such, bioluminescence is a "special" language likely used only in the light organ symbioses, whereas the response to MAMPs is a more general

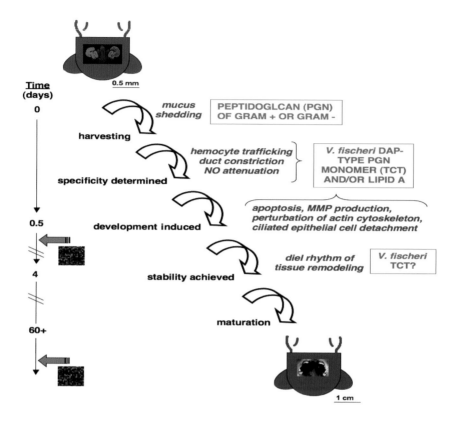

FIGURE 2-7 The series of events that characterize the trajectory of the symbiosis. Solid gray arrows = time points of current microarrays. DAP = diaminopimelic acid; MMP = matrix metalloproteinase; NO = nitric oxide; TCT = tracheal cytotoxin.

language. Numerous molecular, biochemical, and cellular events in the host's superficial epithelium are associated with this process (Figure 2-7). The crypt epithelia respond to their direct interactions with the symbiont with a swelling of the cells and an increase in microvillar density. The underlying signals from the symbiont that mediate these phenotypes remain to be determined.

Deciphering the Molecular Language of Symbiosis

The studies of the squid-vibrio symbiosis described earlier have revealed the dynamic cellular events associated with the onset of the symbiosis. In addition, genomic tools have recently been developed for the system. Specifically, a database of the expressed sequence tags (EST) of the juvenile light organ that contains a set of nearly 14,000 unique clusters has been generated, and these clusters have been spotted onto glass slide arrays (Chun et al., 2006, 2008). The genome of a *V. fischeri* strain isolated from the *E. scolopes* light organ has been fully sequenced (Ruby et al., 2005), and probes representing greater than 90 percent of the open reading frames (ORFs) and small RNAs have been arrayed on an Affymetrix® chip. These host and symbiont resources have paved the way for characterization of the various phases of the symbiosis with an hour-by-hour resolution.

The first array experiments of host responses were carried out at 18 hours following initial exposure to the symbiont (Chun et al., 2008). The experiments were designed to determine the effects of symbiosis on host gene expression. In addition to analyzing the responses to symbiosis with wild-type symbionts, these experiments used mutants of *V. fischeri* to explore the influences of luminescence and autoinducer (AI) production by the symbiont population. A few hundred host genes were reproducibly differentially regulated at 18 hours in response to interactions with the wild-type symbiont. In addition to providing important information on the inner workings of the squid-vibrio system, they offer insight into those characters conserved over animal evolution that typify the colonization of the apical surface of epithelia by gram-negative bacteria. This window into animal-bacterial interactions was made possible by comparisons of the squid-vibrio data with those obtained from the colonization of germ-free vertebrates with their bacterial partners (Hooper et al., 2001; Rawls et al., 2004). One recurring theme was the differential regulation of genes associated with biochemical pathways and cellular behaviors that are key in the bacterial pathogenesis of animal tissues (Figure 2-8). Studies with *V. fischeri* mutants revealed that symbiont bioluminescence is a powerful inducer of host gene regulation, whereas AI is

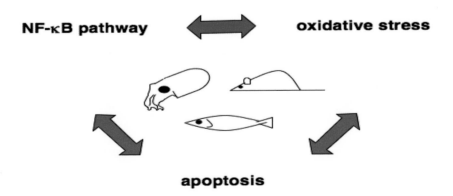

FIGURE 2-8 Some symbiosis characteristics shared among the mouse, zebrafish, and squid.

less so. Moreover, comparisons of the changes in light-organ gene expression in animals colonized with mutants defective in AI production with those exposed to AI pharmacologically revealed that the presence of the bacteria is critical to the responses of host tissues (i.e., they appear to prime the host cells for interaction with AI). This finding is a cautionary tale to those who study the pharmacological effects of bacterial products on animal cells.

A microarray study of the symbiosis in adult animals is currently under way. In these studies, both host and symbiont gene expressions are being examined to define the conversation between the partners over the day-night cycle. We antici-pate that this study will provide insight into how animals maintain symbioses stably over their life cycle.

Summary: The Esperanto of Symbiosis

Research on the squid-vibrio symbiosis has demonstrated that both the host and symbiont signal one another with a biochemical, molecular, and cellular language that is typically associated with mediating bacterial pathogenesis of animal tissues. Because biologists are becoming increasingly aware that mutu-alistic symbioses are far more prevalent than pathogenic ones, these findings in the squid-vibrio system invite us to question our premises about the true nature of host responses to pathogens (Figure 2-9).

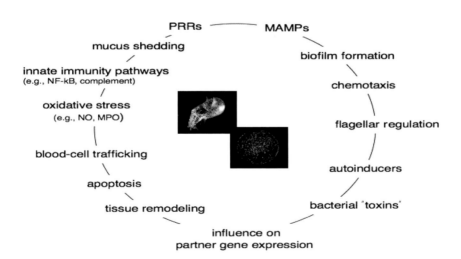

FIGURE 2-9 The language of symbiosis. Some aspects of the squid-vibrio symbiosis that illustrate the similarities with bacterial pathogenesis. MPO = myeloperoxidase-like proteins; NO = nitric oxide; PRR = pattern recognition receptors.

They suggest that the principal selection pressure on the evolution of animal-microbe relationships is for the establishment and maintenance of mutualisms and that pathogens are often "spies" subverting a preexisting conversation with the host. As such, both types of associations involve the same molecular language, and it is the context that defines how the fitness of the host is affected.

REFERENCES

Overview References

Andersson, A. F., and J. F. Banfield. 2008. Virus population dynamics and acquired virus resistance in natural microbial communities. *Science* 320(5879):1047-1050.

Banfield, J. F. 2008a. *Subproject 1: integrating genomics, proteomics, and functional analyses into ecosystem-level studies of natural microbial communities*, http://eps.berkeley.edu/~jill/gtl/subproject_1.htm (accessed July 31, 2008).

————. 2008b. *Geomicrobiology Program University of California, Berkeley*, http://eps.berkeley.edu/~jill/banres.html (accessed July 31, 2008).

Brelles-Mariño, G., and J. M. Ané. 2008. Nod factors and the molecular dialogue in the rhizobia-legume interaction. In *Nitrogen fixation research progress*, edited by G. N. Couto. Hauppauge, NY: Nova Science Publishers, Inc.

Denef, V. J., N. C. VerBerkmoes, M. B. Shah, P. Abraham, M. Lefsrud, R. L. Hettich, and J. F. Banfield. 2009. Proteomics-inferred genome typing (PIGT) demonstrates inter-population recombination as a strategy for environmental adaptation. *Environmental Microbiology* 11(2):313-325.

Kiers, E. T., R. A. Rousseau, S. A. West, and R. F. Denison. 2003. Host sanctions and the legume-rhizobium mutualism. *Nature* 425(6953):78-81.

Lo, I., V. J. Denef, N. C. Verberkmoes, M. B. Shah, D. Goltsman, G. DiBartolo, G. W. Tyson, E. E. Allen, R. J. Ram, J. C. Detter, P. Richardson, M. P. Thelen, R. L. Hettich, and J. F. Banfield. 2007. Strain-resolved community proteomics reveals recombining genomes of acidophilic bacteria. *Nature* 446(7135):537-541.

Makarova, K. S., N. V. Grishin, S. A. Shabalina, Y. I. Wolf, and E. V. Koonin. 2006. A putative RNA-interference-based immune system in prokaryotes: computational analysis of the predicted enzymatic machinery, functional analogies with eukaryotic RNAi, and hypothetical mechanisms of action. *Biology Direct* 1(7): http://www.biology-direct.com/content/1/1/7 (accessed August 28, 2008).

Mojica, F. J., C. Diez-Villasenor, J. Garcia-Martinez, and E. Soria. 2005. Intervening sequences of regularly spaced prokaryotic repeats derive from foreign genetic elements. *Journal of Molecular Evolution* 60(2):174-182.

Riely, B. K., J.-H. Mun, and J.-M. Ané. 2006. Unravelling the molecular basis for symbiotic signal transduction in legumes. *Molecular Plant Pathology* 7(3):197-207.

Taormina, P. J., L. R. Beuchat, and L. Slutsker. 1999. Infections associated with eating seed sprouts: an international concern. *Emerging Infectious Diseases* 5(5):626-634.

Tyson, G. W., and J. F. Banfield. 2008. Rapidly evolving CRISPRs implicated in acquired resistance of microorganisms to viruses. *Environmental Microbiology* 10(1):200-207.

Weerasinghe, R. R., D. M. Bird, and N. S. Allen. 2005. Root-knot nematodes and bacterial Nod factors elicit common signal transduction events in *Lotus japonicus*. *Proceedings of the National Academy of Sciences* 102(8):3147-3152.

Relman References

Bäckhed, F., H. Ding, T. Wang, L. V. Hooper, G. Y. Koh, A. Nagy, C. F. Semenkovich, and J. I. Gordon. 2004. The gut microbiota as an environmental factor that regulates fat storage. *Proceedings of the National Academy of Sciences* 101(44):15718-15723.

Cummings, J. H., and G. T. Macfarlane. 1997. Role of intestinal bacteria in nutrient metabolism. *Journal of Parenteral and Enteral Nutrition* 21(6):357-365.

Dethlefsen, L., M. McFall-Ngai, and D. A. Relman. 2007. An ecological and evolutionary perspective on human-microbe mutualism and disease. *Nature* 449(7164):811-818.

Dethlefsen, L., S. M. Huse, M. L. Sogin, and D. A. Relman. 2008. The pervasive effects of an antibiotic on the human gut microbiota, as revealed by deep 16S rRNA sequencing. *PLoS Biology* 6(11):e280.

Eckburg, P. B., E. M. Bik, C. N. Bernstein, E. Purdom, L. Dethlefsen, M. Sargent, S. R. Gill, K. E. Nelson, and D. A. Relman. 2005. Diversity of the human intestinal microbial flora. *Science* 308(5728):1635-1638.

Folke, C., S. Carpenter, B. Walker, M. Scheffer, T. Elmqvist, L. Gunderson, and C. S. Holling. 2004. Regime shifts, resilience and biodiversity in ecosystem management. *Annual Review of Ecology, Evolution, and Systematics* 35:557-581.

Handelsman, J. 2004. Metagenomics: application of genomics to uncultured microorganisms. *Microbiology and Molecular Biology Reviews* 68(4):669-685.

Huse, S. M., L. Dethlefsen, J. A. Huber, D. M. Welch, D. A. Relman, and M. L. Sogin. 2008. Exploring microbial diversity and taxonomy using SSU rRNA hypervariable tag sequencing. *PLoS Genetics* 4(11):e1000255.

Ley, R. E., M. Hamady, C. Lozupone, P. J. Turnbaugh, R. R. Ramey, J. S. Bircher, M. L. Schlegel, T. A. Tucker, M. D. Schrenzel, R. Knight, and J. I. Gordon. 2008. Evolution of mammals and their gut microbes. *Science* 320(5883):1647-1651.

Mazmanian, S. K., C. H. Liu, A. O. Tzianabos, and D. L. Kasper. 2005. An immunomodulatory molecule of symbiotic bacteria directs maturation of the host immune system. *Cell* 122(1):107-118.

Palmer, C., E. M. Bik, D. B. Digiulio, D. A. Relman, and P. O. Brown. 2007. Development of the human infant intestinal microbiota. *PLoS Biology* 5(7):e177.

Rakoff-Nahoum, S., J. Paglino, F. Eslami-Varzaneh, S. Edberg, and R. Medzhitov. 2004. Recognition of commensal microflora by Toll-like receptors is required for intestinal homeostasis. *Cell* 118(2):229-241.

Rawls, J. F., M. A. Mahowald, R. E. Ley, and J. I. Gordon. 2006. Reciprocal gut microbiota transplants from zebrafish and mice to germ-free recipients reveal host habitat selection. *Cell* 127(2):423-433.

Stappenbeck, T. S., L. V. Hooper, and J. I. Gordon. 2002. Developmental regulation of intestinal angiogenesis by indigenous microbes via Paneth cells. *Proceedings of the National Academy of Sciences* 99(24):15451-15455.

Turnbaugh, P. J., M. Hamady, T. Yatsunenko, B. L. Cantarel, A. Duncan, R. E. Ley, M. L. Sogin, W. J. Jones, B. A. Roe, J. P. Affourtit, M. Egholm, B. Henrissat, A. C. Heath, R. Knight, and J. I. Gordon. 2009. A core gut microbiome in obese and lean twins. *Nature* 457(7728):480-484.

McFall-Ngai References

Chun, C. K., T. E. Scheetz, F. Bonaldo Mde, B. Brown, A. Clemens, W. J. Crookes-Goodson, K. Crouch, T. DeMartini, M. Eyestone, M. S. Goodson, B. Janssens, J. L. Kimbell, T. A. Koropatnick, T. Kucaba, C. Smith, J. J. Stewart, D. Tong, J. V. Troll, S. Webster, J. Winhall-Rice, C. Yap, T. L. Casavant, M. J. McFall-Ngai, and M. B. Soares. 2006. An annotated cDNA library of juvenile *Euprymna scolopes* with and without colonization by the symbiont *Vibrio fischeri*. *BMC Genomics* 7:154.

Chun, C. K., J. V. Troll, I. Koroleva, B. Brown, L. Manzella, E. Snir, H. Almabrazi, T. E. Scheetz, F. Bonaldo Mde, T. L. Casavant, M. B. Soares, E. G. Ruby, and M. J. McFall-Ngai. 2008. Effects of colonization, luminescence, and autoinducer on host transcription during development of the squid-vibrio association. *Proceedings of the National Academy of Sciences* 105(32):11323-11328.

Dale, C., and N. A. Moran. 2006. Molecular interactions between bacterial symbionts and their hosts. *Cell* 126(3):453-465.

Hooper, L. V., M. H. Wong, A. Thelin, L. Hansson, P. G. Falk, and J. I. Gordon. 2001. Molecular analysis of commensal host-microbial relationships in the intestine. *Science* 291(5505):881-884.

IOM (Institute of Medicine). 2006. *Ending the war metaphor: the changing agenda for unraveling the host-microbe relationship.* Washington, DC: The National Academies Press.

Koropatnick, T. A., J. T. Engle, M. A. Apicella, E. V. Stabb, W. E. Goldman, and M. J. McFall-Ngai. 2004. Microbial factor-mediated development in a host-bacterial mutualism. *Science* 306(5699):1186-1188.

McFall-Ngai, M. J. 2007. The squid-vibrio association: a naturally occurring experimental model of animal-bacterial partnerships. In *Gut microbiota and regulation of the immune system*, edited by G. Huffnagle and M. Noverr. Austin, TX: Landes Bioscience Press.

———. 2008. Are biologists in future shock? Symbiosis integrates biology across domains. *Nature Reviews Microbiology* 6(10):789-792.

McFall-Ngai, M. J., and J. I. Gordon. 2005. Experimental models of symbiotic host-microbial relationships: understanding the underpinnings of beneficence and the origins of pathogenesis. In *Evolution of microbial virulence*, edited by H. Seifert and V. DiRita. Washington, DC: ASM Press.

Nicholson, J. K., E. Holmes, J. C. Lindon, and I. D. Wilson. 2004. The challenges of modeling mammalian biocomplexity. *Nature Biotechnology* 22(10):1268-1274.

Nyholm, S. V., and M. J. McFall-Ngai. 2004. The winnowing: establishing the squid-vibrio symbiosis. *Nature Reviews Microbiology* 2(8):632-642.

Rawls, J. F., B. S. Samuel, and J. I. Gordon 2004. Gnotobiotic zebrafish reveal evolutionarily conserved responses to the gut microbiota. *Proceedings of the National Academy of Sciences* 101(13):4596-4601.

Ruby, E. G. 2008. Symbiotic conversations are revealed under genetic interrogation. *Nature Reviews Microbiology* 6(10):752-762.

Ruby, E. G., M. Urbanowski, J. Campbell, A. Dunn, M. Faini, R. Gunsalus, P. Lostroh, C. Lupp, J. McCann, D. Millikan, A. Schaefer, E. Stabb, A. Stevens, K. Visick, C. Whistler, and E. P. Greenberg. 2005. Complete genome sequence of *Vibrio fischeri*: a symbiotic bacterium with pathogenic congeners. *Proceedings of the National Academy of Sciences* 102(8):3004-3009.

Visick, K. L., and E. G. Ruby. 2006. *Vibrio fischeri* and its host: it takes two to tango. *Current Opinion in Microbiology* 9(6):632-638.

3

Pathogen Evolution

OVERVIEW

As Lederberg (2000) observed, the host-microbe relationship is a dynamic equilibrium. Physiological or genetic changes in either partner may prompt commensal microbes to invade the tissue of their host, thereby triggering an immune response that destroys the invaders, but may also injure or kill the host. As they explored this process from the perspectives of pathogen and host, the workshop speakers featured in this chapter proposed a variety of possible evolutionary routes to the host-microbe relationships that underlie infectious diseases.

The chapter's first paper, by Stanley Falkow of Stanford University, considers the nature of bacterial pathogenicity as it has been viewed historically, and as revealed by his research and that of his colleagues at Stanford University. He explains how key discoveries—beginning with Lederberg's fundamental work on bacterial genetics—shaped the developing field of molecular biology, and more specifically, Falkow's nearly 50 years of research on the genetic basis of bacterial pathogenicity.

Using the tools of molecular genetics to study *Salmonella*, Falkow and coworkers have observed how bacteria manipulate host cell functions, how horizontal gene transfer shapes pathogen specialization, and how inherited pathogenicity islands transform commensal bacteria into pathogens. Having screened the entire *Salmonella* genome for genes that are associated with different stages of infection with a microarray-based negative selection strategy, they have identified many pathogen genes expressed in the multistage process of host invasion. Using a mouse model, they have also identified host genes and gene pathways expressed in response to *Salmonella* infection.

Falkow also considers the importance of the microbes he refers to as "commensal pathogens": bacterial species (e.g., *Streptococcus pneumoniae*, *Neisseria meningitidis*, *Haemophilus influenzae* type b, *Streptococcus pyogenes*) that typically inhabit the human nasopharynx without symptom, but sometimes cause disease. Their existence raises a host of scientific questions regarding the relationship between microbial pathogenicity, infectious disease, and immune function—questions that, he argues, should be approached by studying microbial pathogenicity as a biological phenomenon, and not merely from the perspective of its role in causing disease.

Just as there is more to microbial pathogenicity than disease, there is more to infectious disease than the actions of pathogens on host cells and systems. The chapter's second paper, coauthored by Elisa Margolis and workshop speaker Bruce Levin of Emory University, considers the host response to microbial virulence, which, the authors note, does not correspond to simple evolutionary models. They examine why bacteria harm the (mostly human) hosts they need for their survival, offering evidence that "much of the virulence of bacterial infections can be blamed on the seemingly misguided overresponse of the immune defenses."

These immunological failings include responding more vigorously than needed, as occurs in bacterial sepsis; responding incorrectly to a pathogen, as occurs in lepromatous leprosy; or responding to the wrong signals, as occurs in toxic shock syndrome. Margolis and Levin explore these and other examples of the "perversity of the immune system" and consider this view in light of various current hypotheses for the evolution of bacterial virulence. They offer possible explanations as to why natural selection has not tempered immune overresponse to bacterial infections and discuss the implications of their host-response perspective on virulence for the treatment of bacterial infections.

Two additional speakers, Gordon Dougan and Julian Parkhill, of the Wellcome Trust Sanger Institute in Cambridge, United Kingdom, contributed to workshop discussions concerning the evolution of the host-pathogen relationship. Each presenter discussed the evolutionary pathways taken by *Salmonella* serovars to become diverse pathogens. These include *Salmonella enterica* serovar Typhimurium (hereinafter *S. typhimurium*), which infects a wide range of hosts and is a major cause of gastroenteritis in humans, and *S. enterica* serovar *Typhi* (hereinafter *S. typhi*), the human-specific agent of the systemic infection typhoid fever. In humans, *S. typhimurium* infections are generally (but not always; see below) contained within the intestinal epithelium. *S. typhi* evades destruction by the immune system and is transported, via the liver and spleen, to the gall bladder and bone marrow, in which the bacteria can persist (Figure WO-9). Thus, significant numbers of people infected with typhoid—including those asymptomatically infected with *S. typhi*—become chronic carriers of the pathogen and reservoirs of a disease that poses a considerable threat to public health. From the perspective of *S. typhi*, however, this "stealth" strategy is essential to its survival.

Like many human-adapted pathogens, such as *Yersinia pestis*, *Bacillus anthracis*, and *Mycobacterium tuberculosis*, *S. typhi* is monophyletic; that is, it is restricted in terms of genomic variation, Dougan noted. "These human-restricted and recently evolved pathogens entered the human population, like many pathogens, no more than about 30,000 to 40,000 years ago," and thus, he explained, *S. typhi* has coevolved with humans, and at a similar evolutionary rate.

In his presentation, Parkhill presented evidence that, in addition to acquiring genes that confer invasiveness (pathogenicity islands, as described by Falkow), monophyletic pathogens become virulent through loss of function in genes that regulate the expression of virulence factors (e.g., the pertussis toxin in *Bordetella spp.*, as described in detail in Box WO-2). Much of this evidence derives from determining the identity of the few differences among the genomes of monophyletic pathogens, as revealed by comparator genomics.

"We do comparator genomics in the hope that the comparison between the genomes will tell us something about the comparison between the phenotypes," Parkhill said. "We might expect that we can go and look in those genes and find [virulence factors]," he continued, but in the case of *Bordetella spp.*, that did not happen (Box WO-2). Rather, their comparisons revealed that *Bordetella pertussis*, the primary causative agent of whooping cough in humans, evolved toward host restriction and greater virulence by losing function in genes associated with host interaction (thereby narrowing host ranges) and also genes that regulate the expression of virulence factors, such as the pertussis toxin (Parkhill et al., 2003).

Similar events appear to have influenced the evolution of a variety of human, equine, and plant pathogens, Parkhill noted. In the case of *S. typhi*, a large number of pseudogenes (recently inactivated genes, as indicated by the presence of point mutations) have inactivated cell surface proteins and pathogenicity proteins (McClelland et al., 2001). "This is the signature of an organism that has changed its niche," he said. "It has gone from a fecal-oral-transmitting pathogen that is limited to the cells lining the gut [to] become a systemic pathogen. It has lost function. It has inactivated genes that are involved in pathogenicity, genes that were involved in its previous lifestyle."

"Almost certainly, some of these inactivations are selective," he continued. "They are necessary for that change in niche, [such as the inactivation of] type III secreted effector genes that we know are important in the interaction of *S. typhimurium* with its host. We can see that genes that we know are involved in host range determination in *S. typhimurium* have been inactivated." However, he added, "a lot of these changes, we believe, are probably collateral damage. There is a massive event, massive changes, that the organism can't control." Such circumstances produced an evolutionary bottleneck, during which a massive number of pseudogenes became fixed as the pathogen's host range and virulence changed.

The comparative sequencing of *S. typhi* and a second, independent derivative of the ancestor *Salmonella enterica*, *S. enterica* serovar Paratyphi A (hereinafter *S. paratyphi* A), provides further evidence for Parkhill's hypothesis (McClelland et al., 2004). Like *S. typhi*, *S. paratyphi* A has become a systemic pathogen restricted to infecting humans. Each serovar contains approximately 200 pseudo-genes, but only about 30 of them are common to both. Those shared pseudogenes comprise a "list of genes that we thought were important and we thought might be selective for *Salmonella* starting to become an invasive pathogen, [such as] secreted effector proteins, genes involved in host range, and shedding genes, amongst others," Parkhill observed. Moreover, he said, "The interesting thing about most of these common pseudogenes in *typhi* and *paratyphi* A is that they don't have the same inactivating mutation. They have been acquired indepen-dently. That suggests that they are probably selectively required."

The same evolutionary processes have also produced less-virulent patho-gens, Parkhill said. For example, the sequence of the bacterium *Streptococcus thermophilus*, used to ferment yogurt, reveals its descent from an oral human pathogen, *Streptococcus salivarius* (Bolotin et al., 2004). "That suggests—and it seems likely—that people started fermenting yogurt 10,000 years ago on the Russian steppes while spitting into milk to initiate fermentation," he explained. "After a while, they probably realized that this was quite disgusting or they found some really good strains and they propagated them because they made nice yogurt. Basically, what people have been doing with yogurt is a 10,000-year microbiology experiment. What happens if you take a pathogen and adapt it to a new niche—fermenting yogurt—that has not existed before? What happens is, you get a massive increase . . . in pseudogenes, which has knocked out most of the genes that were involved in making this an oral pathogen."

Thus, he concluded, the presence of many pseudogenes in an organism's genome bespeaks recent and precipitous evolutionary change, but not necessarily change toward pathogenicity. Pseudogenes are what remain in the chromosome of an organism that has adapted rapidly to a new niche, Parkhill observed; the loss of those nonfunctional genes occurs much more slowly. "This suggests that a large proportion of the changes we see in these organisms is really due to drift," he said. "There are a few selective changes, but a lot of it is random drift."

Turning to more recent events in evolution in *S. typhi*, Dougan described a sequencing study he and coworkers conducted to compare 200 gene fragments of approximately 500 base pairs, each from 105 globally representative *S. typhi* isolates.[1] In this monophyletic pathogen, they identified only 88 single nucleotide polymorphisms (SNPs), which included at least 15 independent mutations to the same crucial gene encoding a DNA gyrase subunit (Roumagnac et al., 2006).

---

[1] Using various advanced sequencing techniques, Dougan and coworkers have also attempted to classify variation across the entire *S. typhi* genome (Roumagnac et al., 2006). Interestingly, this comparison detected no evidence of genetic recombination, indicating that the species is genetically isolated, Dougan said, adding that this may be a feature of other host-adapted pathogens.

These mutations confer resistance to fluoroquinolone (nalidixic acid) antibiotics, which were introduced in the late 1980s for the treatment of multiantibiotic-resistant *S. typhi* infections.

Using this information, Dougan and colleagues constructed a phylogenetic tree of *S. typhi*. Based on its SNP content, "any new isolate can be unequivocally assigned to the tree," Dougan said. Moreover, "the SNPs actually associate with different types of mutations in different parts of the backbone of the protein, which give rise to different nalidixic acid-resistant clones." Thus, the tree can be used to discriminate among isolates, but also "to stratify the acquisition even down to the point mutation of a drug resistance marker."

Dougan predicted that this method, which he termed DNA-based signature typing, will give rise to a "new era" of field- and clinic-based microbial pathogenesis studies. Researchers will be able to link phenotypes with particular SNP markers present in bacteria isolated from patients, he said; applications could include efforts to identify the genetic basis of enhanced transmission or virulence in emergent pathogen strains, to trace carriers of infectious diseases, and to conduct type-specific vaccine efficacy studies.

In addition to SNPs, Dougan noted another route to antibiotic resistance that appears recently to have been taken by non-typhoidal serovars of *Salmonella*, including *S. typhimurium*. These invasive infections—by pathogens that normally cause gastroenteritis—have become a major cause of morbidity and mortality in African children (Gordon et al., 2008; Graham, 2002). "Most of the children and people in Africa who were dying of salmonellosis, invasive disease, were not dying of *S. typhi*; they were actually dying of the strains that normally cause gastroenteritis, like *S. typhimurium* and *enteritidis*," Dougan observed. Sequences of strains causing non-typhoidal salmonellosis (NTS) proved genetically distinct from *Salmonella* strains (of the same serovars) that cause gastroenteritis in Western populations: they bore plasmids containing two distinct genetic elements that conferred resistance to multiple antibiotics, as well as to quaternary ammonium (a disinfectant; Graham et al., 2000). "It's almost designed by nature to be the perfect solution to man's attempt to treat with antibiotics," Dougan said, as well as with antibiotics such as chloramphenicol.

Dougan warned that these resistance genes could spread rapidly through horizontal transfer to other *Salmonella* strains following the planned introduction of large-scale antibiotic prophylaxis (trimethoprim-sulfonamide) for human immunodeficiency virus (HIV)-infected African children. "We talked about the relationship between commensals and pathogens: they know no boundaries," he observed. "I can't think it's going to be very long [after the introduction of large-scale antibiotic prophylaxis] before we actually trigger the movement of this potential transporter around the population of Africa. I'm very alarmed at this, and I think we need to think a little bit further about how we go about doing that."

Meanwhile, as they attempted to understand the genetic origins of NTS,

Dougan and coworkers discovered that antibody protects against the disease, which disproportionately affects children between four months of age (before which they are protected by maternal antibodies) and two years of age (after which their own immune systems develop effective defense against the pathogen; MacLennan et al., 2008). This finding suggests that vaccines against NTS may be effective in inducing protective antibody in the vulnerable age group.

BACTERIAL PATHOGENICITY:
AN HISTORICAL AND EXPERIMENTAL PERSPECTIVE

Stanley Falkow, Ph.D.[2]
Stanford University

Joshua Lederberg noted in his 1987 essay that "the importance of bacteria as agents of infectious disease was clearly established by 1876, but this motivated little interest in their fundamental biology until about sixty-five years later" (Lederberg, 1987). He was taught, as I was, that bacteria were *Schizomycetes*—asexual primitive plants—so it was hard to think of them as being inherently pathogenic. Salvador Luria said of those times that microbiology was the last stronghold of Lamarckism.

Lederberg, while he was a student, was influenced by several pivotal discoveries in the mid-1940s that paved the way for his subsequent work on bacterial conjugation, including the demonstration of the mechanism of bacterial transformation by Avery, MacLeod, and McCarty (1944) and of bacterial mutagenesis and selection by Luria and Delbrück (1943). Lederberg's discovery of bacterial conjugation permitted investigators for the first time to study microbial genetics and biochemistry. It was a dream come true for the young Lederberg; he recalled that he had worn out the pages of the book on physiological chemistry that he received for his Bar Mitzvah. Josh also realized from the outset that the techniques he was developing might have practical applications for vaccine improvement and also in attaining "an understanding of virulence, a latter-day extension of Pasteur's primitive techniques" (Lederberg, 1987).

Lederberg, with his student Norton Zinder and collaborator Bruce Stocker, discovered in the early 1950s that any piece of bacterial DNA can be incorporated into a bacteriophage genome (Stocker et al., 1953). He understood from this that gene recombination, termed generalized transduction, probably occurred in nature because phages were shown to be the basis for several of the different kinds of known *Salmonella* serotypes.

Lederberg's fundamental studies in bacterial genetics were a major factor for the discoveries, in subsequent years, of messenger RNA, the genetic code, and the work of Jacob and Monod (1961) on gene regulation. This revolutionary body of

---

[2]Robert W. and Vivian K. Cahill Professor of Microbiology and Immunology.

work became the foundation for modern molecular biology and also set the stage for the present-day study of bacterial pathogenicity. Since I have been asked to talk about bacterial pathogenicity in both an historical as well as experimental perspective, perhaps I will be forgiven for using some of my own work as well as some of the work of my Stanford colleagues for the discussion of this topic.

I began my work on the genetic basis of pathogenicity in 1959, working on the typhoid bacillus with Louis S. Baron at the Walter Reed Army Institute of Research. Baron had worked in Lederberg's laboratory (Lou once told me that Lederberg claimed that if an experiment had more than six plates and four pipettes, it was over-designed). My goal at the time can be simply stated: I wanted to know the genetic differences between *Salmonella* and non-pathogenic residents of the bowel such as *Escherichia coli*.

I worked with medical microbiologists who thought, as many still do, that a pathogen is any organism that causes disease. Microbiologists at the time characterized pathogenic bacteria as degenerate forms that have lost their way and that simply grew at the expense of the host, thereby causing damage (disease). I thought, as I said at a seminar at Cold Spring Harbor in 1964, that pathogens have evolved unique genetic traits that made them that way (to which a very famous scientist in the audience replied, "Falkow, no one gives a s—t about typhoid or pathogens. Why don't you work on something important?"). Alas, there was a point to this criticism. While I attempted to show that there were unique pathogenicity genes, I lacked the necessary genetic and molecular experimental tools to make the point.

Instead, I turned to the study of episomes, later to be called plasmids (a term Lederberg coined). Josh and Esther Lederberg discovered the first plasmid, the F factor, which appeared to be a transmissible genetic element that determined the fertility of *E. coli* K-12. Strains harboring F could transfer their genes to other bacteria. The work of William Hayes and later Jacob, Monod, Wollman, and others subsequently refined the biology of bacterial fertility. Soon, other examples of plasmids were reported, including transferable resistance to a number of antibiotic drugs. Infectious multiple-drug resistance was described to the Western world around 1960 by Tsutomu Watanabe, and confirmed by the work of Naomi Datta in England and David Smith and others in the United States. These R plasmids, as they were called, became the focus of a large number of scientists during the mid-1960s. Another scientist's work on plasmids—that of a veterinarian named H. Williams Smith—has not been well appreciated. He demonstrated that some plasmids could transmit bacterial toxins, adhesins, and, to some extent, host specificity, from one bacterial cell to another (Smith and Halls, 1967). Smith used the classic Lederberg approach, using only pipettes, Petri dishes, and simple genetic crosses to make these significant discoveries.

In 1972, Stanley N. Cohen and Herbert Boyer discovered that genes could be cut and spliced by using R plasmids and their derivatives, thus signaling the dis-

covery of gene cloning. This elegant new technique, as well as the development
of DNA sequencing, made it possible to finally study pathogenicity genes.

Redefining Bacterial Pathogenicity Using the Tools of Molecular Genetics

Among the things we have learned about bacterial pathogenicity are these
fundamental characteristics:

- Pathogens are impressive cell biologists. Twenty-five years of accumulated data demonstrate that bacteria manipulate the normal functions of
the host cell in ways that benefit the bacteria (Figure 3-1).
- Horizontal gene transfer via mobile genetic elements has been an extremely
important force in the evolution of bacterial specialization, including that
of pathogens. The genes for many specialized "bacterial" products, like
toxins and adhesins, actually reside on transposons and phages.
- The inheritance of blocks of genes, called pathogenicity islands, is often
the key to the expression of pathogenicity in bacteria.

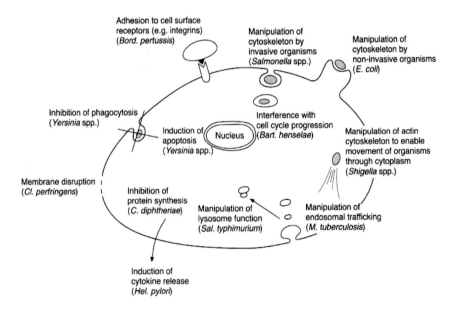

FIGURE 3-1 Pathogenic bacteria interfere with or manipulate for their own benefit the
normal function(s) of the host cell.
SOURCE: Figure reprinted from Wilson et al. (2002) with permission from Cambridge
University Press. Copyright 2002.

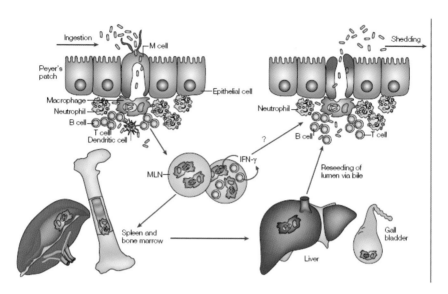

FIGURE 3-2 *Salmonella* infection. IFN-γ = gamma interferon; MLN = mesenteric lymph node.
SOURCE: Monack et al. (2004b).

When we began using the tools of molecular genetics to examine virulence genes and identify their functions, we first tried to isolate particular genes for toxins and other likely virulence products. Today, we take a very different approach, which reflects our understanding of pathogen behavior. For example, unlike commensal bacteria, *Salmonella* breaches the host's epithelial barrier, usually in areas of the intestinal epithelium known as Peyer's patches, and becomes engulfed by phagocytic cells. Instead of being killed, the pathogen replicates there and is then distributed to the liver and the spleen. Eventually, in many cases, the pathogen will be shed by the host, often over long periods of time. These key events in the pathogenesis of infection, in addition to the interaction of the pathogen with the host's innate and adaptive immune systems, are illustrated in Figure 3-2.

The difference between the pathogen *Salmonella* and ancestral, commensal organisms in the bowel is based on the inheritance of pathogenicity islands, which give these bacteria the ability to leave the confines of the colon for locations where other bacteria would be killed. There the evolving pathogen can act free from competition. To identify the genes that enable such incursions into the host, we use a microarray-based negative selection strategy, as shown in Figure 3-3, which allows us to screen the entire *Salmonella* genome for genes that are associated with different stages of infection (Chan et al., 2005).

Using this strategy, one finds that while many genes are expressed within

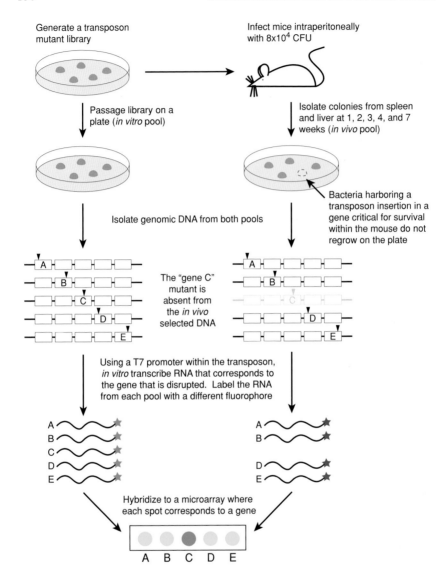

FIGURE 3-3 Microarray-based negative selection strategy. hyb = hybridization; IP = intraperitoneal.
SOURCE: Figure courtesy of Kaman Chan, Ph.D., Stanford University.

the first week of disease, there is a group of genes that are not expressed until the second week of the disease, and even others that are not expressed until the third or fourth week of disease (Figure 3-4; Lawley et al., 2006). These results indicate that particular genes are required for different stages of persistent infection within the mouse. Some of the genes are involved in the ability of *Salmonella* to excrete proteins that kill macrophages during initial infection, while others

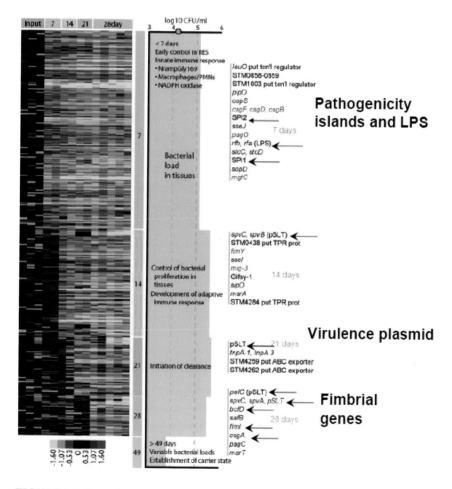

FIGURE 3-4 Time-dependent selection of persistence genes. A yellow box indicates that the persistence gene is absent. A blue or black box indicates that the persistence gene is present. CFU = colony-forming units; LPS = lipopolysaccharide; PMNs = polymorphonuclear leukocytes; prot = protein; RES = reticuloendothelial system.
SOURCE: Modified from Lawley et al. (2006).

have evolved to allow the bacterium to replicate and persist within the vacuoles of macrophages.

Thus, it takes many genes to contribute to the ability of *Salmonella* to make its journey from the mouth to the Peyer's patch in the epithelium of the small intestine, and from there into a macrophage that is spread to other sites. With the technology now available to us, we can now identify all such genes quite readily, but it will take many years before we will be able to determine their exact biological function in *Salmonella*-host interactions.

The Host as a Reporter of Response to Infection

The host can also tell us what happens during acute and persistent bacterial infections. When Lucinda Thompson, Denise Monack, and I examined gene expression in the peripheral blood of infected mice following *Salmonella* infection, we found that 344 genes were induced during acute *Salmonella* infection of naïve mice (Figure 3-5). These genes are associated with interferon signaling, infiltration of neutrophils, and leukocyte extravasation.

We looked at specific gene pathways that were up regulated or down regulated in response to *Salmonella* infection, and we compared these patterns of host response with those that occurred in mice that had been previously immunized with a known *Salmonella typhimurium* vaccine strain, *aroA*. Similar genes were induced in early infection in immunized mice, as in naïve mice, and therefore constitute an "immunity signature" (Figure 3-6). Persistently-infected mice (10 months post-infection) also expressed a characteristic profile of induced genes, as compared with uninfected controls, that included genes involved in antigen presentation, defense responses, and the regulation of various cellular processes. At six months, most animals are moving toward a "normal" pre-infection gene expression profile, but some continued to express infection-associated genes.

Thanks to mouse models developed by Denise Monack et al. (2004a), it is now possible to look at the transmission of *Salmonella* from persistently-infected animals that are asymptomatic, but which excrete the bacterium for up to a year. Naïve mice placed in the same cage as persistently infected animals become infected. Some, but not all, become "supershedders," excreting very high levels of bacteria in their stools and transmitting the infection to their uninfected cage mates (Lawley et al., 2008). Epidemiologists have identified people who are supershedders of certain infectious diseases, as well. Although they are asymptomatic, *Salmonella* supershedders have been found to have a pronounced inflammatory response as compared to other persistently-infected mice (which excrete relatively low levels of bacteria).

The bacteria from supershedders are no more virulent or transmissible than those from other infected mice. It was not clear what factors contributed to the supershedder property. The mice are genetically inbred so presumably an animal becoming a supershedder was governed by non-genetic factors. Mice immuno-

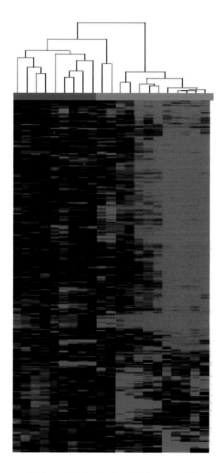

FIGURE 3-5 Response to *Salmonella* infection. Columns under the blue box correspond to separate naïve animals before challenge. Each column under the pink box corresponds to the gene activation in a single animal after wild-type infection (days 4-9 post-challenge). SAM (Statistical Analysis of Microarrays) of naïve animals pre-challenge versus post-challenge showed a total of 334 genes that were significantly (more than 2 fold) induced and 105 that were significantly repressed in infected animals. FDR (False Discovery Rate) = 0.93 percent, 2-fold change.
SOURCE: Unpublished experiments of L. Thompson, D. Monack, and S. Falkow (2007).

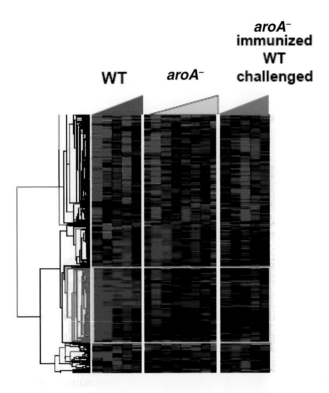

FIGURE 3-6 Response to *Salmonella* infection in 129sv mice following immunization. The host transcriptional response in peripheral blood cells was measured in naïve mice challenged by oral infection with *Salmonella typhimurium* (red). The transcriptional response of peripheral blood cells in animals similarly challenged with a vaccine strain (*aro*A⁻; yellow). The transcriptional response of peripheral cells in animals previously vaccinated with the *aro*A⁻ strain and subsequently challenged with wild type *S. typhimurium* (orange) *aro*A⁻ = *aroA* mutant.
SOURCE: Unpublished experiments of L. Thompson, D. Monack, and S. Falkow (2007).

compromised by steroid treatment were no more likely to become supershedders. However, if mice are given a small, single dose of neomycin—enough to reduce the commensal intestinal flora for less than 48 hours—and then in four days are exposed to *Salmonella*, almost all of them become supershedders (Lawley et al., 2008). These animals go from having low to high levels of intestinal inflammation, and at the same time, they become very efficient transmitters of the disease. One possible interpretation of this observation is that the fine balance between the commensal flora and the innate immune system is disrupted by the elimination of certain members of the normal flora by antibiotic killing and provides a window of opportunity for the establishment of larger populations of *Salmonella* in the bowel.

Pathogens Versus Commensals

The term *pathogen* is derived from a Greek phrase that means, "the birth of pain." A successful pathogen must do the following:

- Acquire virulence genes
- Sense the environment
- Switch virulence genes on and off
- Move to the site of infection
- Become established
- Acquire nutrients
- Survive stress
- Avoid the host's immune system
- Subvert host cytoskeleton and signaling pathways
- Disseminate to other sites and/or hosts (in some cases)
- Be transmitted to a new susceptible host

Successful commensals also do many of these things, but unlike a pathogen, a commensal does not have an inherent ability to cross anatomic barriers or breach host defenses that limit the survival or replication of other microbes. As previously noted, pathogens (but not commensals) choose to live in dangerous places to avoid competition and get nutrients. These invasive properties are essential to the pathogen's survival in nature.

Nevertheless, several members of the human bacterial flora that usually live uneventfully in the human nasopharynx—including *Streptococcus pneumoniae*, *Neisseria meningitidis*, *Haemophilus influenzae* type b, and *Streptococcus pyogenes*—sometimes cause disease. These microbes, which I call "commensal pathogens," have virulence determinants that suggest that they regularly come into intimate contact with elements of the innate and adaptive immune system (Falkow, 2006). Recent history shows that immunization against these pathogens not only prevents human disease but also eliminates the microbes' ability to colonize the human host efficiently. Such commensal pathogens persist in a significant proportion of the human population, the vast majority of whom are asymptomatic carriers.

This raises several questions: Are such microbes pathogens, or are they commensals? Are "virulence" factors often actually a subset of adaptive factors that allow microbes to exploit a particular niche, but not necessarily designed to cause disease? Are they virulence factors, or would a better term be *colonization factors*? I submit that medicine's focus on disease may sometimes distract us from understanding the biology of pathogenicity. Of course, it is important that there be a medical definition of a pathogen that is based on disease. On the other hand, I would argue that disease does not encompass the biological aspects of pathogenicity and the evolution of the host-parasite relationship. Many pathogens cause persistent infections in humans (e.g., *Mycobacterium tuberculosis*, *Treponema pallidum* [the cause of syphilis], *Chlamydia*, *S. typhi*, and *Helicobacter pylori*), and most such infections are asymptomatic. Might these microbes be considered commensal pathogens?

Moreover, as Figure 3-7 suggests, the decline of many infectious diseases in industrialized countries has been accompanied by an increase in immune disorders, such as multiple sclerosis, Crohn's disease, and asthma. The eradication of *Helicobacter* has been associated with an increase in esophageal cancer and, more recently, asthma (Blaser et al., 2008). Perhaps one third or more of the world's population is asymptomatically colonized with *Mycobacterium tuberculosis*. Over the course of the last century, as Western society has eliminated this pathogen through basic public health measures, it seems that we may actually have eliminated a microbe with which our species had a dialog, a microbe that "talked" to the human immune system and kept it primed for defensive action. Perhaps, as *Pogo* cartoonist Walt Kelly has suggested, "we have met the enemy and he is us."

If the nature of microbial pathogenicity is schizophrenic—characterized by inconsistent or contradictory elements—then it is important to study every aspect

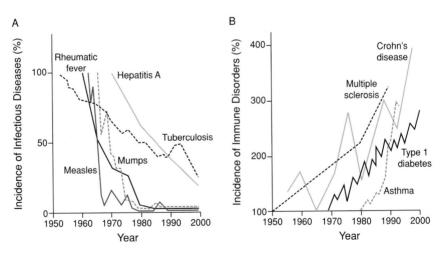

FIGURE 3-7 The decline of many infectious diseases in industrialized countries (A) has been accompanied by an increase in immune disorders (B), United States, 1950 to 2000. SOURCE: Bach (2002).

of its biology, and not be distracted by its role in causing disease. Instead, we must follow the instruction of Thomas Huxley, as expressed in a quotation from Huxley's letter to Charles Kingsley on September 23, 1860, that both Lederberg and I have treasured: "Sit down before fact as a little child, be prepared to give up every preconceived notion, follow humbly wherever and to whatever abysses nature leads, or you shall learn nothing" (Huxley, 1901).

Acknowledgments

Josh Lederberg was a remarkable man with a remarkable intellect. His legacy is yet to be fully appreciated. He influenced me from the outset of my career, and although I told him so in person on several occasions, I think it appropriate to express my gratitude to him here as well.

My remarks at this symposium were summarized by Eileen Choffnes and her staff and formed the basis for this paper. I am grateful to Lucinda Thompson and Denise Monack for permitting me to present here some of their previously unpublished results on the host response in naïve and immunized mice to *Salmonella* infection. I also thank Denise Monack and Trevor Lawley for discussing with me their recent exciting findings on *Salmonella* transmission.

EVOLUTION OF BACTERIAL-HOST INTERACTIONS: VIRULENCE AND THE IMMUNE OVERRESPONSE[3]

Elisa Margolis[4]
Emory University

Bruce R. Levin, Ph.D.[4]
Emory University

While many people may not believe in evolution, for those of us with the great taste and good fortune to work with bacteria, viruses, and single cell fungi, evolution is not a matter of belief, and much less one of faith. Evolution is something we constantly see whether we want to or not. For those who are evolutionary biologists by training, inclination, or aspiration there is an obligation to be more than just witnesses and historians of evolution. We have to provide explanations for the origin and maintenance of all biological phenomena. There can be no exceptions.

Coming up with these explanations and better yet with testable evolutionary

---

[3]This paper was originally published in Baquero, F., C. Nombela, G. H. Cassell, and J. A. Gutiérrez, eds. 2008. *Evolutionary biology of bacterial and fungal pathogens*. Washington, DC: ASM Press. Pp. 3-13. Reprinted with permission from ASM Press.

[4]Department of Biology, Atlanta, GA 30322.

hypotheses is not hard for characters that provide obvious fitness advantages to the organisms that express them. The ascent of resistance following the introduction of antibiotics came as no surprise to evolutionary biologists. In the presence of antibiotics, bacteria that are resistant to their action have an obvious selective advantage relative to their susceptible ancestors. More challenging to account for are situations where it is not clear how the character in question could have evolved by natural selection favoring the individual organisms. While the interactions between parasitic bacteria and their mammalian hosts include many characters that can be explained by natural selection operating at the level of individual bacteria or individual hosts (Burnet and White, 1972), there are many that cannot. Virulence is one of these traits that is hard to account for by simple evolutionary models; why would bacteria harm the hosts they need for their survival?

In this chapter (speculative rant, if you prefer) we focus primarily on aspects of the evolution of the bacterium-host (mostly human) interactions that cannot be readily accounted for by simple, advantage-to-the-individual evolutionary scenarios. We postulate and provide evidence that much of the virulence of bacterial infections can be blamed on the seemingly misguided overresponse of the immune defenses, what is sometimes referred to as "friendly fire" (Levin and Anita, 2001; Whitnack, 1993) or immunopathology (Graham et al., 2005). We consider how this perversity of the immune system fits with current hypotheses for the evolution of virulence, the evolution of the so-called virulence factors, and speculate on the reasons natural selection has failed to or is unable to blunt the immune overresponse to bacterial infections. We conclude with a brief discussion of the implications of this perspective on virulence for the treatment of bacterial infections.

Bacterial Virulence as an Immune Response

We define virulence as the magnitude of the morbidity and the increase in the likelihood of mortality resulting from the colonization and proliferation of bacteria in or on a host. To facilitate our consideration of this virulence and its evolution we use the gross simplification, a cartoon, of the bacterium-host interaction presented in Figure 3-8. Bacteria enter a site, the blue box, where they replicate and establish a population and colonize the host, but in which they do not generate perceptible symptoms. Virulence requires their passage into a second site, the red box, where the presence of bacteria (or their products) can, but need not, cause symptoms, e.g., for a *Streptococcus pneumoniae* bacteremia the blue site is the nasopharynx and the red is the bloodstream. In this model the red site needn't be a different physical location. It could be a different state of the bacteria in the site of their colonization, e.g., for a *Staphylococcus aureus* skin infection, the blue site would be the skin and the red a boil.

In Figure 3-8 as well as in mammals, virulence occurs in two ways, both of which require the bacteria to enter the red, potentially symptomatic site or state:

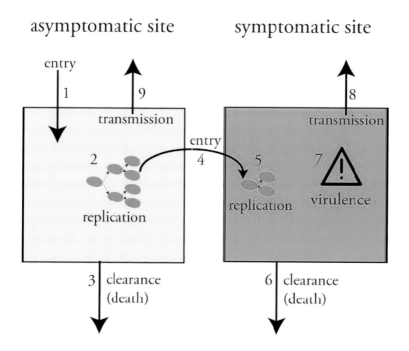

FIGURE 3-8 The artist's conception of the infection process and the host's immune response and overresponse: Blue—site where the presence of bacteria does not result in symptoms—asymptomatic. Red—site/state where the presence of bacteria can result in symptoms.

(i) direct damage to the host tissue is caused by the replication of the bacteria and/or the production of specific products (toxins), or (ii) indirect damage to the host occurs through an inappropriate or overresponse of the immune system. Both types of damage are represented by the "!" within a triangle. In this scheme the immune defenses can prevent virulence in one or more of seven related ways:

1. Limiting the entry of bacteria into the asymptomatic site
2. Limiting the proliferation of the bacteria in the asymptomatic site
3. Increasing the rate of clearance of bacteria and their products from the asymptomatic site
4. Preventing entry of bacteria or their products into the potentially symptomatic site or state
5. Reducing the rate of proliferation of the bacteria within the potentially symptomatic site or state
6. Increasing the rate of clearance of bacteria and their products from the potentially symptomatic site or state

7. Preventing an immune overresponse to the bacteria or their products in the potentially symptomatic site or state

The first three of these immune responses maintain the density of bacteria and concentrations of their products in the asymptomatic site at levels where they are unlikely to spill over or otherwise enter the site or state where they can generate symptoms. Whether they do generate symptoms and the magnitude of those symptoms given passage into the red site or state also depends on how well the immune system limits their densities and the concentrations of their products. In Figure 3-8, the number 8 is the infectious transmission of bacteria promoted by the generation of symptoms, and the number 9 is the transmission of bacteria from the asymptomatic site. All of these enumerated steps in which the immune system limits the virulence of the bacteria can be classified as appropriate responses. However, inappropriate responses, for which we use the term *overresponse* when they lead to host damage, may occur due to defects in one of these steps or as a secondary consequence of mounting an immune response.

In the following we focus primarily on the virulence resulting from the overresponse of the immune system. There are, however, examples of virulence that can be attributed to the direct damage of host tissue by the replication of bacteria or the secretion of their products. Included among these are (i) dental caries, resulting from the acid produced by metabolizing *Lactobacillius acidophilus or Streptococcus mutans* (Gibbons, 1964), (ii) paralysis due to the neurotoxins secreted by *Clostridium botulinum* or *Clostridium tetani* acting on the nerve and motor endplates (Schiavo et al., 1992), and (iii) diarrhea resulting from enterotoxins that inhibit resabsorption of sodium chloride or promote its secretion. Examples of virulence being a direct product of the interaction between bacteria and host cells appear to be rare relative to those in which the morbidity and mortality can be attributed to the indirect damage due to an immune overresponse.

As illustrated in Table 3-1, the morbidity and mortality of bacterial infections can be attributed to the host's immune system operating in one of three inappropriate ways: (i) being more vigorous than needed, (ii) being incorrect for that pathogen, or (iii) responding to the wrong signals. The best-investigated example of the immune system responding too vigorously is bacterial sepsis, where the entry of cytokines and bacteria into the bloodstream brings about widespread blood vessel injury and multiple organ failure (impaired pulmonary, hepatic, or renal function). Here the response to the bacteria is at one level appropriate, as the cytokines released play an important part in attracting neutrophils (immune cells that phagocytose bacteria) to the local infection site, but is also excessive (Kurahashi et al., 1999). The distinction between an inappropriate and appropriate immune response can be seen in the spectrum of illness associated with *Mycobacterium leprae* (Modlin, 2002; Sieling et al., 1999). Hosts that respond to infection predominantly with antibodies and very few CD4 T cells have infectious sites with large macrophages that contain numerous mycobacteria. These macrophages are responsible for the multiple skin lesions and nodules seen in lepromatous leprosy, while a host with T helper 1-type response (high interferon-γ production and low interleukin-4 [IL-4]) has numerous well-formed granulomas with very few mycobacteria that form minor skin lesions. Superantigens provide an example of the immune system responding to an incorrect signal. Superantigens are bacterial products that stimulate a large number of T cells (1–40% of T cells will react) by binding to major histocompatibility complex class II molecules and T cell receptors (beta chain) independently of their specificity for antigens (Rott and Fleischer, 1994). *S. aureus*, *Streptococcus pyogenes*, *Mycoplasma arthritidis*, and *Yersinia pseudotuberculosis* are among the bacteria that produce superantigens. In the case of toxic shock syndrome, the superantigens produced by *S. aureus* induce the indiscriminate and overwhelming activation of T cells leading to the production of cytokines that mediate shock and tissue injury. In all three of these cases the morbidity and mortality of the host can be attributed to an apparently misguided response of the immune system, which we refer to as an overresponse.

The Evolution of Bacterial Virulence as an Immune Response

How does the observation that much of the morbidity and mortality can be attributed to a host overresponse to bacteria help in understanding the evolution of virulence? In a perspective written a decade ago, one of us (Levin, 1996) listed four hypotheses that account for the evolution of virulence: (i) the conventional wisdom, (ii) epidemiological selection, (iii) coincidental evolution, and (iv) short-sighted, within-host evolution. Since that time, although there have been a number of theoretical, experimental, and speculative articles on the evolution of virulence (for a small and admittedly biased sample see Andre and Godelle, 2006; Bonhoeffer and Nowak, 1994; Brown et al., 2006; Bull, 1994; Ebert and Bull, 2003; Ebert and Herre, 1996; Frank, 1996; Grech et al., 2006; Lipsitch et al., 1995, 1996; Lipsitch and Moxon, 1997; Mackinnon and Read, 2004; Regoes et al., 2000), we do not know of studies that have rejected any of these hypotheses and only one adding what may be a new hypothesis: quasispecies evolution (Pfeiffer and Kirkegaard, 2005). This fifth hypothesis may only apply to viruses with high mutation rates and arguably could be subsumed under the broader rubric of within-host evolution. In this section we consider how the observation that morbidity and mortality of bacterial infections can be attributed to the hosts' immune overresponse fits each of these hypotheses for the evolution of the virulence of bacteria.

The Conventional Wisdom

This phrase, which the late John Kenneth Galbraith coined to describe ideas and explanations that are widely accepted as true by the public, was applied by

TABLE 3-1 Some Examples of Virulence Resulting from an Immune Overresponse

Disease	Bacteria	Red site[a]	Bacterium-host interaction[b]	Damage induced by immune response
Pneumonia	Multiple species, e.g., S. pneumoniae, N. meningitidis, H. influenzae	Lungs	Cell wall and other bacterial components → induce proinflammatory cytokines, edema, and fibrin deposition (Bergeron et al., 1998)	Recruitment of fluid and cells into the air spaces of the lungs
Toxic shock syndrome/scarlet fever	Staphylococcus, Streptococcus	Circulatory system, systemic	Superantigens → indiscriminate activation of T cells (McCormick et al., 2001)	Extreme inflammation leading to septic shock
Duodenal ulcers	Helicobacter pylori	Gastric and duodenal mucosa	Bacterial persistence → chronic inflammation (Czinn and Nedrud, 1997)	Mucosal atrophy
Septicemic plague	Yersinia pestis	Circulatory system, systemic	Endotoxin and/or unknown factors → systemic inflammatory response	Acute shock
Cutaneous anthrax	Bacillus anthracis	Skin	Anthrax lethal factor (toxin) → release of TNF-α and IL1β (Firoved et al., 2005)	Tissue necrosis

Petechiae, Waterhouse-Friderichsen syndrome	*Neisseria meningitidis*	Skin and adrenal glands	Endotoxin → release of TNF-α, Blood vessel destruction IL-1, and IL-6 (Klein et al., 1996)	Blood vessel destruction
Rheumatic fever	*Streptococcus pyogenes*	Heart, joints, skin, and brain	M protein → activation of autoreactive (anticardiac myosin) antibodies (Cunningham, 2003)	Antibody-induced damage of heart muscle; immune complex deposited in large joints and skin
Reactive arthritis	*Chlamydia trachomatis*	Joints, eyes, urethra	Bacterial persistence → activation of autoreactive immune cells (Sieper, 2001)	Inflammation of large joints, eyes, and urethra
Tuberculosis	*Mycobacterium tuberculosis*	Lungs	Bacterial presence → release of high levels of TNF-α (Bekker et al., 2000; Kaushal et al., 2002)	Recruitment of fluid and cells into the air spaces of the lungs; necrosis
Meningitis	Multiple species, e.g., *S. pneumoniae, N. meningitidis, H. influenzae*	Meninges	Cell wall and other bacterial components → induce proinflammatory cytokines and chemokines (Braun et al., 1999)	Increased blood-brain barrier permeability; neutrophil recruitment; increased intracranial pressure; brain damage

[a]See Figure 3-8 for an explanation of red site.
[b]TNF-α, tumor necrosis factor alpha.

Bob May and Roy Anderson (May and Anderson, 1983b) to describe the then-prevailing view of the evolution of the virulence. According to that wisdom, virulence is an artifact of the relative novelty of parasite's association with its host. As the relationship between the parasite and host matures, natural selection in either the parasite or host population or both will lead to the extinction of one or the other species or the evolution of symbiosis or mutualism.

While the original theory behind this hypothesis for the evolution of virulence of infections amounts to little more than the adage "don't bite the hand that feeds you," the evidence in support of it was and remains compelling. Many of the bacteria responsible for morbidity and mortality of humans were acquired from other species in the not-so-distant past (after the advent of agriculture), and some are continuously acquired in this way. Included among these zoonotic (and protozoonotic) infections are plague, tuberculosis, Legionnaires' disease, botulism, anthrax, brucellosis, tularemia, Rocky Mountain spotted fever, cholera, and other diarrheal diseases. The bacteria responsible for some of these infections, such as *Mycobacterium tuberculosis*, are transmitted between humans and can be maintained without the animal source. Others such as *Legionella pneumophila* are not. Also consistent with the conventional wisdom is the correlated observations that only a very small minority of the vast numbers of species of bacteria that colonize mammals cause disease.

It may seem that the proposition that the virulence of bacterial infections can be attributed to host immune overresponse fits quite well with this conventional wisdom. To wit, the immune system has not yet had the time to evolve to moderate the response to these novel bacteria and their products and/or these bacteria have not yet evolved into being nice. Eventually, or as it was once referred to, on "equilibrium day," (Levin et al., 2000), mutualism will prevail and the immune overresponse will be tempered.

Epidemiological Selection

The conventional wisdom is an observation rather than a mechanism, an observation that focuses on the interactions between bacteria and the individual hosts they colonize. To fully understand the evolution of commensal and pathogenic bacteria, however, it is necessary to consider their lifestyle outside the host and, in particular, their transmission between hosts. One approach to this more comprehensive picture of the evolution of parasitic microbes has been to draw inferences about the nature and direction of selection from epidemiological models (Levin et al., 1982; Levin and Pimentel, 1981; May and Anderson, 1983a). In accord with this perspective, the fitness of a particular strain of bacteria is given by its basic reproductive number, R_0, the number of secondary infections caused by a single infected individual in a wholly susceptible population of hosts; the higher the value of R_0, the greater the fitness of the bacteria. In these traditional epidemiological models, virulence is only expressed as mortality. Morbidity and other more subtle effects of infections are not directly considered in these epidemiological models.

As long as the transfer to new hosts requires viable hosts, selection will favor bacteria that are not only infectiously transmitted at ever-higher rates but also persist longer in colonized hosts (i.e., are less likely to kill the host). In other words, selection will favor ever-more-benign, symbiotic, or better yet, mutualistic bacteria. Evolution in the host population will also be for reduced virulence; hosts that are less subject to infection-associated morbidity and mortality will be favored. As long as transmission occurs from the blue asyptomatic site (8 in Figure 3-8) rather than the red site (9 in Figure 3-8), these epidemiological models can be seen as the theoretical basis of the conventional wisdom (also see Lenski and May, 1994). If, however, transmission and the morbidity and mortality of the host are coupled so the more virulent bacteria are transmitted at higher rates than the more benign, there is a trade-off between the loss of the host and gain to the bacteria; virulence would be favored in the bacterial population (Ebert and Bull, 2003).

On first consideration it may seem that this transmission and virulence trade-off is inconsistent with the proposition that the morbidity and mortality of the infection is a product of the host's immune overresponse. We suggest this is not necessarily the case. The host overresponse could be a by-product of selection operating on individual bacteria to promote their transmission. While we don't know of overwhelming, quantitative, empirical evidence of this being the case for any pathogenic bacteria (viruses are another matter; see Fenner and Ratcliffe, 1965), this interpretation is supported by reasonable plausibility arguments. Here we consider two of the more compelling of these examples of pathogenic bacteria of humans.

The first is the diarrheal diseases in which humans play a significant role in the transmission process. Because of the massive output of bacteria, diarrhea is likely to increase the density of bacteria in water and food products and thereby the transmission rate of these bacteria. Thus, as long as transmission is promoted by diarrhea, selection in the bacterial population will favor mechanisms that cause diarrhea. In some cases the induction of diarrhea is attributed to what can be seen as immune overresponse. The dysentery bacteria *Shigella flexneri* induces the release of the cytokine interleukin 1 (IL-1) in infected macrophages, which leads to extensive injury of the colon mucosa, which in turn results in fluid and protein loss into the intestinal lumen and the ensuing diarrhea (Hilbi et al., 1997). This hypothesis for the evolution of diarrhea to increase transmission requires that the transmission advantage more than makes up for the loss in transmission due to host mortality. To our knowledge, there are no quantitative empirical studies demonstrating that this trade-off obtains for any diarrheal disease.

The second example is plague. Albeit not yet as well documented as the oft-told mother of all trade-off stories, myxoma and the Australian rabbits (Fenner, 1965), the emerging tale of the evolution of the virulence of the plague bacillus,

Yersinia pestis, has parallels to that story. There is compelling evidence that this flea-transmitted pathogen evolved from a not very virulent enteric, oral-fecal transmitted *Yersinia* relatively recently by the acquisition of a couple of plasmids and a few chromosomal genes (Achtman et al., 1999; Carniel, 2003). Since fleas acquire these bacteria from the blood of rodents, the density of bacteria in circulating blood would be directly associated with the likelihood of their transmission to other rodents (or humans). Also directly associated with this density of bacteria in the blood is sepsis, the virulent manifestation of *Y. pestis* infections. Elisabeth Carniel (personal communication) has suggested that the capacity to generate lethal sepsis is not just a by-product of the proliferation of bacteria in the blood, but may be selected for in the bacterial population. Although the cost-benefit calculation has not been made, it may be that the rate of transmission of the bacteria is augmented by their killing infected rodents, as fleas move to new hosts when their original host dies. For both diarrheal diseases and plague, the virulence resulting from the host overresponse is associated with transmission. Clearly more empirical work would be necessary to confirm the existence of a trade-off between bacterial transmission and an immune overresponse and the postulated exploitation of this overresponse for the epidemiological advantage of the parasite.

Coincidental Evolution

In accord with this hypothesis there is no advantage to the bacteria to make the host sick and certainly no advantage for the host to be ill; virulence is a consequence of the bacteria being in the wrong host or in a wrong site in the right host (Levin and Svanborg Eden, 1990) (the arrow above 7 in Figure 3-8). The bacterial products responsible for the morbidity and/or mortality of the host, virulence determinants, evolved in response to selection for some function other than virulence.

Reasonable candidates for coincidental virulence due to an immune overresponse are diseases associated with *Helicobacter pylori*. These bacteria colonize and maintain populations in the stomachs of the majority of humans for most of their lives without generating symptoms and appear to have done so since prehistoric times (Falush et al., 2003). However, it wasn't until Marshall and Warren (1984) presented evidence that a curved bacteria we now know as *H. pylori* was an etiologic agent for gastric and peptic ulcers that this seemingly commensal bacteria was elevated to the status of pathogen. This distinction was further enhanced by evidence that *H. pylori* was also associated with gastric cancers (Moss and Blaser, 2005; Tatematsu et al., 2005). *H. pylori* colonization can result in a chronic inflammatory state that is generated when the host responses (such as the release of IL-8 and other chemokines, the attraction of neutrophils, macrophages, and the local stimulation of T cells) fails to clear the bacteria and lymphoid aggregates form in the lamina propria of the stomach and duodenum.

This continued stimulation of the immune and inflammatory cells (termed chronic atrophic gastritis) results in the destruction of the gastric epithelium, formation of peptic ulcers, and increased risk for gastric cancers. Presumably, but not yet formally demonstrated, the induction of the inflammatory response and the subsequent diseases provides no advantage to *H. pylori* in a colonized host or its transmission to new hosts. In this sense, the virulence of *H. pylori* in colonized humans is coincidental.

While they are commonly described as pathogens, especially in grant proposals and by people suffering from the symptoms they can generate, a number of bacteria responsible for morbidity and mortality in humans also have good credentials as commensals. Like *H. pylori* they are carried asymptomatically by many and cause disease in few. Included among the more prominent of these commensal pathogens for humans are *S. aureus*, *Haemophilus influenzae*, *S. pneumoniae*, and *Neisseria meningitidis*. From an evolutionary perspective, invasive disease seems to be the wrong thing for these bacteria to do—dead ends. The sites of their virulence, blood and meninges, are certainly not good for their transmission to new hosts by their normal route, through respiratory droplets. The rare virulence of these commensal bacteria can be accounted for by an immune overresponse in these sites (Bergeron et al., 1998; Braun et al., 1999). The occasional movement of bacteria into a site where they can cause disease (the red in Figure 3-8) may be due to chance or coincidental evolution or as we argue below may be a consequence of within-host evolution of the bacterial population.

Within-Host Evolution

In accord with this hypothesis, the virulence of bacteria is the product of selection favoring more pathogenic members of a population colonizing an individual host (Levin and Bull, 1994). The advantage gained by the bacteria by generating symptoms in a colonized host is restricted to that host and may be to its disadvantage in its transmission to a new host; this evolution is short-sighted. A mutant commensal bacterium with the capacity to establish and maintain populations in normally sterile sites, cells, or tissues could be favored within a colonized host because in those sites there is less competition for nutrients and/or those mutant bacteria are somewhat protected from the host immune defenses.

Although we can make a good case and even cite evidence for the virulence of some viruses, such as poliovirus and Coxsackievirus, being the product of within-host evolution (Gay et al., 2006; Levin and Bull, 1994), for bacteria the best we can do at this stage is present arguments founded on plausibility and consistency with observations (see, for example, Meyers et al., 2003). Central to these arguments are the results of studies with mice and rats demonstrating that the bacteria responsible for invasiveness (blood infection) are commonly derived from one of very few cells (Meynell, 1957; Moxon and Murphy, 1978; Pluschke et al., 1983; Rubin, 1987). One possible explanation for these observations is that the bacteria responsible for the blood infections are the products of single, mutant cells with an enhanced capacity to invade and proliferate in blood.

While supporting the within-host evolution hypothesis for virulence, these observations are also consistent with the coincidental evolution hypothesis: that, by chance alone, only one or a few cells establish blood infections can be attributed to very small holes in the host's defenses through which only one or very few bacteria traverse the arrow above 7 in Figure 3-8. Although the coincidental and within-host hypotheses could be distinguished by demonstrating that the bacteria establishing a blood infection have an inherited propensity for the invasion of blood, to our knowledge there are no published studies that have done this test. However, whether the invasiveness of the blood or other normally sterile sites is coincidental or due to within-host evolution, the virulence of bacteria in these sites can be attributed to a host's immune overresponse.

The Evolution of Virulence Determinants

Not all bacteria or even all members of the same species of bacteria capable of colonizing mammals are responsible for disease. One explanation for why some bacteria cause disease and others do not is what have become known as virulence factors or virulence determinants, the expression of which are, by definition, essential for that bacteria to cause disease in (or on) colonized hosts (Finlay and Falkow, 1989). Included among these are characters that facilitate adhesion to host cells, evade the host constitutive and inducible immune defenses, and produce toxins. Appropriately, much of contemporary bacteriology is devoted to understanding the molecular biology, genetics, evolutionary origin, and mode of action of virulence determinants as a way to understand bacterial diseases and ideally prevent or treat them. While virulence determinants (factors) are almost certainly the products of adaptive evolution in bacterial populations, not so clear are the selection pressures responsible for their evolution and maintenance. Are they favored because of virulence, i.e., the morbidity and mortality of the host promotes the colonization, persistence, and infectious transmission of bacteria that express these determinants? Are virulence factors by-products of selection for other functions, e.g., their expression provides protection against grazing protozoa (Wildschutte et al., 2004) and/or facilitates competition with other microbes? Or is the virulence attributed to these factors an inadvertent by-product of their normal function in a host, a primitive character that will be lost on or before equilibrium day. While these hypotheses may be mutually exclusive for any specific bacterium-host and virulence factor, they are clearly not so collectively. Whether they evolve in response to selection for virulence or not, some of these virulence factors are responsible for triggering the immune overresponse.

Why Does the Immune System Overrespond?

In the preceding, we have portrayed the host immune system as misguided, overresponding in ways that cause rather than prevent the morbidity and mortality of a bacterial infection. From the perspective of evolutionary biology, however, "misguided" is hardly an explanation. Colonization by bacteria is not a rare event but rather something mammals confront all the time, and overresponding in a way that results in their morbidity and mortality would almost certainly be selected against. In their review of "immunopathogy," Graham and colleagues postulated a number of reasons for this transgression of the immune response (Graham et al., 2005). Here we offer our perspective on this issue.

As we see it, there are two general classes of explanations for the maintenance of an overresponse of the immune system. (i) While infectious disease may be a major source of morbidity and mortality (Haldane, 1949), disease-mediated selection can be relatively weak, and extensive amounts of time would be required to evolve mechanisms to modulate the immune response to specific bacterial infections. (ii) Functional constraints on the immune system limit the ability of natural selection to totally prevent and maybe even partially mitigate an immune overresponse to bacterial infections.

(a) Even if selection universally favors tempering the immune overresponse to infections, and the favored genotypes could be generated (which we question below [b]), the time required for temperance to evolve could be considerable, especially if the overresponse is specific for particular bacteria and/or their products. This is due to two factors. (a) At its maximum the intensity of selection for modulating the immune overresponse to an infection would equal the fraction of the population with that infection. It would be substantially lower if the symptoms of the infection were not expressed in all colonized hosts, were rarely lethal or sterilizing, or were primarily manifest after reproductive years or if the magnitude of the reduction of the overresponse of the favored genotype was less than absolute. For most of the diseases listed in Table 3-1 virulence is a rare occurrence in colonized hosts (less than 1%), and therefore the intensity of selection against an immune overresponse would be relatively weak. (b) It can take a considerable amount of time for a rare beneficial mutant to ascend to substantial frequencies. For example, if the selection for a reduced overresponse is operating on genotypes at a single locus (the best case), the initial frequency of a favored allele is 10^{-3}, the favored genotype has a 1% selective advantage, and there is no dominance, it would take 1,381 generations (more than 20,000 years for humans) for that gene to reach a gene frequency of 50%. If the favored genotype is recessive, the corresponding number of generations would be 100,491 (Crow and Kimura, 1971).

What about the role of the bacteria in the evolution of a more temperate immune system? As a consequence of their vastly shorter generation times, haploid genomes, and propensity to receive genes and pathogenicity islands by horizontal transfer, it seems reasonable to assume that bacteria would have an edge in an evolutionary arms race with their mammal hosts. We suggest, however, that this edge contributes little if anything to the slowing pace at which mammalian evolution could modulate the immune overresponse. Although there maybe situations where virulence is positively correlated with the infectious transmission of bacteria, in most of these cases the morbidity and mortality associated with their transmission is not to the bacteria's advantage and may be to their disadvantage. Even greater transmission of these bacteria would be possible if the hosts were not debilitated or killed as a result of diarrhea or if the bacteremias required for vector-borne transmission did not result in sepsis. In this interpretation evolution in the bacteria population would not oppose the evolution of a more temperate host immune system. Of all the examples considered in this chapter, the only one in which evolution in the bacterial population might favor an immune overresponse is Carniel's suggestion that by killing their host, *Y. pestis* acquires a transmission advantage.

(ii) While the above realities of the ecology and genetics of natural selection may be part of the answer to the question of why evolution has not eliminated the immune system's overresponse to bacterial (and other) infections, we suggest it is not the most important reason. We conjecture that the primary reason mammalian evolution has not tempered and perhaps cannot temper the immune overresponse to bacterial and other infections is functional constraints that limit the extent to which the immune system can be modified. The immune system has roles other than clearing bacterial infections. It has been postulated that these other roles dominated the evolution of the mammalian immune system (Burnet, 1970). These different roles as well as the extraordinary diversity of organisms colonizing mammals, bacteria, viruses, fungi, and worms of various ilks and the variety of sites of colonization impose different and potentially conflicting demands on the immune defenses, phenomena referred to as antagonistic pleiotropy. An appealing hypothesis for the immunopathogy known as allergies is an overresponse of those elements of the immune system that in less-pristine times would otherwise be occupied with the control of helminth infections (Wilson and Maizels, 2004).

There is a fine line between responding (1-6 in Figure 3-8) and overre-sponding (7 in that figure), which may be difficult for the systems regulating the immune response to perceive, much less avoid. As suggested by Frank (Andre et al., 2004), the intensity of an immune response may be determined by a trade-off between increasing the strength and rapidity of an immune defense and the virulence from an immune system overresponse.

Is there evidence in support of these two hypotheses for why evolution has not eliminated the virulence resulting from the immune overresponse? Not much—at least not yet. We suggest, however, that some of the considerable amount of inherited variability in the susceptibility to infectious disease in human populations (Bellamy and Hill, 1998; Bellamy et al., 2000; Segal and Hill, 2003; Sorensen et al., 1988) can be interpreted as support for these hypotheses. To be sure, there is good and even overwhelming evidence that some of this variation is maintained by disease-mediated balancing or frequency-dependent selection, but this is not the case for all or even the majority of it. We suggest that much of the standing genetic variation in disease susceptibility in human populations is a reflection of the myopia and limitations of natural selection: (i) the relative weakness of selection for modulating the immune overresponse and (ii) even more, the impotency of natural selection due to the constraints on the immune system—antagonistic pleiotropy. Genetic variation that is not or is poorly perceived by natural selection will build up and persist (Crow and Kimura, 1971).

Implications

While the morbidity and mortality of most bacterial infections can be attributed to an immune overresponse, virtually all of our efforts to treat these infections are directed at controlling the proliferation and clearing the bacteria, primarily with antibiotics. This approach has been and continues to be effective, but not completely so. Antibiotic treatment commonly fails, and patients die or remain ill for extended periods. Resistance of the pathogen to the antibiotics employed for treatment is only one of the reasons for this failure and for some infections is not the major one, at least not yet (Levin and Rozen, 2006; Yu et al., 2003).

The obvious alternative approach to treating infections is to reduce the morbidity and prevent the mortality by modulating the immune system's overresponse. There have been attempts to do just that for the treatment of bacteria-mediated sepsis. Clinical trials have evaluated the use of glucocorticoids (Bone et al., 1987), drugs designed to neutralize endotoxins (Ziegler et al., 1991), tumor necrosis factor α (Fisher et al., 1996), and IL-1β (Fisher et al., 1994), but none of these treatments was effective. The most successful trials in humans to date have been with a component of the natural anticoagulant system, activated protein C, which has substantial anti-inflammatory properties along with being a potent anticoagulant (reduces the formation of clots that are responsible for organ failure in late stages of sepsis) (Fourrier, 2004). In addition, new agents redirect the immune response and hold promise as effective future therapies for sepsis, such as IL-12 (O'Suilleabhain et al., 1996) and antibodies against complement (C5a) (Czermak et al., 1999). However, understanding the specifics of the immune overreaction and the intricacies of the feedback mechanisms that control an immune response is necessary for therapies to be directed at enhancing or inhibiting the patient's immune response.

At this time, taken at large, the success of these immune modulating methods in preventing the morbidity and mortality of bacterial infections can at the very best be described as modest. However, in maintaining the speculative nature of this rant, and desiring an optimistic conclusion, we suggest that as we learn more about the regulation of the immune response and develop procedures to monitor as well as administer regulatory immune molecules in real time, these methods will become increasingly effective for the treatment of bacterial infection.

Acknowledgments

We thank Elisabeth Carniel for sharing her ideas about the evolution of the virulence of *Y. pestis*. We are grateful to Jim Bull and Harris Fienberg for insightful comments and suggestions. B. R. L. acknowledges his continuous gratitude to Fernando Baquero, for inspiration, ideas, never-ending whimsy, support, and friendship. This endeavor was supported by a grant from the NIH, AI40662 (B. R. L.), and an NIH Training Grant (E. M.).

REFERENCES

Overview References

Bolotin, A., B. Quinquis, P. Renault, A. Sorokin, S. D. Ehrlich, S. Kulakauskas, A. Lapidus, E. Goltsman, M. Mazur, G. D. Pusch, M. Fonstein, R. Overbeek, N. Kyprides, B. Purnelle, D. Prozzi, K. Ngui, D. Masuy, F. Hancy, S. Burteau, M. Boutry, J. Delcour, A. Goffeau, and P. Hols. 2004. Complete sequence and comparative genome analysis of the dairy bacterium *Streptococcus thermophilus*. *Nature Biotechnology* 22(12):1554-1558.

Gordon, M. A., S. M. Graham, A. L. Walsh, L. Wilson, A. Phiri, E. Molyneux, E. E. Zijlstra, R. S. Heyderman, C. A. Hart, and M. E. Molyneux. 2008. Epidemics of invasive *Salmonella enterica* serovar Enteritidis and *S. enterica* serovar Typhimurium infection associated with multidrug resistance among adults and children in Malawi. *Clinical Infectious Diseases* 46(7):963-969.

Graham, S. M. 2002. Salmonellosis in children in developing and developed countries and populations. *Current Opinion in Infectious Diseases* 15(5):507-512.

Graham, S. M., E. M. Molyneux, A. L. Walsh, J. S. Cheesbrough, M. E. Molyneux, and C. A. Hart. 2000. Nontyphoidal *Salmonella* infections of children in tropical Africa. *Pediatric Infectious Disease Journal* 19(12):1189-1196.

Lederberg, J. 2000. Infectious history. *Science* 288(5464):287-293.

MacLennan, C. A., E. N. Gondwe, C. L. Msefula, R. A. Kingsley, N. R. Thomson, S. A. White, M. Goodall, D. J. Pickard, S. M. Graham, G. Dougan, C. A. Hart, M. E. Molyneux, and M. T. Drayson. 2008. The neglected role of antibody in protection against bacteremia caused by nontyphoidal strains of *Salmonella* in African children. *Journal of Clinical Investigation* 118(4):1553-1562.

McClelland, M., K. E. Sanderson, J. Spieth, S. W. Clifton, P. Latreille, L. Courtney, S. Porwollik, J. Ali, M. Dante, F. Du, S. Hou, D. Layman, S. Leonard, C. Nguyen, K. Scott, A. Holmes, N. Grewal, E. Mulvaney, E. Ryan, H. Sun, L. Florea, W. Miller, T. Stoneking, M. Nhan, R. Waterston, and R. K. Wilson. 2001. Complete genome sequence of *Salmonella enterica* serovar Typhimurium LT2. *Nature* 413(6858):852-856.

McClelland, M., K. E. Sanderson, S. W. Clifton, P. Latreille, S. Porwollik, A. Sabo, R. Meyer, T. Bieri, P. Ozersky, M. McLellan, C. R. Harkins, C. Wang, C. Nguyen, A. Berghoff, G. Elliott, S. Kohlberg, C. Strong, F. Du, J. Carter, C. Kremizki, D. Layman, S. Leonard, H. Sun, L. Fulton, W. Nash, T. Miner, P. Minx, K. Delehaunty, C. Fronick, V. Magrini, M. Nhan, W. Warren, L. Florea, J. Spieth, and R. K. Wilson. 2004. Comparison of genome degradation in paratyphi A and typhi, human-restricted serovars of *Salmonella enterica* that cause typhoid. *Nature Genetics* 36(12):1268-1274.

Parkhill, J., M. Sebaihia, A. Preston, L. D. Murphy, N. Thomson, D. E. Harris, M. T. Holden, C. M. Churcher, S. D. Bentley, K. L. Mungall, A. M. Cerdeno-Tarraga, L. Temple, K. James, B. Harris, M. A. Quail, M. Achtman, R. Atkin, S. Baker, D. Basham, N. Bason, I. Cherevach, T. Chillingworth, M. Collins, A. Cronin, P. Davis, J. Doggett, T. Feltwell, A. Goble, N. Hamlin, H. Hauser, S. Holroyd, K. Jagels, S. Leather, S. Moule, H. Norberczak, S. O'Neil, D. Ormond, C. Price, E. Rabbinowitsch, S. Rutter, M. Sanders, D. Saunders, K. Seeger, S. Sharp, M. Simmonds, J. Skelton, R. Squares, S. Squares, K. Stevens, L. Unwin, S. Whitehead, B. G. Barrell, and D. J. Maskell. 2003. Comparative analysis of the genome sequences of *Bordetella pertussis, Bordetella parapertussis* and *Bordetella bronchiseptica. Nature Genetics* 35(1):32-40.

Roumagnac, P., F. X. Weill, C. Dolecek, S. Baker, S. Brisse, N. T. Chinh, T. A. Le, C. J. Acosta, J. Farrar, G. Dougan, and M. Achtman. 2006. Evolutionary history of *Salmonella* typhi. *Science* 314(5803):1301-1304.

Falkow References

Avery, O. T., C. M. MacLeod, and M. McCarty. 1944. Studies on the chemical nature of the substance inducing transformation of pneumococcal types. *Journal of Experimental Medicine* 79(2):137-158.

Bach, J. F. 2002. The effect of infections on susceptibility to autoimmune and allergic diseases. *New England Journal of Medicine* 347(12):911-920.

Blaser, M. J., Y. Chen, and J. Reibman. 2008. Does *Helicobacter pylori* protect against asthma and allergy? *Gut* 57(5):561-567.

Chan, K., C. C. Kim, and S. Falkow. 2005. Microarray-based detection of *Salmonella enterica* serovar Typhimurium transposon mutants that cannot survive in macrophages and mice. *Infection and Immunity* 73(9):5438-5449.

Falkow, S. 2006. Is persistent bacterial infection good for your health? *Cell* 124(4):699-702.

Huxley, L. 1901. *Life and letters of Thomas Henry Huxley,* Vol. I. New York: D. Appleton and Company. P. 235.

Jacob, F., and J. Monod. 1961. Genetic regulatory mechanisms in the synthesis of proteins. *Journal of Molecular Biology* 3:318-356.

Lawley, T. D., K. Chan, L. J. Thompson, C. C. Kim, G. R. Govoni, and D. M. Monack. 2006. Genome-wide screen for *Salmonella* genes required for long-term systemic infection of the mouse. *PLoS Pathogy* 2(2):e11.

Lawley, T. D., D. M. Bouley, Y. E. Hoy, C. Gerke, D. A. Relman, and D. M. Monack. 2008. Host transmission of *Salmonella enterica* serovar Typhimurium is controlled by virulence factors and indigenous intestinal microbiota. *Infection and Immunity* 76(1):403-416.

Lederberg, J. 1987. Genetic recombination in bacteria: a discovery account. *Annual Review of Genetics* 21:23-46.

Luria, S. E., and M. Delbrück. 1943. Mutations of bacteria from virus sensitivity to virus resistance. *Genetics* 28(6):491-511.

Monack, D. M., D. M. Bouley, and S. Falkow. 2004a. *Salmonella typhimurium* persists within macrophages in the mesenteric lymph nodes of chronically infected *Nramp1$^{+/+}$* mice and can be reactivated by IFNγ neutralization. *Journal of Experimental Medicine* 199(2):231-241.

Monack, D. M., A. Mueller, and S. Falkow. 2004b. Persistent bacterial infections: the interface of the pathogen and the host immune system. *Nature Reviews Microbiology* 2(9):747-765.

Smith, H. W., and S. Halls. 1967. Studies on *Escherichia coli* enterotoxin. *Journal of Pathology and Bacteriology* 93(2):531-543.

Stocker, B. D., N. D. Zinder, and J. Lederberg. 1953. Transduction of flagellar characters in *Salmonella. Journal of General Microbiology* 9(3):410-433.

Wilson, M., R. McNab, and B. Henderson, eds. 2002. *An introduction to cellular microbiology.* Cambridge, United Kingdom: Cambridge University Press.

Margolis and Levin References

Achtman, M., K. Zurth, G. Morelli, G. Torrea, A. Guiyoule, and E. Carniel. 1999. Yersinia pestis, the cause of plague, is a recently emerged clone of Yersinia pseudotuberculosis. *Proc. Natl. Acad. Sci. USA* 96:14043-14048.

Andre, J. B., and B. Godelle. 2006. Within-host evolution and virulence in microparasites. *J. Theor. Biol.* 241:402-409.

Andre, J.-B., S. Gupta, S. Frank, and M. Tibayrenc. 2004. Evolution and immunology of infectious diseases: what new? An E-debate. *Infect. Genet. Evol.* 4:69-75.

Bekker, L. G., A. L. Moreira, A. Bergtold, S. Freeman, B. Ryffel, and G. Kaplan. 2000. Immuno-pathologic effects of tumor necrosis factor alpha in murine mycobacterial infection are dose dependent. *Infect. Immun.* 68:6954-6961.

Bellamy, R., and A. V. Hill. 1998. Genetic susceptibility to mycobacteria and other infectious pathogens in humans. *Curr. Opin. Immunol.* 10:483-487.

Bellamy, R., N. Beyers, K. P. McAdam, C. Ruwende, R. Gie, P. Samaai, D. Bester, M. Meyer, T. Corrah, M. Collin, D. R. Camidge, D. Wilkinson, E. Hoal-Van Helden, H. C. Whittle, W. Amos, P. van Helden, and A. V. Hill. 2000. Genetic susceptibility to tuberculosis in Africans: a genome-wide scan. *Proc. Natl. Acad. Sci. USA* 97:8005-8009.

Bergeron, Y., N. Ouellet, A. M. Deslauriers, M. Simard, M. Olivier, and M. G. Bergeron. 1998. Cytokine kinetics and other host factors in response to pneumococcal pulmonary infection in mice. *Infect. Immun.* 66:912-922.

Bone, R. C., C. J. Fisher, Jr., T. P. Clemmer, G. J. Slotman, and C. A. Metz. 1987. Early methyl-prednisolone treatment for septic syndrome and the adult respiratory distress syndrome. *Chest* 92:1032-1036.

Bonhoeffer, S. A., and M. A. Nowak. 1994. Mutation and the evolution of virulence. *Proc. R. Soc. London B* 258:133-140.

Braun, J. S., R. Novak, K. H. Herzog, S. M. Bodner, J. L. Cleveland, and E. I. Tuomanen. 1999. Neuroprotection by a caspase inhibitor in acute bacterial meningitis. *Nat. Med.* 5:298-302.

Brown, N. F., M. E. Wickham, B. K. Coombes, and B. B. Finlay. 2006. Crossing the line: selection and evolution of virulence traits. *PLoS Pathog.* 2:e42.

Bull, J. J. 1994. Virulence. *Evolution* 48:1423-1437.

Burnet, F. 1970. *Immunological Surveillance.* Pergamon Press, Oxford, United Kingdom.

Burnet, F. M., and D. O. White. 1972. *Natural History of Infectious Diseases.* Cambridge University Press, Cambridge, United Kingdom.

Carniel, E. 2003. Evolution of pathogenic *Yersinia*: some lights in the dark. *Adv. Exp. Med.* 529:3-12.

Crow, J. F., and M. Kimura. 1971. *An Introduction to Population Genetics Theory*, 1st ed. Harper & Row, New York, NY.

Cunningham, M. W. 2003. Autoimmunity and molecular mimicry in the pathogenesis of post-streptococcal heart disease. *Front Biosci.* 8:S533-S543.

Czermak, B. J., V. Sarma, C. L. Pierson, R. L. Warner, M. Huber-Lang, N. M. Bless, H. Schmal, H. P. Friedl, and P. A. Ward. 1999. Protective effects of C5a blockade in sepsis. *Nat. Med.* 5:788-792.

Czinn, S. J., and J. G. Nedrud. 1997. Immunopathology of Helicobacter pylori infection and disease. *Springer Semin. Immunopathol.* 18:495-513.

Ebert, D., and J. J. Bull. 2003. Challenging the trade-off model for the evolution of virulence: is virulence management feasible? *Trends Microbiol.* 11:15-20.

Ebert, D., and E. A. Herre. 1996. The evolution of parasitic diseases. *Parasitol. Today* 12:96-101.

Falush, D., T. Wirth, B. Linz, J. K. Pritchard, M. Stephens, M. Kidd, M. J. Blaser, D. Y. Graham, S. Vacher, G. I. Perez-Perez, Y. Yamaoka, F. Megraud, K. Otto, U. Reichard, E. Katzowitsch, X. Wang, M. Achtman, and S. Suerbaum. 2003. Traces of human migrations in Helicobacter pylori populations. *Science* 299:1582-1585.

Fenner, F., and F. N. Ratcliffe. 1965. *Myxomatosis*. Cambridge University Press, Cambridge, United Kingdom.

Finlay, B. B., and S. Falkow 1989. Common themes in microbial pathogenicity. *Microbiol. Rev.* 53:210-230.

Firoved, A. M., G. F. Miller, M. Moayeri, R. Kakkar, Y. Shen, J. F. Wiggins, E. M. McNally, W. J. Tang, and S. H. Leppla. 2005. Bacillus anthracis edema toxin causes extensive tissue lesions and rapid lethality in mice. *Am. J. Pathol.* 67:1309-1320.

Fisher, C. J., Jr., G. J. Slotman, S. M. Opal, J. P. Pribble, R. C. Bone, G. Emmanuel, D. Ng, D. C. Bloedow, and M. A. Catalano. 1994. Initial evaluation of human recombinant interleukin-1 receptor antagonist in the treatment of sepsis syndrome: a randomized, open-label, placebo-controlled multicenter trial. *Crit. Care Med.* 22:12-21.

Fisher, C. J., Jr., J. M. Agosti, S. M. Opal, S. F. Lowry, R. A. Balk, J. C. Sadoff, E. Abraham, R. M. Schein, and E. Benjamin. 1996. Treatment of septic shock with the tumor necrosis factor receptor:Fc fusion protein. The Soluble TNF Receptor Sepsis Study Group. *N. Engl. J. Med.* 334:1697-1702.

Fourrier, F. 2004. Recombinant human activated protein C in the treatment of severe sepsis: an evidence-based review. *Crit. Care Med.* 32:S534-S541.

Frank, S. A. 1996. Models of parasite virulence. *Q. Rev. Biol.* 7(1):37-78.

Gay, R. T., S. Belisle, M. A. Beck, and S. N. Meydani. 2006. An aged host promotes the evolution of avirulent coxsackievirus into a virulent strain. *Proc. Natl. Acad. Sci. USA* 103:13825-13830.

Gibbons, R. J. 1964. Bacteriology of dental caries. *J. Dent. Res.* 43(Suppl):1021-1028.

Graham, A. L., J. E. Allan, and A. F. Read. 2005. Evolutionary causes and consequences of immuno-pathology. *Annu. Rev. of Ecol. Evol. Syst.* 36:373-397.

Grech, K., K. Watt, and A. F. Read. 2006. Host-parasite interactions for virulence and resistance in a malaria model system. *J. Evol. Biol.* 19:1620-1630.

Haldane, J. B. S. 1949. Disease and evolution. *Ric. Sci.* 19:68-76.

Hilbi, H., A. Zychlinsky, and P. J. Sansonetti. 1997. Macrophage apoptosis in microbial infections. *Parasitology* 115(Suppl):S79-S87.

Kaushal, D., B. G. Schroeder, S. Tyagi, T. Yoshimatsu, C. Scott, C. Ko, L. Carpenter, J. Mehrotra, Y. C. Manabe, R. D. Fleischmann, and W. R. Bishai. 2002. Reduced immunopathology and mortality despite tissue persistence in a Mycobacterium tuberculosis mutant lacking alternative sigma factor, SigH. *Proc. Natl. Acad. Sci. USA* 99:8330-8335.

Klein, N. J., C. A. Ison, M. Peakman, M. Levin, S. Hammerschmidt, M. Frosch, and R. S. Heyderman. 1996. The influence of capsulation and lipooligosaccharide structure on neutrophil adhesion molecule expression and endothelial injury by Neisseria meningitidis. *J. Infect. Dis.* 173:172-179.

Kurahashi, K., O. Kajikawa, T. Sawa, M. Ohara, M. A. Gropper, D. W. Frank, T. R. Martin, and J. P. Wiener-Kronish. 1999. Pathogenesis of septic shock in Pseudomonas aeruginosa pneumonia. *J. Clin. Invest.* 104:743-750.

Lenski, R. E., and R. M. May. 1994. The evolution of virulence in parasites and pathogens: reconciliation between two competing hypotheses. *J. Theor. Biol.* 169:253-265.

Levin, B. R. 1996. The evolution and maintenance of virulence in microparasites. *Emerg. Infect. Dis.* 2:93-102.

Levin, B. R., and R. Antia. 2001. Why we don't get sick: the within-host population dynamics of bacterial infections. *Science* 292:1112-1125.

Levin, B. R., and J. J. Bull. 1994. Short-sighted evolution and the virulence of pathogenic microorganisms. *Trends Microbiol.* 2:76-81.

Levin, B. R., and C. Svanborg Eden. 1990. Selection and evolution of virulence in bacteria: an ecumenical excursion and modest suggestion. *Parasitology* 100:S103-S115.

Levin, B. R., and D. E. Rozen. 2006. Non-inherited antibiotic resistance. *Nat. Rev. Microbiol.* 4:556-562.

Levin, S. A., and D. Pimentel. 1981. Selection of intermediate rates of increase in parasite host systems. *Am. Nat.* 117: 308-315.

Levin, B. R., A. C. Allison, H. J. Bremermann, L. L. Cavalli-Sforza, B. C. Clarke, R. Frentzel-Beymem, W. D. Hamilton, S. A. Levin, R. M. May, and H. R. Thieme. 1982. Evolution of parasite systems (group report), p. 212-243. *In* R. M. Anderson and R. M. May (ed.), *Population Biology of Infectious Diseases.* Springer, Berlin, Germany.

Levin, B. R., V. Perrot, and N. Walker. 2000. Compensatory mutations, antibiotic resistance and the population genetics of adaptive evolution in bacteria. *Genetics* 154:985-997.

Lipsitch, M., and E. R. Moxon. 1997. Virulence and transmissibility of pathogens: what is the relationship? *Trends Microbiol.* 5:31-37.

Lipsitch, M., E. A. Herre, and M. A. Nowak. 1995. Host population structure and the evolution of parasite virulence: a "law of diminishing returns." *Evolution* 49:743-748.

Lipsitch, M., S. Siller, and M. A. Nowak. 1996. The evolution of virulence in pathogens with vertical and horizontal transmission. *Evolution* 50:1729-1741.

Mackinnon, M. J., and A. F. Read. 2004. Virulence in malaria: an evolutionary viewpoint. *Philos. Trans. R. Soc. London B* 359:965-986.

Marshall, B. J., and J. R. Warren. 1984. Unidentified curved bacilli in the stomach of patients with gastritis and peptic ulceration. *Lancet* 1:1311-1315.

May, R. M., and R. M. Anderson. 1983a. Epidemology and genetics in the coevolution of parasite and hosts. *Proc. R. Soc. London B* 219:281-313.

————. 1983b. Parasite-host coevolution, pp. 186-206. *In* D. J. Futuyama and M. Slatkin (ed.), *Coevolution* Sinauer, Sunderland, MA.

McCormick, J. K., J. M. Yarwood, and P. M. Schlievert. 2001. Toxic shock syndrome and bacterial superantigens: an update. *Annu. Rev. Microbiol.* 55:77-104.

Meyers, L. A., B. R. Levin, A. R. Richardson, and I. Stojiljkovic. 2003. Epidemiology, hypermutation, within-host evolution and the virulence of Neisseria meningitidis. *Proc. Biol. Sci.* 270:1667-1677.

Meynell, G. G. 1957. The applicability of the hypothesis of independent action to fatal infections in mice given Salmonella typhimurium by mouth. *J. Gen. Microbiol.* 16:396-404.

Modlin, R. L. 2002. Learning from leprosy: insights into contemporary immunology from an ancient disease. *Skin Pharmacol. Appl. Skin Physiol.* 15:1-6.

Moss, S. F., and M. J. Blaser. 2005. Mechanisms of disease: inflammation and the origins of cancer. *Nat. Clin. Pract. Oncol.* 2:90-97 (quiz 1 p. following 113).

Moxon, E. R., and P. A. Murphy. 1978. *Haemophilus influenzae* bacteremia and meningitis resulting from the survival of a single organism. *Proc. Nat. Acad. Sci. USA* 75:1534-1536.

O'Suilleabhain, C., S. T. O'Sullivan, J. L. Kelly, J. Lederer, J. A. Mannick, and M. L. Rodrick. 1996. Interleukin-12 treatment restores normal resistance to bacterial challenge after burn injury. *Surgery* 120:290-296.

Pfeiffer, J. K., and K. Kirkegaard. 2005. Increased fidelity reduces poliovirus fitness and virulence under selective pressure in mice. *PLoS Pathog.* 1:e11.

Pluschke, G., A. Mercer, B. Kusecek, A. Pohl, and M. Achtman. 1983. Induction of bacteremia in newborn rats by Escherichia coli K1 is correlated with only certain O (lipopolysaccharide) antigen types. *Infect. Immun.* 39:599-608.

Regoes, R. R., M. A. Nowak, and S. Bonhoeffer. 2000. Evolution of virulence in a heterogeneous host population. *Evolution* 54:64-71.

Rott, O., and B. Fleischer. 1994. A superantigen as virulence factor in an acute bacterial infection. *J. Infect. Dis.* 169:1142-1146.

Rubin, L. G. 1987. Bacterial colonization and infection resulting from multiplication of a single organism. *Rev. Infect. Dis.* 9:488-493.

Schiavo, G., F. Benfenati, B. Poulain, O. Rossetto, P. Polverino de Laureto, B. R. DasGupta, and C. Montecucco. 1992. Tetanus and botulinum-B neurotoxins block neurotransmitter release by proteolytic cleavage of synaptobrevin. *Nature* 359:832-835.

Segal, S., and A. V. Hill. 2003. Genetic susceptibility to infectious disease. *Trends Microbiol.* 11:445-448.

Sieling, P. A., D. Jullien, M. Dahlem, T. F. Tedder, T. H. Rea, R. L. Modlin, and S. A. Porcelli. 1999. CD1 expression by dendritic cells in human leprosy lesions: correlation with effective host immunity. *J. Immunol.* 162:1851-1858.

Sieper, J. 2001. Pathogenesis of reactive arthritis. *Curr. Rheumatol. Rep.* 3:412-418.

Sorensen, T. I., G. Nielson, P. Anderson, and T. Teasdale. 1988. Genetic and environmental influences on premature death in adult adoptees. *N. Engl. J. Med.* 318:727-732.

Tatematsu, M., T. Tsukamoto, and T. Mizoshita. 2005. Role of Helicobacter pylori in gastric carcinogenesis: the origin of gastric cancers and heterotopic proliferative glands in Mongolian gerbils. *Helicobacter* 10:97-106.

Whitnack, E. 1993. Sepsis, pp. 770-778. *In* M. Schaechter, G. Medhoff, and B. I. Eistenstein (ed.), *Mechanisms of Microbial Disease*, 2nd ed. Williams and Wilkins, Baltimore, MD.

Wildschutte, H., D. M. Wolfe, A. Tamewitz, and J. G. Lawrence. 2004. Protozoan predation, diversifying selection, and the evolution of antigenic diversity in Salmonella. *Proc. Natl. Acad. Sci. USA* 101:10644-10649.

Wilson, M. S., and R. M. Maizels. 2004. Regulation of allergy and autoimmunity in helminth infection. *Clin. Rev. Allergy Immunol.* 26:35-50.

Yu, V. L., C. C. Chiou, C. Feldman, A. Ortqvist, J. Rello, A. J. Morris, L. M. Baddour, C. M. Luna, D. R. Syndman, M. Ip, W. C. Ko, M. B. Chedid, A. Andremont, and K. P. Klugman. 2003. An international prospective study of pneumococcal bacteremia: correlation with in vitro resistance, antibiotics administered, and clinical outcome. *Clin. Infect. Dis.* 37:230-237.

Ziegler, E. J., C. J. Fisher, Jr., C. L. Sprung, R. C. Straube, J. C. Sadoff, G. E. Foulke, C. H. Wortel, M. P. Fink, R. P. Dellinger, N. N. Teng, et al. 1991. Treatment of gram-negative bacteremia and septic shock with HA-1A human monoclonal antibody against endotoxin. A randomized, double-blind, placebo-controlled trial. The HA-1A Sepsis Study Group. *N. Engl. J. Med.* 324:429-436.

4

Antibiotic Resistance:
Origins and Countermeasures

OVERVIEW

From Joshua Lederberg's appreciation of the microbial world's immense and fluid genetic resources came his recognition that humans, despite their dominion over "higher" forms of life, remain prey to microscopic predators. After a few decades in which it appeared that human ingenuity, in the form of antibiotics, had outwitted the pathogens—but during which Lederberg and others warned of our distinct disadvantage in an escalating "arms race" with infectious microbes—antibiotic-resistant bacterial strains, or "superbugs," have now become ubiquitous. As Stanley Cohen of Stanford University observes in his contribution to this chapter, "It seems quite remarkable that despite the enormous progress made in the treatment of infectious disease during Joshua Lederberg's lifetime, the ominous microbial threat discussed by Lederberg on multiple occasions continues." The papers collected in this chapter explore the evolutionary origins of the antibiotic resistance phenomenon, take its measure as a present and future threat to public health, and propose scientific approaches to addressing it, including investigating environmental reservoirs of antibiotic resistance, identifying sources of novel antibiotics, and developing alternatives to conventional antibiotic therapies.

In the chapter's first paper, workshop speaker Julian Davies, of the University of British Columbia, reviews the history of the development of antibiotic resistance, beginning in the early twentieth century. "Although we have gained considerable understanding of the biochemical and genetic bases of antibiotic resistance," he writes, "we have failed dismally to control the development of antibiotic resistance, or to stop its transfer among bacterial strains." This failure is perhaps understandable given the ubiquity of resistance genes in the environment—a vast collection of genes referred to by Davies and others as the "resistome"—and the many opportunities for genetic exchange that both nature and man have availed microbes. Thus, as Davies notes, "any drug usage, no matter how well controlled, inevitably leads to the selection of multidrug-resistant pathogens."

Davies describes the spectrum of resistance mechanisms and places them in an evolutionary context, from their function in "virgin" (antibiotic-free) environments to their potential for transmission and reassortment with other resistance elements in human-created environments, such as wastewater treatment systems. He also considers the function of the natural bioactive compounds from which many clinical antibiotics are derived and, in particular, the low-dose effects of natural antibiotics, which appear to differ significantly from the therapeutic effects of clinical antibiotics administered at high doses.

In the next contribution to this chapter, Cohen contends that antibiotic resistance has become a global public health threat "largely as a consequence of the conventional approach used to treat infections, which is to attack the pathogen in the hope that the host will not be harmed by the drug." He proposes a different approach to dealing with microbial pathogens: by interfering with the cooperative relationship that many of them have with their hosts. Host cells, he notes, furnish many invading pathogens with genes and gene products necessary for pathogen propagation and transmission.

Cohen describes the strategy by which he and coworkers have identified several such pathogen-exploited host genes and discusses their prospects as therapeutic targets. "There is hope that devising antimicrobial therapies that target host cell genes exploited by pathogens, rather than—or in addition to—targeting the pathogens themselves may help in the effort to thwart the microbial threat," he concludes. However, he also acknowledges that it remains to be determined whether such host-oriented therapies will, as expected, be less vulnerable to circumvention by pathogen mutations.

The final paper in this chapter, by Jo Handelsman of the University of Wisconsin, describes research in her laboratory on two topics relevant to antibiotic resistance: the use of metagenomics to discover (potentially transferrable) resistance functions in soil bacteria; and the potential for manipulating endogenous microbial communities so that they will defend their hosts against pathogens, thus eliminating the need for antibiotic therapy.

In order to understand the process of host invasion from the perspective of the commensal microbes involved in such events, Handelsman and coworkers are characterizing the composition, dynamics, and functions of model endogenous communities. Their studies of the microbial community of the gypsy moth gut have yielded intriguing evidence that commensal bacteria interact in ways that can influence host health, sometimes in surprising ways—including collaborating with invaders to kill their hosts. Handelsman's group has also found that treatment with antibiotics alters the response of their model microbial community to pathogens. These findings and observations have led to further inquiries into the nature of community robustness and the genetic attributes of invading microbes that permit them to overcome a robust community.

ANTIBIOTIC RESISTANCE AND THE FUTURE OF ANTIBIOTICS

Julian Davies, Ph.D.[1]
University of British Columbia

Naturally occurring small molecules, and their synthetic and semisynthetic derivatives, have been used as the foundation of infectious disease therapy since the late 1930s. Following the introduction of the "wonder drugs" penicillin and streptomycin, dozens of novel and derived bioactive compounds have been developed and used for the treatment of microbial maladies in humans, animals, and plants. Figure 4-1 shows a brief history of antibiotic development from the pre-antibiotic era to the present. The use of these important therapeutic agents for nonhuman applications, such as animal feed additives, started in the early 1950s and has expanded enormously. More than half of the total annual production volumes of all antimicrobials today are employed for nontherapeutic use as growth promotants and prophylactics in the food animal and aquaculture industries and for many other agricultural purposes.

In 2000, more than 25 million pounds of antibiotics were manufactured in the United States alone. Over a half century, this equates to about 1 million metric tons! Considering the fact that Russia, China, and India each currently produces more antibiotics than the United States, the amounts of these compounds made worldwide is very significant.[2] From an ecological point of view, the dispersion of these bioactive compounds into the environment has created increasing pressure for the selection of antibiotic-resistant microbes throughout the entire biosphere. This selection pressure is clearly most crucial in specific therapeutic situations, such as hospitals and their associated intensive care units, but it is evident that antibiotic-resistant bacteria are now ubiquitous! The increasing use of antibiotics, since the 1950s, for both appropriate and inappropriate applications, has led almost simultaneously to increases in antibiotic-resistant bacteria, and the spectrum of resistance determinants has grown correspondingly. Little wonder that antibiotic resistance is a continuing and major threat concomitant with efforts to cure human disease.

A series of epidemics of resistant organisms have marked the antibiotic era: penicillin-resistant *Staphylococcus aureus*, methicillin-resistant *Staphylococcus*

---

[1]Emeritus professor of microbiology and immunology.
[2]Accurate production figures are difficult to obtain.

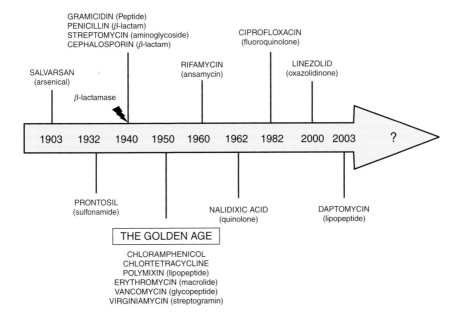

FIGURE 4-1 Major classes of antimicrobials and the year of their discovery.

aureus (MRSA),[3] vancomycin-intermediate *Staphylococcus aureus* (VISA),[4] drug-resistant *Vibrio cholerae*, multidrug-resistant (MDR)[5] and extensively drug-resistant (XDR)[6] *Mycobacterium tuberculosis* (hereinafter, MDR- and XDR-TB), CTX-M resistant *Escherichia coli* and *Klebsiella pneumoniae*, *Clostridium dif-*

---

[3]MRSA is a type of *S. aureus* that is resistant to antibiotics called β-lactams. β-lactam antibiotics include methicillin and other more common antibiotics such as oxacillin, penicillin, and amoxicillin (for more information, see http://www.cdc.gov/ncidod/dhqp/ar_MRSA_ca_public.html#2).

[4]VISA and vancomycin-resistant *S. aureus* (VRSA) are specific types of antimicrobial-resistant staphylococcal bacteria. Although most staphylococci are susceptible to the antimicrobial agent vancomycin, some have developed resistance to vancomycin. Infections caused by VISA and VRSA isolates cannot be successfully treated with vancomycin because these organisms are no longer responsive to vancomycin. However, to date, all VISA and VRSA isolates have been susceptible to other Food and Drug Administration (FDA)-approved drugs (for more information, see http://www.cdc.gov/ncidod/dhqp/ar_visavrsa_FAQ.html).

[5]MDR-TB is tuberculosis that is resistant to at least two of the best anti-tuberculosis drugs, isoniazid and rifampin. These drugs are considered first-line drugs and are used to treat all individuals with tuberculosis (for more information, see http://www.cdc.gov/tb/pubs/tbfactsheets/mdrtb.htm).

[6]XDR-TB is a relatively rare type of MDR-TB. XDR-TB is defined an *M. tuberculosis* isolate that is resistant to isoniazid and rifampin plus any fluoroquinolone and at least one of three injectable second-line drugs (i.e., amikacin, kanamycin, or capreomycin; for more information see http://www.cdc.gov/tb/pubs/tbfactsheets/mdrtb.htm).

ficile, and many others. Reports of new outbreaks of these so-called "superbugs" in the popular press are regular events.

The CTX-M family of extended-spectrum β-lactamases is of particular interest and concern. These enzymes inactivate the extended-spectrum (third-generation) cephalosporins of the cefotaxime class that were first introduced for the treatment of infections caused by gram-negative organisms in the 1990s. The increasing use of these antibiotics led to the appearance of resistant strains in several countries. As indicated by the frequent reports of the increasing prevalence of extended-spectrum β-lactamases, this resistance mechanism is now endemic in hospitals and the community throughout the world (Livermore et al., 2007; Queenan and Bush, 2007). Moreover, families of CTX-M enzymes that differ in amino acid sequence and catalytic activity have been reported in almost every country. Isolates of pathogenic *Enterobacteriaceae* carrying these resistance determinants, in particular *K. pneumoniae,* are essentially untreatable. In a number of hospitals and among certain populations, such as the military, the appearance of multidrug-resistant *Acinetobacter baumannii* isolates carrying a CTX-M enzyme has been of particular concern: this is an emerging pathogen capable of heightened mortality and morbidity (Peleg et al., 2008).

Antibiotic-resistant infections are commonplace. Human and nonhuman use of antimicrobial agents has guaranteed this status quo; bacteria and other microbes evolve rapidly to adapt to diverse environments and novel stresses and remain alive. Nonetheless, in industrialized nations "old" antibiotics, penicillin and the sulfonamides, are effective most of the time treating in routine outpatient. Until recently the pharmaceutical industry has survived by developing novel agents (usually by synthetic chemical modification of existing compounds) in order to counter new types of infections; this continuing arms race has permitted modern medicine to keep up, in some measure, with bacterial genetics. Occasionally, structurally novel compounds have been isolated that offer a narrow window of therapeutic success, but this has become less common and there is no panacea yet.

Given the seriousness of the current situation, a number of well-considered proposals have been mooted with the objective of controlling resistance development and restoring and maintaining the efficacy of treatments against infectious diseases (Norrby et al., 2005; Spellberg et al., 2008). But any drug usage, no matter how well controlled, inevitably leads to the selection of drug-resistant pathogens.

Antibiotic Resistance in the Environment

Since the first report of a penicillin-destroying enzyme (penicillinase) in a bacterial strain (Abraham and Chain, 1940), antibiotic resistance traits have been found in many environmental bacteria isolates (Table 4-1).

It is generally believed that pristine environments represent the major reser-

TABLE 4-1 Reports on Antibiotic Resistance Genes Isolated from the Environment

Year	Report
1972	Identification of β-lactamases in soil actinomycetes
1974	Identification of aminoglycoside-modifying enzymes in soil bacteria
1988	Identification of *Citrobacter spp.* and *Kluyvera spp.* as origins of extended-spectrum β-lactamases
2001	Identification of *gyrA* allelism in soil isolates that provides such isolates with "natural" fluoroquinolone resistance
2004	Identification of resistance genes in the soil metagenome
2006	Identification of the environmental "resistome" that conveys multidrug resistance in soil isolates
2006	Identification of the "intrinsic" resistome of pathogens (gene knockouts)
2008	Identification of the environmental "subsistome"—a population of bacteria that degrades antibiotics

voir of resistance genes to be acquired by bacterial pathogens. The identification of the "resistome" expanded the range of antibiotic resistance determinants and soil-derived actinomycete strains involved (D'Costa et al., 2006). More recently, the demonstration that a proportion of the microbes in the environment are capable of metabolizing antibiotics provided additional evidence for the existence of enzymatic mechanisms for the biodegradation of antibiotics that enable bacteria to subsist on antibiotic substrates (the "subsistome"); this work has revealed the presence of a great diversity of antibiotic-modifying enzymes in nature, and these enzymes provide the basis for a wide range of antibiotic resistance mechanisms (Dantas et al., 2008). In addition, it has been shown that bacterial genomes possess a significant number of genes that, if over-expressed in other hosts, would be candidate antibiotic resistance genes (intrinsic resistance; Tamae et al., 2008). In these latter cases, it has not yet been demonstrated that these genes are bona fide resistance genes; however, they have the potential to be. The existence of a universe of latent resistance genes in pathogens, commensals, and environmental organisms provides further evidence that resistance genes are common in nature; resistance is everywhere genotypically, if not phenotypically.

In an ideal world the identification of a new antibiotic resistance phenotype might provide the basis for "early-warning" measures in clinical situations and could guide the development of preventive measures that should be taken to avoid dissemination of the determinant. However, this is difficult to achieve in practice, given that resistance genes may spread rapidly and become well established in the population before they are detected definitively. In addition, it is debatable how much of the environmental and intrinsic resistance is significant in clinical terms. The strategy certainly could have worked in the case of CTX-M determinants, since their putative source was first identified in environmental *Kluyvera spp.* and later in enteric pathogens.

TABLE 4-2 Mechanisms of Resistance, 2008[a]

• Increased efflux	• Decreased influx[b]
• Enzymatic inactivation	• Sequestration
• Target modification[b]	• Target by-pass
• Target repair/protection	• Target amplification[b]
• Biofilm formation[b]	• Intracellular localization

[a]All of the mechanisms are acquired by horizontal gene transfer.
[b]Also acquired by mutation.

As has been known since the first use of antibiotics in research and the clinic, random mutation (spontaneous or induced) gives rise to antibiotic resistance. The mutations can occur in the genes encoding drug targets, drug export, or other mechanisms (Table 4-2). These mutations (usually point mutations) are pleiotropic;[7] it could be argued, conversely, that resistant mutants are frequently the result of point mutations selected by nutritional or other pressures. Indeed, the pleiotropic effects may influence a wide variety of other functions. Even resistant strains that are detected in the environment (i.e., the resistome or subsistome) may, in reality, be the result of multifunctional activities, such as nutritional balances. Experiments performed in our laboratory demonstrate that antibiotic-resistant mutants may have multiple phenotypes that could act as means of resistance selection depending on the nutrients available in the environment. Which came first?

Cryptic resistance genes that are present in bacteria or bacterial populations may be revealed by metagenomic cloning and expression (Riesenfeld et al., 2004), although it will be necessary to employ a wide range of expression hosts, in addition to *E. coli*, to identify functional resistance cloned from diverse bacterial genera. We do not yet know the mechanisms for gene "pick-up" from environmental sources, nor in what way such genes are "tailored" for efficient expression in heterologous hosts, such as bacterial pathogens. Environmental bacteria vary greatly in their G+C[8] content and codon usage; converting the transferred genes into functional genes in new hosts must require extensive "tailoring" by mutation and recombination. Native promoters need to be altered to permit transcription and translation in the new hosts. A good example of the way in which "new" genes may be acquired and expressed is provided by the omnipresent integron-mediated acquisition and expression system. In principle, any open reading frame can be inserted as a cassette into an attachment site associated with an integron recombinase and become a functional gene (Figure 4-2).

The integron cassette promoter is strong and very effective in a variety of

---

[7]Producing more than one effect; having multiple phenotypic expressions (see http://www.merriam-webster.com/dictionary/pleiotropic).

[8]guanine-cytosine.

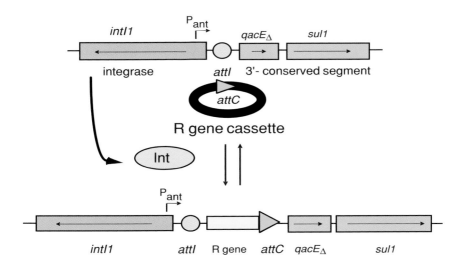

FIGURE 4-2 Integron mechanism of gene capture. Integron-mediated gene capture and the model for cassette exchange. Outline of the process by which circular antibiotic resistance gene cassettes (*antR*) are repeatedly inserted at the specific *attI* site in a class 1 integron downstream of the strong promoter P_c.*intI1*, integrase-encoding gene; Int = integrase IntI1. SOURCE: Patrice Courvalin, Institut Pasteur, Paris, France.

microbial hosts. However, the acquisition process is not well understood, and essentially all aspects of the origins and evolution of antibiotic resistance genes remain unsolved. Integron-encoded gene cassettes are widespread in all natural environments (marine and terrestrial), with the vast majority of such genes being of unknown function (Koenig et al., 2008). Although primarily of gram-negative bacterial origin, there have been reports of resistance integrons in gram-positive bacteria (Nandi et al., 2004). Gene exchange between bacterial genera is an evolutionary axiom, but it is still necessary to solve the dilemma of access to the resistance gene sources.

Antibiotic Resistance in Urban Environments

Urban areas, constantly exposed to the large variety of antibiotics that are commonly used in the hospital and the community, have considerable reservoirs of resistance. In large cities, thousands of people receive antibiotic treatment every day (e.g., in hospitals, in nursing homes, and at home), and accordingly, many antibiotics are released as part of the waste stream of the sanitary sewer system into wastewater. The "hotspots" are generally considered to be hospital-associated, but given that the main disposal route is through sewers, it is very likely that wastewater treatment plants (WWTPs) are also hotspots, with significant concentrations of antibiotic-resistant microbes containing multiple resistance genes from a wide variety of antibiotics and myriad potential vectors.

That WWTPs provide ideal environments for gene exchange and gene acquisition has been confirmed by studies of antibiotic resistance plasmids isolated from WWTP bacterial cultures (Tennstedt et al., 2005). Sequencing of purified plasmid DNAs demonstrated that very complex genomes are formed, as evidenced by the combinations of resistance genes, transfer factors, transposases, integrons and their associated integrases, bacteriophage remnants, and other potential mobile elements. Thus, a considerable level of genetic mixing and matching occurs, with the consequence that novel combinations of antibiotic resistance determinants may be discharged into WTTP effluents (Schlüter et al., 2007). It is also possible that virulence genes in the form of pathogenicity islands are acquired by plasmids.

While the roles of such newly formed resistance elements may be a matter of speculation there is absolutely no doubt that they are produced, and could contribute to, the gene pool responsible for increasing antibiotic resistance in the human community.

Antibiotic Hormesis

Antibiotics have long been known to have multiple effects on target cells. For many years, clinicians and a growing number of microbiologists and biochemists have reported that sub-inhibitory concentrations of commonly used antibiotics may affect microbial cell growth, morphology, structure, adhesion, virulence, and a variety of other phenotypes (Davies et al., 2006). In the majority of studies the changes induced were advantageous to the microbes. For example, a number of so-called antibiotics induce the formation of biofilms that permit microbial communities to survive under adverse conditions; others may promote swarming and motility, perhaps enhancing nutrient accessibility; even protein synthesis inhibitors may cause changes in cell-wall structure and function. Recently, detailed studies of transcriptional and proteomic changes (rather than phenotypic ones) have confirmed the wide range of responses of microbes to bioactive small molecules.

Almost all antibiotics have side effects that in many cases require careful dosage monitoring; these effects often occur at sub-inhibitory concentrations. The recent demonstration that these compounds exert the phenomenon of hormesis[9]

---

[9]Hormesis defines a dose-response activity of antibiotics (and other agents). It usually refers to a positive (beneficial) effect at low concentrations and a negative (toxic or inhibitory) effect at higher concentrations (Yim et al., 2006).

provides an explanation for the dual activities of microbially-derived natural products (Davies, 2006). The differences are due to transcriptional effects that are concentration-dependent. At low concentrations, antibiotics cause strong stimulation of transcription from specific groups of promoters. At higher concentrations (near-inhibitory), the transcription patterns change, as illustrated in Figure 4-3. Such dose-response dependence has been demonstrated by the use of promoter-reporter constructs or microarray analyses in the laboratory. We have proposed that these hormetic effects of antibiotics account for the differences in activity of low-molecular-weight compounds in their natural environment (soils, etc.) compared to their role in the treatment of infectious diseases.

In pristine environments, complex microbial populations can be assumed to exist in some form of homeostatic equilibrium that change as a result of fluctuations in nutritional resources. Stability in the population is probably maintained by cell-to-cell signaling that modulates their metabolic activity. When soil isolates of bacteria producing useful bioactive compounds (such as antibiotics) are

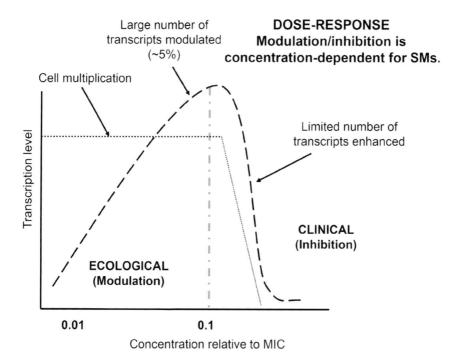

FIGURE 4-3 Concentration dependence of transcription modulation by antibiotics. MIC = minimal inhibitory concentrations; SMs = small molecules.
SOURCE: Adapted from Yim et al. (2006).

transplanted to a completely different environment (a laboratory), and exposed to potent mutagens and fermentation on unusual substrates to very high densities under artificial conditions, the yields of small molecules are often amplified considerably, generating sufficiently large amounts of the desired compounds to permit their testing as antibiotics at elevated (inhibitory) concentrations.

Antibiotics and Other Anthropomorphisms

The microbiological literature is spiced with comments such as "the bacteria have to make a decision" or "the microbe has made a choice." These statements are patently ridiculous. Microbes are genetically programmed to respond to different external environments; they do not make choices. Perhaps the most pervasive of such comments refers to the activity of microbially-produced small molecules. Selman Waksman's seminal work on the discovery of potent compounds such as neomycin and streptomycin led him to define them as antibiotics. He later realized that this definition was based on laboratory studies of soil microbes, grown in media containing exotic substrates and tested for their activities against human pathogens, not environmental bacteria. Waksman (1961) subsequently made a different judgment:

> The existence of microbes that have the capacity to produce antibiotics in artificial culture cannot be interpreted as signifying that such phenomena are important in controlling microbial populations in nature. . . . Unless one accepts the argument that laboratory environments are natural, one is forced to conclude that antibiotics play no part in modifying or influencing living processes that normally occur in nature.

We may disagree with his conclusion, in the light of current knowledge, but must concur with the distinction between the laboratory and complex natural environments.

Another "misinterpretation" is that all of these compounds kill bacteria. On the contrary, it is well known that most of the antimicrobial agents used therapeutically do not kill other bacteria—they only inhibit the growth of the target organism. This is an important aspect of antimicrobial therapy: the drug retards or inhibits bacterial growth and virulence and allows the human immune system to eliminate the weakened pathogens.[10]

The Global Microbiome

Bacteria are the most abundant living organisms on this planet and, given that they are essential to the maintenance of all other living organisms, they are

---

[10]For this reason, treatment with compounds with a cidal action is favored for patients who are immunosuppressed or in cases of critical infections caused by highly virulent organisms.

the most important. This is becoming increasingly evident as results of studies of the human microbiome demonstrate that the vast microbial population of the human gastrointestinal tract varies drastically with changes in age, diet, and disease (Bäckhed et al., 2005).

The microbial world is immense in number and diversity; specific details are subject to debate, but estimates indicate that there are between 10,000 to 50,000 taxa per gram in different soil types (Quince et al., 2008), while the bacterial density may be as large as 10^9 total bacteria per gram of soil. These are communities, not isolated organisms! In many environments, in the human intestinal tract for example, a small number of dominant phyla may constitute as much as 90 percent of the total population. Are these populations in constant conflict (the "war metaphor"), or do they undergo flux and coexist as controlled communities? It seems most logical that the latter is true, in which case some form of signaling process would be required to modulate the mixed populations under differing conditions. We believe that the signaling is primarily chemical, involving a variety of naturally occurring, low-molecular-weight compounds; after all, this is the predominant form of signaling in higher organisms. Of course, other forms of interactions, such as electrical connections between bacteria, are not excluded; the latter notion is supported by the demonstration of nanowires linking bacterial cells in biofilms (Gorby et al., 2006).

There is increasing evidence that the bioactive molecules generally classified as antibiotics play roles as chemical signals. Based on estimates of the putative small molecule biosynthetic clusters predicted from the nucleotide sequences of bacterial genomes, most soil and other environmental microbes have the genetic capacity to produce a number of bioactive compounds that are active at very low concentrations. Since many soil bacteria have the capacity to produce up to 20 different bioactive compounds, the notion of community structures controlled by small molecules becomes obvious. Welcome to the world of chemical biology!

Although these compounds may have antibiotic activity at high and often nonphysiological concentrations, their functions as chemical signals or modulators, as with the well-known quorum-sensing autoinducers, may be paramount in the environment. Even in the case of microcins[11] or bacteriocins,[12] which are natural peptides having potent in vitro activity as antibiotics, one can question whether specific bacterial killing in mixed bacterial populations is their natural function. As mentioned earlier, we and others have shown that at low concentrations (several log units below the determined minimum inhibitory concentration), these highly potent molecules modulate transcription in target bacteria without

---

[11]Microcins are essentially bacteriocins that contain a smaller number of amino acids.

[12]Bacteriocins are bacterially produced, small, heat-stable peptides that are active against other bacteria and to which the producer has a specific mechanism of immunity. Bacteriocins can have a narrow or a broad target spectrum (see http://www.nature.com/nrmicro/journal/v3/n10/glossary/nrmicro1273_glossary.html).

any deleterious effects (inhibition or killing); each antibiotic induces specific groups of genes, depending on the type of compound and its concentration. Until the concentrations of microcins and antibiotics in the gut (and other environments) are determined accurately, we will not be able to make any predictions about the potential roles of these molecules. It is probable that they also modulate functions in the epithelial cells lining the gastrointestinal (GI) tract as well as in the bacterial population. The signaling is inter-generic!

It all comes down to the question of concentration; the activities of all bioactive small molecules are dependent on the dose used. It is possible to study the range of responses in the laboratory, but at present this is not possible in complex natural environments. Current studies with antibiotics and other bioactive molecules suggest that synergy may be favored over antagonism in bacterial populations in nature. As observed in an earlier Institute of Medicine Forum workshop summary report *Ending the War Metaphor* (IOM, 2006), bacteria are not necessarily involved in extensive warfare in the GI tract; rather they use the bioactive small molecules to sense and tune into each other in a state of coexistence in microbial communities. It could be that the considerable distress suffered by many people with the ingestion of oral antibiotics is due to the fact that the presence of additional signaling compounds in the medicines creates mixed "signals" in the gut. Increasing knowledge of the natural biological activities of microbes may provide novel methods for their detection and isolation; in all likelihood, they will continue to be the source of agents with indispensable therapeutic value—until inevitably they develop resistance, of course.

The Pre-Antibiotic Era and Antibiotic Virgin Lands

Finally, coming back to the environmental origins of antibiotic resistance, there have been a number of studies in which archived bacteria, isolated before the extensive use of antibiotics, were examined for the presence of plasmids and resistance genes. In a seminal study by V. Hughes and N. Datta (1983), a collection of bacteria isolated from the 1920s (the Murray Collection) were found to carry plasmids, but no resistance genes; a number of the plasmids were shown to be transferable by conjugation. In another study by D. H. Smith (1967) using a collection of *E. coli* strains lyophilized in 1946 (some from Joshua Lederberg), R-factors were found in one strain; subsequent conjugation studies showed inter-strain transfer of resistance to tetracycline and streptomycin. Thus, one can conclude that plasmids were present, but resistance determinants were rare. These were all human isolates. It might be worthwhile to have another look at these collections using modern molecular biology techniques. The uropathogenic strains of *E. coli* in the Murray Collection have been found to contain pathogenicity islands typical of modern isolates (Dobrindt and Davies, unpublished).

Studies have also been done with bacteria isolated from native communities that have had no apparent contact with other humans and no exposure to antibiotics. The first such study, conducted in 1969, demonstrated that the indigenous population in the Solomon Islands had intestinal bacteria carrying R-factors that mediated resistance to streptomycin and tetracycline; the streptomycin resistance was due to a streptomycin phosphotransferase found in resistant strains in the United States (Gardner et al., 1969). More recent investigations examining the inhabitants of remote villages in South America found that as many as 90 percent of the population harbored multidrug-resistant strains (as detected in stool samples); Class I integrons were also identified (Bartoloni et al., 2009; Pallecchi et al., 2007). Resistance to tetracycline, ampicillin, sulfatrimethoprim, chloramphenicol, and streptomycin was frequent. These latter studies had the benefit of using sensitive detection technologies that were not available previously. None of the work examined the presence of resistance genes in associated environmental samples such as soils or in animals. These results imply that antibiotic resistance is at least maintained in the absence of antibiotic exposure. However, there is the possibility that occasional use of antibiotics, or that rare contact with humans who have used antibiotics, triggers the selection of populations of resistant bacteria in naïve populations. It may also be that therapeutic use of plant products by native populations selected for integron-encoded resistance elements, since the Class I integrons carry efflux genes with a wide substrate range.

Other studies have examined the bacterial populations of wild animals (birds, mammals, rodents, fish, reptiles, etc.) with similar results: resistance genes are present in all species tested. The extent to which these animals encountered human detritus is not known, but they clearly play a role in the global dissemination of antibiotic resistance. Seagulls and geese have close contact with humans and migrate over long distances! Suffice it to say, antibiotic resistance plasmids, integrons, transposons, and so forth, are widespread in bacteria. The natural resistome is ubiquitous.

Conclusions

Antibiotics are indispensable in the treatment of infectious and other diseases. Their discovery was essentially the result of human intervention, the consequence of a laboratory phenomenon that developed into an industry. Unfortunately, although antibiotic production and use have been of enormous medical and commercial value in transforming the therapy of infectious diseases in the last 60 years, the extensive use of antimicrobial agents during this period has not led to the eradication of any one bacterial pathogen. Moreover, pharmaceutical companies are discontinuing their infectious disease programs, largely for economic reasons. Discoveries of new, broad-spectrum antibiotics are very rare.

The widespread development of antibiotic resistance in hospitals and the community makes it imperative that the search for new, novel bioactive compounds be maintained.

The pharmaceutical industry has screened hundreds of thousands of compounds for antibiotic activity over the last 50 years. Many antimicrobial agents were discovered that did not appear to be useful or were not competitive for one reason or another. They might have had undesirable side effects and/or an unacceptable toxicity profile. Could some of these agents be used effectively for a short time in cases of life-threatening infections with multidrug-resistant pathogens? In circumstances where the currently available antibiotics are not effective, might some of the rejected compounds be resurrected for use at sub-inhibitory concentrations, alone or in synergistic combination with known compounds so that toxic side effects would be minimized?

National and international agencies have been seeking ways to maintain appropriate levels of work on antibiotic discovery (and recovery) while coping with the problem of antibiotic resistance (Norrby et al., 2005; Spellberg et al., 2008). The Infectious Disease Society of America and the Federation of European Microbiological Societies have made a number of rational recommendations, but their implementation appears to have been less successful than had been hoped. Questions still abound. What definitive actions can be taken? Containment procedures to date have successfully curtailed the spread of some viral infections. Any appearance of bird flu, for instance, has been met with rapid and drastic action and the disease has not yet reached a significant number of the human population. Should the same tactics be employed with the appearance of all new antibiotic resistance determinants? Could the CTX-M β-lactamase pandemic have been arrested had the first identified outbreak been contained by strict isolation measures? Is there an active and effective early-warning system for new resistant pathogens worldwide? What agencies would be the most effective in exercising this oversight?

Surely governments, the pharmaceutical industry, and academia can cooperate in actions that will curtail resistance development in pathogens and support efforts to provide a continuing supply of potent antimicrobial agents. Such a concerted effort is urgent. As Joshua Lederberg said, "[I]n that natural evolutionary competition, there is no guarantee that we will find ourselves the survivor" (Lederberg, 1988).

MICROBIAL DRUG RESISTANCE:
AN OLD PROBLEM IN NEED OF NEW SOLUTIONS

Stanley N. Cohen, M.D.[13]
Stanford University

Barring geno-suicide, the human dominion is challenged by only pathogenic microbes, for whom we remain the prey; they the predator.

Joshua Lederberg, May 1993

As anyone who knew Joshua Lederberg is aware, he was not prone to hyperbole or overstatement, so this pronouncement is sobering. The confrontation between microbes and humanity that Lederberg referred to has had a long history—from the biblical Plague of the Philistines described in the Book of Samuel, to the first documented instance of plague, Justinian's Plague, in the sixth century A.D., through a series of major pandemics that decimated Western European populations in the centuries that followed. In modern times, perhaps the greatest plague was that of pandemic influenza, which occurred during the winter of 1918-1919.

The challenge from microbes is multifaceted. An early instance of biological warfare occurred during the bubonic plague epidemic known as the Black Death during the mid-fourteenth century. During the Siege of Caffa (now the Ukranian city of Feodosija) in 1346, the invading Mongols, who were suffering from Plague, catapulted their comrades' corpses over the walls of Caffa in the hope that the rotting corpses would poison the city's water supply and destroy its defenders (Wheelis, 2002).[14]

Human intervention against microbial disease commenced with the practice of variation—the subcutaneous inoculation of smallpox-susceptible persons with material taken from a pustule of an individual afflicted with smallpox—an approach that can be traced to ancient folk practices in China, Turkey, and Africa (Riedel, 2005). A modern understanding of infectious disease began with the visualization of microbes by van Leeuwenhoek in 1683. In the 1870s, Louis Pasteur and Robert Koch first advanced the germ theory, which focused on the causal role of microbes in infectious disease. Half a century later, Griffith (1928) showed that disease-producing properties could be transferred between bacteria by a substance they produced. Fifteen years after that observation came the dramatic discovery by Avery, MacLeod, and McCarty that the genetic information

---

[13]Kwoh-Ting Li Professor in the School of Medicine; Professor of Genetics and Professor of Medicine.

[14]The Black Death killed an estimated 43 million people worldwide, including 25 million people in Europe (one-third of the population); devastated commerce; and led to social upheavals (Bishop, 2003).

responsible for such disease-producing properties resides in DNA, rather than in proteins, as had previously been suspected (Avery et al., 1944). This discovery set the stage for the seminal experiments of Lederberg and Edward Tatum on conjugation and recombination in bacteria, which depended significantly on extrachromosomal elements carried by these microbes. In an article published in the journal *Physiological Reviews* in October 1952, Lederberg invented the name "plasmids" for these extrachromosomal elements, and subsequent studies from multiple laboratories showed that plasmids were circles of DNA that could be passed among bacteria conjugally by hair-like structures called pili (for a review, see Cohen, 1993).

As Lederberg was conducting his investigations of bacterial conjugation and recombination, antibiotics were introduced for the clinical treatment of infectious diseases. The use of penicillin saved many lives near the close of World War II, and it was widely anticipated at the time that antibiotic use would soon lead to the worldwide eradication of bacterial infections. This hope quickly dissolved with the appearance of antibiotic resistance. Multidrug-resistant bacteria were observed first in Japan in the late 1950s and soon were detected also in other countries throughout the world. Multidrug-resistant bacteria quickly emerged as a medical problem because the drug resistance genes they carry could pass horizontally via conjugation to other bacteria—some of which were more pathogenic than the original plasmid host—as well as linearly to the progeny of the resistant cells. The spread of bacterial resistance to antimicrobial drugs has spawned years of research on mechanisms underlying such resistance. These investigations have shown that resistance to antimicrobials has become common worldwide among bacterial populations largely as a consequence of the conventional approach used to treat infections, which is to attack the pathogen in the hope that the host will not be harmed by the drug. However, notwithstanding this hope, antibacterial and antiviral drugs clearly do have toxic effects on the host; moreover, those bacteria that acquire mutations that result in insensitivity to the antimicrobial can continue to reproduce while most individuals in the bacterial population are killed or their growth inhibited. This leads to the take-over of bacterial populations by resistant microbes. Another limitation of the conventional approach to antimicrobial therapy, in light of the increasing threat of bioterrorism, is the potential for intentional alteration of pathogens so as to render them drug-resistant.

From the early stages of his career, Joshua Lederberg was interested in learning how drug-resistant bacterial populations evolve. With his first wife, Esther Lederberg, he developed a beautifully simple, effective, and inexpensive procedure called replica plating to investigate this: a sterile piece of velvet attached to a cylinder is touched to a plate on which bacterial colonies are growing; the velvet is then touched to multiple sterile growth plates, which in the Lederbergs' experiment contained antibiotics not present on the original plate. Using this method, the Lederbergs found that some bacteria present on the original plate contained mutations that resulted in resistance to an antibiotic that the bacteria had not been exposed to (Lederberg and Lederberg, 1952). Based on these findings and on subsequent investigations of mutations that lead to antibiotic resistance, research by the scientific community has focused on heritable resistance. It is worth noting, however, that the Lederbergs left open the possibility that resistance to antibiotic exposure or other temporary environmental challenges that disappear after several generations of bacterial growth under nonselective conditions might also occur (Lederberg and Lederberg, 1952). Such adaptive resistance, which results from antibiotic-induced altered expression of nonmutated genes, is in fact an additional problem that has in recent years been the subject of increasing scientific attention.

Host-Oriented Therapeutics

Host-microbe interactions generally are viewed as adversarial, and research on such interactions has largely focused on enhancing the ability of the host to halt invasion by pathogens (Beutler, 2004; Ishii et al., 2008). Pathogens can also exploit the normal functions of host cells for propagation, and the cellular genes exploited by pathogens (CGEPs) are potential targets for antimicrobial therapies that may be less subject to mutational circumvention by pathogens. Moreover, certain host cell genes are required by multiple pathogens, providing the potential for broad-spectrum therapeutic effects from the targeting of a single gene or gene product.

Consider the example of viral propagation and the pathogenesis of viral disease, which require that the infecting virus bind to host cells, that the viral genome be internalized, and that its genes be expressed and its proteins processed. To produce disease, the virus must then replicate, produce a pathological event, undergo morphogenesis, and be released in order to infect other cells of the host. All of these events require the functions of host cells and recruitment of these functions by the virus. At least some of these host cell functions are implicated in the pathogenesis of diseases caused by bacterial pathogens. Similarly, the effects of microbial toxins are also dependent on host functions: anthrax toxicity, for example, requires the recruitment of host proteins to enable uptake and processing of the toxin, as shown in Figure 4-4.

When I raise the prospect of developing antimicrobial therapies that target host cell functions recruited by pathogens, there typically are questions about whether interference with such host functions will result in unacceptable drug toxicity. Possible toxicity is an issue that must of course be addressed with any medication, and for infectious diseases, this is true whether a host function or the pathogen itself is the primary target of the therapy. However, drugs that treat disease by targeting normal cell functions routinely are used for all other types of illness, from high blood pressure to neuropsychiatric disorders. The goal of treatment has been to find agents that differentially—but not necessar-

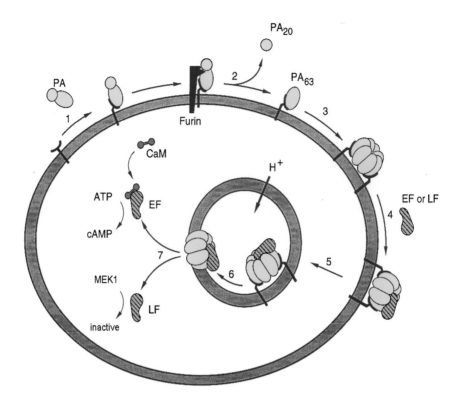

FIGURE 4-4 The road to anthrax toxicity. (1) Anthrax protective antigen (PA) binds to receptors on the surface of the host cell, and (2) is cleaved in two by furin. (3) The remaining 63-kilodalton subunit (PA_{63}) heptamerizes. (4) The heptamer interacts with the edema factor (EF) or lethal factor (LF). (5) It then moves the pathogenic qualities into the cells and (6) it is taken up by endosomes. (7) LF and EF are released into the cytoplasm as a result of acidification to become toxic.
SOURCE: Adapted from Collier and Young (2003).

ily exclusively—affect the pathogenic process. There is no inherent novelty to host-oriented therapy. Rather, only in the infectious disease arena are therapies historically not host oriented.

Finding host genes recruited by pathogens is practical using gene inactivation approaches. However, as mammalian cells normally contain two copies of each gene, inactivation of both copies ordinarily is necessary to produce biological effects. Whereas such homozygous gene inactivation can be accomplished readily when the gene is known and the DNA sequence is available, the use of random inactivation of genes in a mammalian cell population as a tool for the discovery of gene function is a more formidable task. Some years ago, Limin Li and I reported a retrovirus-based strategy for accomplishing this (Li and Cohen, 1996) using viruses having an RNA genome (Figure 4-5).

When such viruses, which are called retroviruses, infect cells, a DNA copy of their genomes is inserted into the chromosome of the infected cell. In large cell populations, inserts can occur in nearly every gene. The inserted DNA copy of the retrovirus genome, which in Figure 4-5 is termed a Gene Search Vector (GSV), inactivates one of the two copies of the chromosomal gene into which it has inserted. The GSV contains a regulated antisense promoter that initiates transcription extending outward from the insert into the adjacent chromosomal DNA segment. This generates an RNA transcript that is complementary to that chromosomal DNA sequence. As the transcript initiated in the GSV is complementary also to mRNA coming from the second copy of the chromosomal gene containing the GSV insertion, the second copy of the chromosomal gene is also inactivated (silenced). The GSV-treated cell population can then be screened, for example, for individual cells that survive the lethal effects of a pathogen or a toxin, and genes containing GSV insertions can be identified and characterized. The overall procedure has been referred to as Random Homozygous Knock Out (RHKO) (Li and Cohen, 1996). The biological effects of gene inactivation observed during

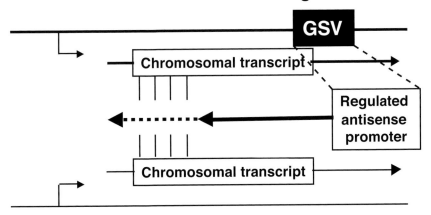

FIGURE 4-5 Diagrammatic representation of the site of chromosomal insertion of a lentiviral GSV. A regulated promoter located in the GSV is used to generate antisense transcripts that functionally inactivate both copies of the chromosomal gene containing the insertion.

the screening can be confirmed by inactivating the same chromosomal gene in naïve cells—using antisense approaches, small interfering RNA, or homologous recombination.

The RHKO strategy enables the identification of genes and genetic pathways required for pathogenicity. Functional homozygous gene inactivation also has proved practical using expressed sequence tag (EST) libraries (Lu et al., 2003) or small interfering RNA libraries (Cherry et al., 2005; Dunn et al., 2004; Hao et al., 2008). However, for EST-based and RNA interference (RNAi)-based methods, the scope of gene inactivation is limited by the contents of the RNAi or EST sequences used to create the library. To the best of my knowledge, the RHKO strategy currently is the only way to comprehensively inactivate mammalian genes randomly.

CGEPs Identified Using Gene Inactivation Approaches

The first gene isolated by the RHKO strategy—tumor susceptibility gene 101 (Tsg101)—was identified using a tumor susceptibility screen (Li and Cohen, 1996). Further study showed that Tsg101 encodes a variant of ubiquitin conjugase enzymes (Koonin and Abagyan, 1997; Ponting et al., 1997) and, importantly, that it has a key role in endocytic trafficking (Babst et al., 2000; Bishop and Woodman, 2001). An initial clue that Tsg101 gene might be involved in endocytic functions had come from two-hybrid analysis for identification of other cellular proteins that interact with Tsg101; one of the Tsg101-binding proteins detected in such experiments was Hrs (Lu et al., 2003), which was known to have a role in the trafficking of receptors to the cell surface (for a review, see Raiborg and Stenmark, 2002). The trafficking process is mediated by multicomponent protein complexes, which have been termed ESCRT (Endosomal Sorting Complexes Required for Transport) complexes, and Tsg101 is an ESCRT complex protein (Babst et al., 2000; Katzmann et al., 2001). Viral budding is topologically similar to receptor trafficking to the cell surface, and virus-encoded proteins can interact with Tsg101 (VerPlank et al., 2001) and other components of ESCRT complexes and usurp ESCRT-complex functions to enable viral egress from cells (Garrus et al., 2001; Goff et al., 2003; Martin-Serrano et al., 2001). Mutations in Tsg101 can interfere with the interaction between Tsg101 and virus-encoded egress proteins (Garrus et al., 2001; Goff et al., 2003; Martin-Serrano et al., 2001) and, consequently, can inhibit viral release. Recent evidence indicates that externally-administered agents such as cyclic peptides can inhibit interaction of Tsg101 with its human immunodeficiency virus (HIV) partner, the egress-promoting Gag protein, and thus interfere with Gag-promoted release of virus-like particles at an inhibitor concentration that does not detectably affect the trafficking of cellular receptors (Tavassoli et al., 2008). Antibodies that interact with Tsg101 on the cell surface have also been found to interfere with virus release (Bonavia et al., 2008).

Collectively, such findings support the notion that host-targeting therapies may prove useful in combating infectious pathogens.

The Search for CGEPs Continues: Anthrax and Tuberculosis

Assay development is crucial to the search for CGEPs. Assays can be as simple as the selection of survivors to pathogen exposure, or can be more complex, for example, involving the identification of cells expressing a "reporter" protein that has been linked to a CGEP-regulated gene. To identify CGEPs involved in the lethal effects of anthrax toxin, we used an EST-based antisense approach to host gene inactivation and an assay that identified cells surviving the lethal effects of anthrax toxin. These investigations have led to the discovery that ARAP3, a phosphoinoside-binding protein previously implicated in membrane vesicle trafficking and cytoskeletal organization, and LRP6, a lipoprotein-receptor-related protein also present on cell surfaces, affect internalization and the consequent lethality of anthrax toxin (Lu et al., 2004; Wei et al., 2006). LRP6 was previously known to be a co-receptor for signaling by the Wnt protein, which has an important role in embryogenesis, differentiation, colorectal cancer, adipogenesis, and osteogenesis (for a review, see He et al., 2004). Reconstitution of ARAP3 or LRP6 deficiency in naïve cells by using antisense methods or small interfering RNAs (siRNAs) has confirmed the role of these genes in anthrax toxin lethality, and confocal imaging studies have shown that internalization of the cell surface-binding component of anthrax toxin (Figure 4-4) is impaired in cells deficient in ARAP3 function or LRP6 function (Lu et al., 2004; Wei et al., 2006). Antibodies generated against LRP6 were found to offer partial protection of cells from toxin-induced death (Wei et al., 2006).

Similar studies to identify CGEPs for *Mycobacterium tuberculosis* currently are under way in my laboratory, in collaboration with the laboratory of Carl Nathan at the Weil School of Medicine at Cornell University. Assays have been established to identify macrophages whose ability to survive infection by *M. tuberculosis* is altered by host gene inactivation. Preliminary experiments have identified host genes that putatively modulate macrophage killing by *M. tuberculosis* and have revealed that some of these genes have pathways in common.

Conclusion

It seems quite remarkable that despite the enormous progress made in the treatment of infectious diseases during Joshua Lederberg's lifetime, the ominous microbial threat discussed by Lederberg on multiple occasions continues. A simple fact underlies this threat: for every pathogen-encoded protein or nucleic acid targeted by a therapeutic agent, there is the potential for a target-altering mutation that can render the therapy ineffective. Thus far, the strategy for coping with drug-resistant microbes has been to discover or design new pathogen-targeting drugs, but there is concern that the time of mutational resistance to even "drugs of last resort" is approaching.

There is hope that devising antimicrobial therapies that target host cell genes exploited by pathogens, rather than—or in addition to—targeting the pathogens themselves, may help in the effort to thwart the microbial threat. Here, I have reviewed some of the evidence indicating that such CGEPs can be identified by genetic screening strategies and that the pathways encoded by CGEPs can offer approachable targets for antimicrobial therapies. Whether therapeutic measures aimed at interfering with pathogen exploitation of host function will be less subject to circumvention by pathogen mutations has yet to be determined. However, host-oriented therapies clearly should not select for host mutations that increase susceptibility to a pathogen, and microbial mutations that enable the pathogen to propagate by exploiting alternative host cell functions may prove to be less common than mutations that simply modify the pathogen site targeted by a drug.

Acknowledgments

I thank multiple colleagues in my laboratory at Stanford University who have participated in studies reviewed here, and the U.S. National Institutes of Health (NIH), the Defense Advanced Products Research Agency (DARPA),[15] and the Defense Threat Reduction Agency (DTRA)[16] for research support.

EXPANDING THE MICROBIAL UNIVERSE: METAGENOMICS AND MICROBIAL COMMUNITY DYNAMICS

Jo Handelsman, Ph.D.[17]
University of Wisconsin

In the real world, microbial processes occur within the context of a community. Most of our understanding of microorganisms is derived from bacteria in pure culture. It is critical, if we are to manage microbes to improve human health, that we develop a complementary understanding of microbial behavior in complex communities. In this essay, I will discuss studies in two different areas of microbial community research in my laboratory: antibiotic resistance and host-microbe interactions. I will begin with a brief overview of our work on the discovery of antibiotic resistance genes in communities of soil bacteria, and proceed to a more detailed discussion of a model system that we have developed to study endogenous microbial communities and their role in host health and disease.

---

[15]An agency of the U.S. Department of Defense.
[16]An agency of the U.S. Department of Defense.
[17]Howard Hughes Medical Institute Professor, Department of Bacteriology.

The Metagenomics of Resistance

Where does antibiotic resistance come from? We know quite a lot about the clinical sources of antibiotic resistance as it affects humans, and we know a bit about the agricultural reservoirs of resistance, particularly in livestock systems. We really know nothing, however, about the source of antibiotic resistance, except that it must come from the environment, because the genes that confer antibiotic resistance vastly predate the use of antibiotics.

So, where do these genes come from? Soil is a likely source for many of them, because most of the antibiotics that we use today are derived from soil microbes (see the contribution to this chapter by Julian Davies). Soil is an extremely species-rich environment. Our models and statistical analyses suggest that a single gram of soil may contain between 5,000 and 40,000 species (conservatively defined) of microbes (Schloss and Handelsman, 2006). In addition to being the richest source of new microbes we have on Earth, soil is also one of the most difficult environments in which to study microbes, because the vast majority of them—probably more than 99.9 percent—cannot currently be cultured. We know these uncultured organisms only by their molecular signatures, which indicate that they are very different from the species that we can culture (see contributions by David Relman in Chapter 2 and Jonathan Eisen in Chapter 5).

Metagenomics was developed in order to study the collective genomes of an assemblage of organisms as a single entity, or a metagenome; the name comes from the Greek, *meta* meaning "transcendent" or "overarching." My group takes a particular approach to such studies, which can be described as functional metagenomics (in contrast to the sequence-driven metagenomic studies, discussed by workshop speaker Jill Banfield, as described in the workshop overview and in Chapter 2). The method we use is very simple and, on the face of it, quite crude. We extract DNA directly from the soil; most of what we obtain comes from Bacteria and Archaea. We clone the extracted DNA in *Escherichia coli,* and then screen the resulting library for an expressed activity, such as antibiotic production or antibiotic resistance.

We have searched for genes that confer resistance to various antibiotics. We have found genes for all three types of enzymes known to modify aminoglycoside antibiotics (e.g., kanamycin) in our libraries. We have also founds genes that confer resistance to β-lactam antibiotics (e.g., penicillin and cephalosporin) in a library derived from soil sampled at a pristine site in Alaska: an island in the middle of the fast-moving, high-volume Tanana River, very far from human activity (except for our visits, and we do not bring antibiotics with us). The β-lactam-resistant clones isolated from this location represent a broad spectrum of resistance mechanisms. Some clones contain gene sequences resembling those of known resistance genes from clinical isolates; these genes express enzymes (such as β-lactamases, pencillinases, and carbapenemases) that degrade β-lactam antibiotics. Other clones contain genes that diverge deeply from known β-lactam resistance genes from cultured organisms; we think these genes may confer resistance through a regulatory function.

One of the most interesting clones that we obtained from the Alaskan soil encodes the first known example of a hybrid β-lactamase: two different kinds of β-lactamases fused into a single protein. One end confers resistance to the penicillin-like compounds, and the other end confers resistance to the cephalosporin-like compounds; together, they confer very broad resistance to β-lactam antibiotics (Allen et al., 2009). This is a gene that we hope to never see in clinically important pathogens, but based on what we know about the transfer of antibiotic resistance determinants, this is entirely possible. We should, therefore, continue to examine and anticipate the broad range of antibiotic resistance genes in our environment. Essential in this exploration is that we are mindful of antibiotic resistance in communities and that we account for both culturable and nonculturable members of those communities since either type of organism could provide an important source of resistance to human pathogens.

Phalanx or Traitors?

The second part of this essay concerns the nature of the microbial communities that comprise the normal gastrointestinal flora of animals. Do these commensal communities protect their hosts from disease, or do they encourage the disease process?

My group is interested in commensal gut communities for a variety of reasons. We know from many studies[18] in mammals, as well as in invertebrates, that the normal gut flora influences host physiology. There is evidence that endogenous microbial communities contribute to obesity, diabetes, and high cholesterol (Ley et al., 2006); there is also evidence that the microbiota is essential to the health of the host. We also know that most pathogens live as commensals and then—under the right conditions—they become invasive, and thereby pathogenic. We are very interested in that switch, because if we want to use microbial communities to protect us from disease (e.g., through the use of probiotics or through the deliberate manipulation of microbial community dynamics), we need to understand how the opposite sometimes happens. We study the role of microorganisms in communities that inhabit the midguts of gypsy moth and cabbage white butterfly larvae. These are good model systems, as the larvae are easy to rear and dissect and their gut microbial communities can be readily and naturally manipulated via host feeding. Moreover, the intestinal epithelium of these insects is surprisingly similar to that of humans and other mammals, in terms of both its anatomy and

---

[18]See Bregman and Kirsner (1965), Fleming et al. (1986), Hirayama and Rafter (1999), Hooper (2004), Hooper and Gordon (2001), Hooper et al. (2001, 2002), Johannsen et al. (1990), Krueger et al. (1982), Ley et al. (2006), and Rath et al. (1996).

immunological function. Using the larval model systems, we have explored the role of microbial communities in health and disease of the host.

Who Is There?

We use both culture-based and culture-independent methods to examine the composition of communities. We find uncultured organisms that are readily recognizable as members of phylogenetic groups that have many culturable members, but these particular organisms cannot themselves be cultured, for largely unknown reasons.

Table 4-3 shows the diversity of bacterial phylotypes that we have identified in the gypsy moth system. Most of these are members of two phyla: the Proteobacteria and the Firmicutes. Over time and under different feeding regimes, we have observed changes in community composition at the species level, but rarely at the phylum level.

Are Signals Exchanged in the Gut Community?

Many disease processes induced by bacteria are dependent on microbial communication through small molecules. Some bacterial species use a mechanism, called quorum sensing, to gauge the density of their populations (Engebrecht and Silverman, 1984; Fuqua et al., 1994; Miller and Bassler, 2001). Our first question was: Does quorum sensing occur in the larval midgut?

Quorum sensing is mediated by small molecules called homoserine lactones. Some thought it was impossible that these molecules—and therefore quo-

TABLE 4-3 Phylogeny of Cultured and Uncultured Bacteria from Third Instar Gypsy Moth Midguts Feeding on an Artificial Diet

Phylotype	Division	Genus	Species
1[a]	low G+C gram positive	*Enterococcus spp.*	*E. faecalis*
2[a]	low G+C gram positive	*Staphylococcus spp.*	*S. lentus*
3[a]	low G+C gram positive	*Staphylococcus spp.*	*S. cohnii*
4[a]	low G+C gram positive	*Staphylococcus spp.*	*S. xylosus*
5[a]	γ-Proteobacterium	*Enterobacter spp.*	
6[a]	γ-Proteobacterium	*Pseudomonas*	*P. putida*
7[a]	γ-Proteobacterium	*Pantoea spp.*	*P. agglomerans*
8[b]	low G+C gram positive	*Enterococcus spp.*	
9[b]	γ-Proteobacterium	*Enterobacter spp.*	
10[b]	α-Proteobacterium	*Agrobacterium spp.*	

[a]Cultured.
[b]Uncultured.
SOURCE: Adapted from Broderick et al. (2004).

rum sensing—could be viable in the lepidopteran (moth and butterfly) midgut, because it is an extremely alkaline environment, typically reaching pH 10 to 12, a pH expected to destroy lactones. Nevertheless, we found that quorum sensing did indeed occur in the gut.

To determine whether a molecular signal was being exchanged between cells in the bacterial community, we used a simple luminescence-based reporter system, described in Figure 4-6. This reporter system enabled us to visualize the impact of homoserine lactone signal exchange among cells within the larval gut. We transformed bacterial cells in the midgut with biosensors that emit light only if they receive quorum-sensing signals sent by another cell. We detected luminescence in whole, living, larvae: clear evidence that the members of the gut bacterial community are communicating with each other in the highly alkaline gut.

What is the biological role of these signals in the insect midgut? We asked whether quorum-sensing signals that are required for *Pseudomonas aeruginosa* infection in other systems are also required in the cabbage white larval gut system; as Figure 4-7 illustrates, the answer was a resounding "yes." We then demonstrated that we could chemically inhibit the larval gut quorum-sensing system using a known quorum-sensing inhibitor synthesized in the laboratory of our colleague at the University of Wisconsin, Helen Blackwell. Both approaches indicated that quorum-sensing is required for the virulence of *P. aeruginosa* in the larval gut. Furthermore, these results suggest that the gut microbes are, in fact, acting as a community and are not merely coexisting.

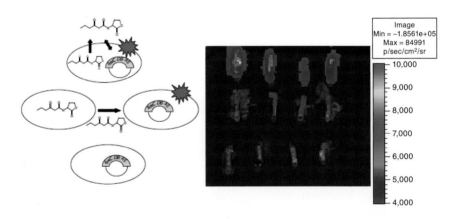

FIGURE 4-6 Detection of quorum-sensing activity and signal exchange in the guts of cabbage white butterfly larvae. Bioluminescence was detected in the individual guts of larvae fed *Pantoea* pSB401 (top row), *Pantoea* mixed with *Pantoea panI*::Tn5 pSB401 (middle row) and *Pantoea panII*::TN5 pSB401 (bottom row).
SOURCE: Borlee et al. (2008).

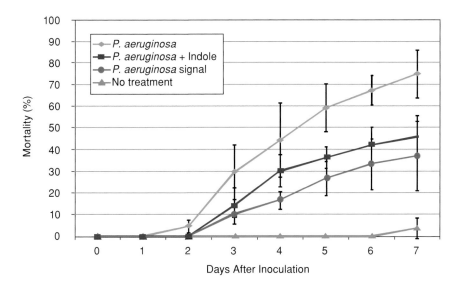

FIGURE 4-7 Mortality of cabbage white butterfly larvae fed *P. aeruginosa* strains and the quorum-sensing analog indole inhibitor. Treatments include *P. aeruginosa* PAO1, *P. aeruginosa* PAO1 and indole inhibitor, *P. aeruginosa* PAO1-JP2 (*lasI rhlI N*-acyl-L-lomoserine lactone-deficient mutant), and no *P. aeruginosa* PAO1 control. Values represent the mean mortality as a percentage of 18 larvae per treatment replicated in five independent experiments. Error bars are the standard errors.
SOURCE: Borlee et al. (2008).

Does the Community Affect the Health of the Host?

Bacillus thuringiensis (Bt) is a well-characterized pathogen of gypsy moth and, in fact, of all members of the order Lepidoptera. The bacterium produces rhomboid protein crystals that are highly toxic only to lepidopterans. The Bt toxin is known to resemble bacterial pore-forming toxins that affect mammals; thus, we are interested in exploring the interaction of Bt, its toxin, and the microbial community in which they function as a model for mammalian gut disease.

Bt has been used widely for the last 50 years to control insect pests that cannot be deterred or eradicated by other means, and in organic agriculture in lieu of synthetic pesticides. Interestingly, despite this broad usage, resistance to Bt has not developed in the field. Over the last decade, crop plants have been genetically engineered to express the Bt toxin gene and cultivated according to strict guidelines designed to prevent the development of resistance to the toxin. Although Bt toxin-expressing crops are now grown on a massive scale in the United States, resistance has not become a problem. Why this is the case, even though antibiotic resistance is rampant, is not known.

Bt was discovered nearly 100 years ago, and both the bacterium and the interaction between its toxin and the cells of the lepidopteran gut epithelium—where the toxin forms pores—have been very well studied. However, the pathway by which pore formation (which damages the intestinal epithelium) leads to the animal's death is not known. It has been assumed that either bacteremia caused by Bt or starvation (because pore formation is associated with reduced feeding) is the proximal cause of death.

Neither of those putative mechanisms satisfied one of my graduate students, Nichole Broderick, whose research had shown that some compounds known to enhance insects' sensitivity to Bt toxin also affect the bacterial community of the gut. As a result, she developed the hypothesis that the gut microbial community acts as a protection—a phalanx—against infection by Bt, and she predicted that eliminating the gut bacteria would enhance Bt activity.

To test this hypothesis, we treated insects with increasing concentrations of an antibiotic cocktail that we knew would kill all of the bacteria that we could detect in the gut (Broderick et al., 2006). However, antibiotic treatment did not enhance Bt killing; instead, the more antibiotic the larvae received in their diet, the greater the larval survival following exposure to Bt (Figure 4-8). That result

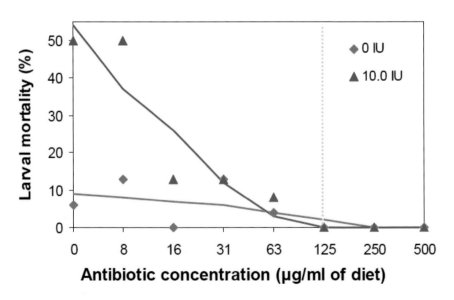

FIGURE 4-8 Gypsy moth larvae reared on antibiotics are not susceptible to Bt. The vertical line indicates the concentration of antibiotic at which no bacteria are detected in the midgut. IU = international units.
SOURCE: Broderick et al. (2006).

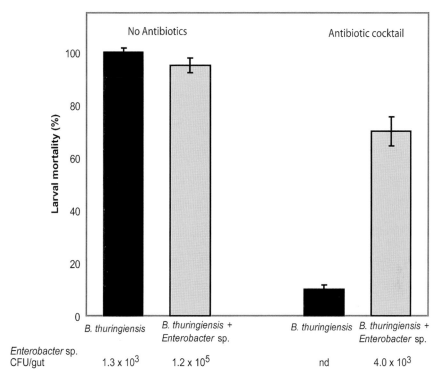

FIGURE 4-9 Restoration of *B. thuringiensis* toxicity by an *Enterobacter spp.* after elimination of the detectable gut flora and *B. thuringiensis* activity by antibiotics. *Lymantria dispar* larvae were reared until the third instar on a sterile artificial diet amended with 500 μg each of penicillin, gentamicin, rifampin, and streptomycin per milliliter (antibiotic cocktail). Each bar represents the mean mortality ± the standard error of the mean for 48 larvae (four replications with 12 larvae each). Values at the bottom represent the sizes of the populations of the *Enterobacter spp.* as detected by culture. nd = not detected. SOURCE: Broderick et al. (2006).

led us to the traitor hypothesis: if Bt is inactive in the absence of the endogenous microbial community, then perhaps community members actually collaborate with Bt in a multispecies infection. If so, one would predict that if we introduced members of the gut bacterial community into antibiotic-treated larvae that lacked a gut flora and were resistant to Bt, sensitivity to Bt would be restored.

We tested this hypothesis by rearing larvae on high levels of the antibiotic cocktail (Figure 4-9), letting the antibiotic clear, then feeding the larvae with *Enterobacter*, a gram-negative bacterium that we have found consistently in the gypsy moth gut; we then assessed the susceptibility of these larvae to Bt

(Broderick et al., 2004). In control experiments with larvae raised without antibiotics, Bt and Bt plus *Enterobacter* both kill nearly 100 percent of the larvae (Broderick et al., 2006). Antibiotics reduced killing by Bt, and the addition of *Enterobacter* prior to Bt exposure restored the killing. This suggests that *Enterobacter* and Bt work together to kill the insects.

Conclusion

Our work with the lepidopteran gut system is at an early stage. We have learned that the microbial assemblage it contains is relatively simple, comprising two phyla that are also found among the human gut microbiota. Chemical signals are exchanged between bacterial cells in the gut, and when this signaling process is inhibited chemically or genetically, the pathogenesis of *P. aeruginosa* is attenuated.

The microbial community affects host health. When the pathogen Bt is introduced, the normally benign gut microbiota mediate pathogenesis. In the absence of the normal gut microbiota, Bt does not induce killing and reintroduction of normal gut residents restores killing. This system might provide a model for studying the common phenomenon that commensal microbes can act as pathogens under the right conditions.

The insect gut system and the soil metagenomic analysis of antibiotic resistance genes both reveal the importance of accounting for communities in the analysis of microbial behavior and genetic potential in biological systems.

REFERENCES

Davies References

Abraham, E. P., and E. Chain. 1940. An enzyme from bacteria able to destroy penicillin. *Nature* 146(3713):837.

Bäckhed, F., R. E. Ley, J. L. Sonnenburg, D. A. Peterson, and J. I. Gordon. 2005. Host-bacterial mutualism in the human intestine. *Science* 307(5717):1915-1920.

Bartoloni, A., L. Pallecchi, H. Rodriguez, C. Fernandez, A. Mantella, F. Bartalesi, M. Strohmeyer, C. Kristiansson, E. Gotuzzo, F. Paradisi, and G. M. Rossolini. 2009. Antibiotic resistance in a very remote Amazonas community. *International Journal of Antimicrobial Agents* 33(2):125-129.

Dantas, G., M. O. A. Sommer, R. D. Oluwasegun, and G. M. Church. 2008. Bacteria subsisting on antibiotics. *Science* 320(5872):100-103.

Davies, J. 2006. Are antibiotics naturally antibiotics? *Journal of Industrial Microbiology and Biotechnology* 33(7):496-499.

Davies, J., G. B. Spiegelman, and G. Yim. 2006. The world of subinhibitory antibiotic concentrations. *Current Opinion in Microbiology* 9(5):1-9.

D'Costa, V. M., K. M. McGrann, D. W. Hughes, and G. D. Wright. 2006. Sampling the antibiotic resistome. *Science* 311(5759):374-377.

Gardner, P., D. H. Smith, H. Beer, and R. C. Moellering, Jr. 1969. Recovery of resistance (R) factors from a drug-free community. *Lancet* 2(7624):774-776.

Gorby, Y. A., S. Yanina, J. S. McLean, K. M. Rosso, D. Moyles, A. Dohnalkova, T. J. Beveridge, I. S. Chang, B. H. Kim, K. S. Kim, D. E. Culley, S. B. Reed, M. F. Romine, D. A. Saffarini, E. A. Hill, L. Shi, D. A. Elias, D. W. Kennedy, G. Pinchuk, K. Watanabe, S. Ishii, B. Logan, K. H. Nealson, and J. K. Fredrickson. 2006. Electrically conductive bacterial nanowires produced by *Shewanella oneidensis* strain MR-1 and other microorganisms. *Proceedings of the National Academy of Sciences* 103(30):11358-11363.

Hughes, V. M., and N. Datta. 1983. Conjugative plasmids in bacteria of the "pre-antibiotic" era. *Nature* 302(5910):725-726.

IOM (Institute of Medicine). 2006. *Ending the war metaphor: the changing agenda for unraveling the host-microbe relationship.* Washington, DC: The National Academies Press.

Koenig, J. E., Y. Boucher, R. L. Charlebois, C. Nesbo, O. Zhaxybayeva, E. Bapteste, M. Spencer, M. J. Joss, H. W. Stokes, and W. F. Doolittle. 2008. Integron-associated gene cassettes in Halifax Harbour: assessment of a mobile gene pool in marine sediments. *Environmental Microbiology* 10(4):1024-1038.

Lederberg, J. 1988. Medical science, infectious disease and the unity of humankind. *Journal of the American Medical Association* 260(5):684-685.

Livermore, D. M., R. Canton, M. Gniadkowski, P. Nordmann, G. M. Rossolini, G. Arlet, J. Ayala, T. M. Coque, I. Kern-Zdanowicz, F. Luzzaro, L. Poirel, and N. Woodford. 2007. CTX-M: changing the face of ESBLs in Europe. *Journal of Antimicrobial Chemotherapy* 59(2):165-174.

Nandi, S., J. J. Maurer, C. Hofacre, and A. O. Summers. 2004. Gram-positive bacteria are a major reservoir of Class 1 antibiotic resistance integrons in poultry litter. *Proceedings of the National Academy of Sciences* 101(18):7118-7122.

Norrby, S. R., C. E. Nord, R. Finch, and the European Society of Clinical Microbiology and Infectious Diseases. 2005. Lack of development of new antimicrobial drugs: a potential serious threat to public health. *Lancet Infectious Diseases* 5(2):115-119.

Pallecchi, L., C. Lucchetti, A. Bartoloni, F. Bartalesi, A. Mantella, H. Gamboa, A. Carattoli, F. Paradisi, and G. M. Rossolini. 2007. Population structure and resistance genes in antibiotic-resistant bacteria from a remote community with minimal antibiotic exposure. *Antimicrobial Agents and Chemotherapy* 51(4):1179-1184.

Peleg, A., H. Seifert, and D. L. Paterson. 2008. *Acinetobacter baumannii*: emergence of a successful pathogen. *Clinical Microbiology Reviews* 21(3):538-582.

Queenan, A. M., and K. Bush. 2007. Carbapenemases: the versatile β-lactamases. *Clinical Microbiology Reviews* 20(3):440-458.

Quince, C., T. P. Curtis, and W. T. Sloan. 2008. The rational exploration of microbial diversity. *The ISME Journal* 2(10):997-1006.

Riesenfeld, C. S., R. M. Goodman, and J. Handelsman. 2004. Uncultured soil bacteria are a reservoir of new antibiotic resistance genes. *Environmental Microbiology* 6(9):981-989.

Schlüter, A., R. Szczepanowski, A. Pühler, and E. M. Top. 2007. Genomics of IncP-1 antibiotic resistance plasmids isolated from wastewater treatment plant provides evidence for a widely accessible drug resistance gene pool. *FEMS Microbiology Reviews* 31(4):449-477.

Smith, D. H. 1967. R factor infection of *Escherichia coli* lyophilized in 1946. *Journal of Bacteriology* 94(6):2071-2072.

Spellberg, B., R. Guidos, D. Gilbert, J. Bradley, H. W. Boucher, W. M. Scheld, J. G. Bartlett, J. Edwards, and the Infectious Diseases Society of America. 2008. The epidemic of antibiotic-resistant infections: a call to action for the medical community from the Infectious Diseases Society of America. *Clinical Infectious Diseases* 46(2):155-164.

Tamae, C., A. Liu, K. Kim, D. Sitz, J. Hong, E. Becket, A. Bui, P. Solaimani, K. P. Tran, H. Yang, and J. H. Miller. 2008. Determination of antibiotic hypersensitivity among 4,000 single-gene-knockout mutants of *Escherichia coli*. *Journal of Bacteriology* 190(17):5981-5988.

Tennstedt, T., R. Szczepanowski, I. Krahn, A. Pühler, and A. Schlüter. 2005. Sequence of the 68,869 bp IncP-1alpha plasmid pTB11 from a waste-water treatment plant reveals a highly conserved backbone, a Tn*402*-like integron and other transposable elements. *Plasmid* 53(3):218-238.

Yim, G., H. Huimi Wang, and J. Davies. 2006. The truth about antibiotics. *International Journal of Medical Microbiology* 296(2-3):163-170.

Waksman, S. A. 1961. The role of antibiotics in nature. *Perspectives in Biology and Medicine* 4(3):271-286.

Cohen References

Avery, O. T., C. M. MacLeod, and M. McCarty. 1944. Studies on the chemical nature of the substance inducing transformation of pneumococcal types. *Journal of Experimental Medicine* 79:137-158.

Babst, M., G. Odorizzi, E. J. Estepa, and S. D. Emr. 2000. Mammalian tumor susceptibility gene 101 (TSG101) and the yeast homologue, Vps23p, both function in late endosomal trafficking. *Traffic* 1(3):248-258.

Beutler, B. 2004. Toll-like receptors and their place in immunology. Where does the immune response to infection begin? *Nature Reviews Immunology* 4(7):498.

Bishop, N., and P. Woodman. 2001. Tsg101/mammalian vps23 and mammalian vps28 interact directly and are recruited to vps4-induced endosomes. *Journal of Biological Chemistry* 276(15):11735-11742.

Bishop, R. A. 2003. *The history of bubonic plague.* Graduate School of Biomedical Sciences, University of Texas Health Science Center at Houston, http://dpalm. med.uth.tmc.edu/courses/BT2003/BTstudents2003_files%5CPlague2003.htm (accessed July 31, 2008).

Bonavia, A., D. Santos, D. Bamba, L. Li, J. Gu, M. Kinch, and M. Goldblatt. 2008. Recruitment of the TSG101/ESCRT-1 machinery in host cells by influenza virus: implications for broad spectrum therapy. Abstract W9-2 for the 27th Annual Meeting of the American Society for Virology, July 12-16, 2008, Ithaca, NY. P. 91.

Cherry, S., T. Doukas, S. Armknecht, S. Whelan, H. Wang, P. Sarnow, and N. Perrimon. 2005. Genome-wide RNAi screen reveals a specific sensitivity of IRES-containing RNA viruses to host translation inhibition. *Genes and Development* 19(4):445-452.

Cohen, S. N. 1993. Bacterial plasmids: their extraordinary contribution to molecular genetics. *Gene* 135(1-2):67-76.

Collier, R. J., and J. A. T. Young. 2003. Anthrax toxin. *Annual Review of Cell and Developmental Biology* 19:45-70.

Dunn, S. J., I. H. Khan, U. A. Chan, R. L. Scearce, C. L. Melara, A. M. Paul, V. Sharma, F.-Y. Bih, T. A. Holzmayer, P. A. Luciw, and A. Abo. 2004. Identification of cell surface targets for HIV-1 therapeutics using genetic screens. *Virology* 321(2):260-273.

Garrus, J. E., U. K. von Schwedler, O. W. Pornillos, S. G. Morham, K. H. Zavitz, H. E. Wang, D. A. Wettstein, K. M. Stray, M. Cote, R. L. Rich, D. G. Myszka, and W. I. Sundquist. 2001. Tsg101 and the vacuolar protein sorting pathway are essential for HIV-1 budding. *Cell* 107(1):55-65.

Goff, A., L. S. Ehrlich, S. N. Cohen, and C. A. Carter. 2003. Tsg101 control of human immunodeficiency virus type 1 Gag trafficking and release. *Journal of Virology* 77(17):9173-9182.

Griffith, F. 1928. The significance of pneumococcal types. *Journal of Hygiene* 27:113-159.

Hao, L., A. Sakurai, T. Watanabe, E. Sorensen, C. A. Nidom, M. A. Newton, P. Ahlquist, and Y. Kawaoka. 2008. *Drosophila* RNAi screen identifies host genes important for influenza virus replication. *Nature* 454(7206):890-893.

He, X., M. Semenov, K. Tamai, and X. Zeng. 2004. LDL receptor-related proteins 5 and 6 in wnt/beta-catenin signaling: arrows point the way. *Development* 131(8):1663-1677.

Ishii, K. J., S. Koyama, A. Nakagawa, C. Coban, and S. Akira. 2008. Host innate immune receptors and beyond: making sense of microbial infections. *Cell Host and Microbe* 3(6):352-363.

Katzmann, D. J., M. Babst, and S. D. Emr. 2001. Ubiquitin-dependent sorting into the multivesicular body pathway requires the function of a conserved endosomal protein sorting complex, escrt-i. *Cell* 106(2):145-155.

Koonin, E. V., and R. A. Abagyan. 1997. TSG101 may be the prototype of a class of dominant negative ubiquitin regulators. *Nature Genetics* 16(4):330-331.

Lederberg, J. 1952. Cell genetics and hereditary symbiosis. *Physiological Reviews* 32(4):403-430.

———. 1993. AIDS pandemic provokes alarming reassessments of infectious disease. *The Scientist* 7(14):11.

Lederberg, J., and E. M. Lederberg. 1952. Replica plating and indirect selection of bacterial mutants. *Journal of Bacteriology* 63(3):399-406.

Li, L., and S. N. Cohen. 1996. Tsg101: a novel tumor susceptibility gene isolated by controlled homozygous functional knockout of allelic loci in mammalian cells. *Cell* 85(3):319-329.

Lu, Q., L. W. Hope, M. Brasch, C. Reinhard, and S. N. Cohen. 2003. TSG101 interaction with HRS mediates endosomal trafficking and receptor down-regulation. *Proceedings of the National Academy of Sciences* 100(13):7626-7631.

Lu, Q., W. Wei, P. E. Kowalski, A. C. Chang, and S. N. Cohen. 2004. EST-based genome-wide gene inactivation identifies ARAP3 as a host protein affecting cellular susceptibility to anthrax toxin. *Proceedings of the National Academy of Sciences* 101(49):17246-17251.

Martin-Serrano, J., T. Zang, and P. D. Bieniasz. 2001. HIV-1 and Ebola virus encode small peptide motifs that recruit Tsg101 to sites of particle assembly to facilitate egress. *Nature Medicine* 7(12):1313-1319.

Ponting, C. P., Y. D. Cai, and P. Bork. 1997. The breast cancer gene product Tsg101: a regulator of ubiquitination? *Journal of Molecular Medicine* 75(7):467-469.

Raiborg, C., and H. Stenmark. 2002. Hrs and endocytic sorting of ubiquitinated membrane proteins. *Cell Structure and Function* 27(6):403-408.

Riedel, S. 2005. Edward Jenner and the history of smallpox and vaccination. *Proceedings (Baylor University Medical Center)* 18(1):21-25.

Tavassoli, A., Q. Lu, J. Gam, H. Pan, S. J. Benkovic, and S. N. Cohen. 2008. Inhibition of HIV budding by a genetically selected cyclic peptide targeting the Gag–Tsg101 interaction. *ACS Chemical Biology* 3(12):757-764.

VerPlank, L., F. Bouamr, T. J. LaGrassa, B. Agresta, A. Kikonyogo, J. Leis, and C. A. Carter. 2001. Tsg101, a homologue of ubiquitin-conjugating (E2) enzymes, binds the L domain in HIV type 1 Pr55(Gag). *Proceedings of the National Academy of Sciences* 98(14):7724-7729.

Wei, W., Q. Lu, G. J. Chaudry, S. H. Leppla, and S. N. Cohen. 2006. The LDL receptor-related protein LRP6 mediates internalization and lethality of anthrax toxin. *Cell* 124(6):1141-1154.

Wheelis, M. 2002. Biological warfare at the 1346 siege of Caffa. *Emerging Infectious Diseases* 8(9):971-975.

Handelsman References

Allen, H. K., L. A. Moe, J. Rodbumrer, A. Gaarder, and J. Handelsman. 2009. Functional metagenomics reveals diverse beta-lactamases in a remote Alaskan soil. *The ISME Journal* 3(2):243-251.

Borlee, B. R., G. D. Geske, C. J. Robinson, H. E. Blackwell, and J. Handelsman. 2008. Quorum-sensing signals in the microbial community of the cabbage white butterfly larval midgut. *The ISME Journal* 2(11):1101-1111.

Bregman, E., and J. B. Kirsner. 1965. Amino acids of colon and rectum. Possible involvement of diaminopimelic acid of intestinal bacteria in antigenicity of ulcerative colitis colon. *Proceedings of the Society for Experimental Biology and Medicine* 118:727-731.

Broderick, N. A., K. F. Raffa, R. M. Goodman, and J. Handelsman. 2004. Census of the bacterial community of the gypsy moth larval midgut by using culturing and culture-independent methods. *Applied Environmental Microbiology* 70(1):293-300.

Broderick, N. A., K. F. Raffa, and J. Handelsman. 2006. Midgut bacteria required for *Bacillus thuringiensis* insecticidal activity. *Proceedings of the National Academy of Sciences* 103(41): 15196-15199.

Engebrecht, J., and M. Silverman. 1984. Identification of genes and gene products necessary for bacterial bioluminescence. *Proceedings of the National Academy of Sciences* 81(13):4154-4158.

Fleming, T. J., D. E. Wallsmith, and R. S. Rosenthal. 1986. Arthropathic properties of gonococcal peptidoglycan fragments: implications for the pathogenesis of disseminated gonococcal disease. *Infection and Immunity* 52(2):600-608.

Fuqua, W. C., S. C. Winans, and E. P. Greenberg. 1994. Quorum sensing in bacteria: the LuxR-LuxI family of cell density-responsive transcriptional regulators. *Journal of Bacteriology* 176(2): 269-275.

Hirayama, K., and J. Rafter. 1999. The role of lactic acid bacteria in colon cancer prevention: mechanistic considerations. *Antonie van Leeuwenhoek* 76(1-4):391-394.

Hooper, L. V. 2004. Bacterial contributions to mammalian gut development. *Trends in Microbiology* 12(3):129-134.

Hooper, L. V., and J. I. Gordon. 2001. Commensal host-bacterial relationships in the gut. *Science* 292(5519):1115-1118.

Hooper, L. V., M. H. Wong, A. Thelin, L. Hansson, P. G. Falk, and J. I. Gordon. 2001. Molecular analysis of commensal host-microbial relationships in the intestine. *Science* 291(5505):881-884.

Hooper, L. V., T. Midtvedt, and J. I. Gordon. 2002. How host-microbial interactions shape the nutrient environment of the mammalian intestine. *Annual Review of Nutrition* 22:283-307.

Johannsen, L., L. A. Toth, R. S. Rosenthal, M. R. Opp, F. Obal Jr., A. B. Cady, and J. M. Kreuger. 1990. Somnogenic, pyrogenic, and hematologic effects of bacterial peptidoglycan. *American Journal of Physiology* 258(1 Pt. 2):R182-R186.

Krueger, J. M., J. R. Pappenheimer, and M. L. Karnovsky. 1982. Sleep-promoting effects of muramyl peptides. *Proceedings of the National Academy of Sciences* 79(19):6102-6106.

Ley, R. E., P. J. Turnbaugh, S. Klein, and J. I. Gordon. 2006. Microbial ecology: human gut microbes associated with obesity. *Nature* 444(7122):1022-1023.

Miller, M. B., and B. L. Bassler. 2001. Quorum sensing in bacteria. *Annual Review of Microbiology* 55:165-199.

Rath, H. C., H. H. Herfarth, J. S. Ikeda, W. B. Grenther, T. E. Hamm Jr., E. Balish, J. D. Tauroq, R. E. Hammer, K. H. Wilson, and R. B. Sartor. 1996. Normal luminal bacteria, especially *Bacteroides* species, mediate chronic colitis, gastritis, and arthritis in HLA-B27/human β_2 microglobulin transgenic rats. *Journal of Clinical Investigation* 98(4):945-953.

Schloss, P. D., and J. Handelsman. 2006. Toward a census of bacteria in soil. *PLoS Computational Biology* 2(7):e92.

5

Infectious Disease Emergence: Past, Present, and Future

OVERVIEW

Emerging infections, as defined by Stephen Morse of Columbia University in his contribution to this chapter, are infections that are rapidly increasing in incidence or geographic range, including such previously unrecognized diseases as HIV/AIDS, severe acute respiratory syndrome (SARS), Ebola hemorrhagic fever, and Nipah virus encephalitis. Among his many contributions to efforts to recognize and address the threat of emerging infections, Lederberg co-chaired the committees that produced two landmark Institute of Medicine (IOM) reports, *Emerging Infections: Microbial Threats to Health in the United States* (IOM, 1992) and *Microbial Threats to Health* (IOM, 2003), which provided a crucial framework for understanding the drivers of infectious disease emergence (Box WO-3 and Figure WO-13). As the papers in this chapter demonstrate, this framework continues to guide research to elucidate the origins of emerging infectious threats, to inform the analysis of recent patterns of disease emergence, and to identify risks for future disease emergence events so as to enable early detection and response in the event of an outbreak, and perhaps even predict its occurrence.

In the chapter's first paper, Morse describes two distinct stages in the emergence of infectious diseases: the introduction of a new infection to a host population, and the establishment within and dissemination from this population. He considers the vast and largely uncharacterized "zoonotic pool" of possible human pathogens and the increasing opportunities for infection presented by ecological upheaval and globalization. Using hantavirus pulmonary syndrome and H5N1 influenza as examples, Morse demonstrates how zoonotic pathogens gain access to human populations. While many zoonotic pathogens periodically infect humans, few become adept at transmitting or propagating themselves, Morse observes. Human activity, however, is making this transition increasingly easy by creating efficient pathways for pathogen transmission around the globe. "We know what is responsible for emerging infections, and should be able to prevent them," he concludes, through global surveillance, diagnostics, research, and above all, the political will to make them happen.

The authors of the chapter's second paper, workshop presenter Mark Woolhouse and Eleanor Gaunt of the University of Edinburgh, draw several general conclusions about the ecological origins of novel human pathogens based on their analysis of human pathogen species discovered since 1980. Using a rigorous, formal methodology, Woolhouse and Gaunt produced and refined a catalog of the nearly 1,400 recognized human pathogen species. A subset of 87 species have been recognized since 1980—and are currently thought to be "novel" pathogens. The authors note four attributes of these novel pathogens that they expect will describe most future emergent microbes: a preponderance of RNA viruses; pathogens with nonhuman animal reservoirs; pathogens with a broad host range; and pathogens with some (perhaps initially limited) potential for human-human transmission.

Like Morse, Woolhouse and Gaunt consider the challenges faced by novel pathogens to become established in a new host population and achieve efficient transmission, conceptualizing Morse's observation that "many are called but few are chosen" in graphic form, as a pyramid. It depicts the approximately 1,400 pathogens capable of infecting humans, of which 500 are capable of human-to-human transmission, and among which fewer than 150 have the potential to cause epidemic or endemic disease; evolution—over a range of time scales—drives pathogens up the pyramid. The paper concludes with a discussion of the public health implications of the pyramid model, which suggests that ongoing global ecological change will continue to produce novel infectious diseases at or near the current rate of three per year.

In contrast to other contributors to this chapter, who focus on what, why, and where infectious diseases emerge, Jonathan Eisen, of the University of California, Davis, considers how new functions and processes evolve to generate novel pathogens. Eisen investigates the origin of microbial novelty by integrating evolutionary analyses with studies of genome sequences, a field he terms "phylogenomics." In his essay, he illustrates the results of such analyses in a series of "phylogenomic tales" that describe the use of phylogenomics to predict the function of uncharacterized genes in a variety of organisms, and in elucidating the genetic basis of a complex symbiotic relationship involving three species.

Knowledge of microbial genomes, and the functions they encode, is severely limited, Eisen observes. Among 40 phyla of bacteria, for example, most of the available genomic sequences were from only three phyla; sequencing of Archaea and Eukaryote genomes has proceeded in a similarly sporadic manner. To fill these gaps in our knowledge of the "tree of life," his group has begun an initiative called the Genomic Encyclopedia of Bacteria and Archaea. Eisen describes this effort and advocates the further integration of information on microbial phylogeny, genetic sequence, and gene function with biogeographical data, in order to produce a "field guide to microbes."

The chapter's final paper, by Peter Daszak of the Consortium for Conservation Medicine, Wildlife Trust, makes the leap from knowing how infectious diseases emerge to predicting where, and under what circumstances, an emergent disease event is likely to occur. Daszak presents several examples of his group's efforts to build predictive approaches to infectious disease emergence based on a thorough understanding of the underlying ecology. These include constructing a model to predict relative risks for Nipah virus reemergence in Malaysia, where a 1999 outbreak devastated a thriving pig farming industry; identifying likely sources by which West Nile virus could spread to Hawaii, the Galapagos, and Barbados; and determining likely reservoirs of H5N1 influenza for specific geographic locations worldwide.

Daszak's group constructed a database of emerging infectious disease "events" first reported in human populations between 1940 and 2004, which they have used to examine correspondences between events and ecological variables, such as human population density and wildlife diversity, in a geographical context. These analyses have revealed "hotspots" for infectious disease emergence. Daszak discusses the implications of hotspot location for global infectious disease surveillance, and describes how he and coworkers have used their knowledge of hotspots to target surveillance for Nipah virus in India, and also to discover a virus with zoonotic potential in Bangladesh.

EMERGING INFECTIONS: CONDEMNED TO REPEAT?

Stephen S. Morse, Ph.D.[1]
Columbia University

We have all learned about the importance of infectious diseases throughout history, including the Plague of Justinian (541-542), the first known pandemic on record (McNeill, 1976), and the Black Death in the fourteenth century. Stanley Falkow, who is included in this volume, has extensively studied *Yersinia pestis*, the responsible organism, and given us important insights into its pathogenesis. Another devastating disease that was once much feared is smallpox, which is said to have killed more people than all the wars in history. The eradication of smallpox was therefore a triumph of public health. Ironically, smallpox has the unique property of being the only species to date that human beings have intentionally

---

[1]Professor of epidemiology and founding director of the Center for Public Health Preparedness at the Mailman School of Public Health.

driven to extinction. While we have unintentionally driven so many species to extinction, it is nice to know we can actually intentionally do some good. Cholera was, of course, a very big concern in the nineteenth century and remains a concern today, especially in places like Bangladesh, as Gerald Keusch of Boston University and a member of the Forum can affirm.

The 1918 influenza pandemic is one of our paradigms of a nightmare emerging infectious disease event. It may very well have been the greatest natural disaster in the early days of the twentieth century. The "official" mortality estimates keep rising as investigators keep finding data from further away, in developing countries and more remote places. But that pandemic is thought to have accounted for about 50 million or more deaths, depending on how you want to count it, and is obviously a matter of great concern.

Despite that, we have had years of complacency about infectious diseases, partly for reasons already discussed—the antibiotic era, immunizations, improved public health measures—all of which have led to the fact that we now live longer and tend to die later of chronic diseases. Unfortunately, this has not been true everywhere. It has not been true in many developing countries. Infectious diseases remain the major causes of morbidity and mortality in much of the world.

But in this paper, I would like to concentrate on emerging infections, the ones that are not previously recognized and that seem to appear suddenly and almost mysteriously—if you will, *The Andromeda Strain* (Crichton, 1969). Figure WO-7 graphically shows a number of examples. Of course, there are also forgotten infections that reappear. We sometimes call those "reemerging infections." I tend to think of most of the "reemerging" infections as reminding us that many infectious diseases in our highly mechanized modern societies, with the standard of living we enjoy, have been pushed to the margins, but have never been entirely eliminated. So when public health measures are relaxed or are abandoned because of lack of money or complacency—complacency being a very big problem—you then see forgotten infections reappearing. An example is diphtheria in the former Soviet Union and Eastern Europe in the early 1990s when those countries no longer had the money to maintain their immunization programs. It reminds us that many of these diseases may be forgotten, but they are not gone.

HIV/AIDS is, of course, the infection that got our attention initially and made it possible at least to think about shaking ourselves out of the growing complacency about infectious diseases. HIV infection and AIDS, starting from obscurity, rose to become a leading cause of death in the United States by 1993 (Figure 5-1). There are recent reports dating HIV to the early twentieth century, but it didn't appear to take off until mid-century. You can find a molecular example of HIV in Zaire in 1969, but that is almost a one-off, and then there were reports of a few cases in the 1970s in Africa, if anyone had been paying attention. Then suddenly, in the early 1980s, it appeared in the United States and took off like the proverbial rocket to overtake all other causes of death in healthy young people. Of course, this is the same age group killed in the 1918 flu, but also the

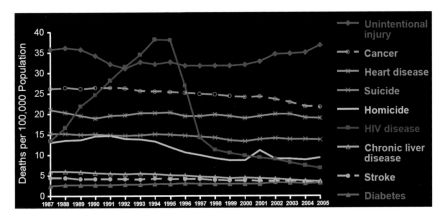

FIGURE 5-1 Leading causes of death in young adults, United States, 1987-2005. Red line: Rise of HIV infection to become leading cause of death.
SOURCE: CDC (2008).

very people we generally expect to have the best survival rate. They have survived childhood and we expect that they ought to be fine. As shown in Figure 5-1, all the other causes of death were unchanged during that period.

HIV was therefore quite a surprise. When you think about it, this does seem rather like *The Andromeda Strain*. We had thousands of years of experience with infections, some of them historically recorded in some detail. Some of these are still unidentified, and we still argue about what they were. But a disease that actually kills by undermining the immune system directly was a novel mechanism of pathogenesis. How often does one find a new mechanism of pathogenesis in an infectious disease, considering the thousands of years of experience that we have had? I think it was quite remarkable.

Since its peak (around 1995), the HIV/AIDS death rate in the young adult population in the United States has dropped (Figure 5-1), thanks largely to the fact that a few effective drugs were finally developed, including in particular the protease inhibitors. As a result, the trend reached a plateau and has recently been going down. HIV/AIDS is now a treatable disease, with many lives saved among those who can afford the medication. But it also worries me that this fortunate situation may not last very long. Inevitably, antiviral resistance has already been identified in some patients. Another concern is that some of the younger people have now become quite complacent about this disease, not knowing the devastation that many of us witnessed in the 1980s, before it could be effectively treated. We are seeing young people now regarding this with less seriousness than they should.

So there we are, facing complacency again. If there is a bottom line to the theme of the Forum on Microbial Threats, it is that we cannot afford to be complacent anymore.

What are emerging infections? I always like informally to define emerging infections as those that would knock a *really* important story off the front page of the newspaper, whether the runaway bride or the Texas polygamy case, at least for a day or two. However, I do have a more formal definition: those infections that are rapidly increasing in incidence or geographic range. In some cases, these are novel, previously unrecognized diseases. But, as I am going to show you, many of them are not *The Andromeda Strain*. They do not come from space. Actually, in many cases, they have already existed in nature. Very often, anthropogenic causes—often as unintended consequences of things we do—are important in the emergence of these infections.

There are many examples. You can pick your favorite: Ebola in 1976; hantavirus pulmonary syndrome, which I will discuss briefly in a moment; Nipah, which Peter Daszak addressed at the workshop (and his group has done some excellent work on this); SARS; and, of course, influenza, which still continues to surprise us.

You could think of the many events shown in Figure 5-2 as "a thousand points of light" (or at least those of you who are old enough to remember the first President Bush). But these are really a lot of little fires all over the world, most of which we did not spot in time before they became big brush fires or even wildfires. That includes many examples, such as West Nile virus entering the United States in 1999, the enteropathogenic *Escherichia coli* (made famous by the "Jack in the Box" case[2]), and a number of others, including SARS, of course.

I have divided the process of disease emergence into two steps, for analysis: (1) what I call introduction, where these "*Andromeda*-like" infections are coming from; and (2) establishment and dissemination, which (fortunately for us) is much harder for most of these agents to achieve. The basic lesson there is that many may be called, but few are chosen.

In this two-step process, as you all know, the opportunities are increasing thanks to ecological changes and globalization, which gives the microbes great opportunities to travel along with us, and to travel very quickly. Even medical technologies have played an inadvertent role in helping to disseminate emerging infections.

I will spend most of my time talking about what seems to be the most mysterious step—and I hope we can demystify it a bit here—and that is the introduction of a "new" infection. What we now know is that many of these infections, exotic as they may seem, are often zoonotic. Some of them do not do very much, and

---

[2]In 1993, four children died and hundreds became ill after eating undercooked hamburger patties contaminated by *E. coli* bacteria at Jack in the Box restaurants (see http://www.about-ecoli.com/ecoli_outbreaks/view/jack-in-the-box-e-coli-outbreak).

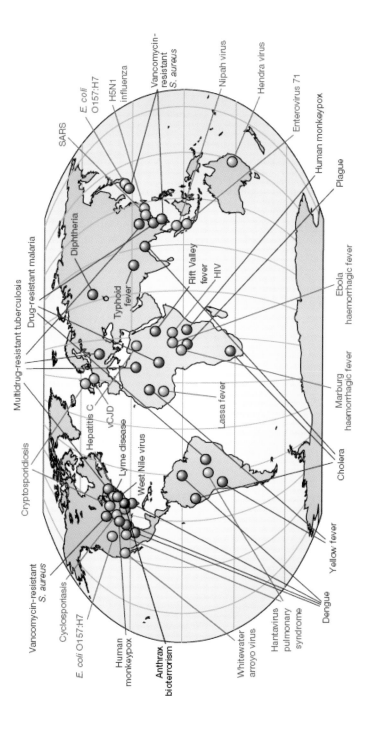

FIGURE 5-2 Global examples of emerging and reemerging infectious diseases, some of which are discussed in the main text. Red represents newly emerging diseases; blue, reemerging or resurging diseases; black, a "deliberately emerging" disease.
SOURCE: Reprinted from Morens et al. (2004) with permission from Macmillan Publishers Ltd. Copyright 2004.

may cause no infection at all; while others may cause a truly dramatic infection, like Ebola.

So that zoonotic pool, if I may use that term, is not fully chlorinated, and it is a rich source of potential emerging pathogens. There is so much biodiversity out there, including a tremendous biodiversity of microbes. Some of that biodiversity—we do not know how much, even now—is still untapped.

Changes in the environment may increase the frequency of contact with a natural host carrying an infection, and therefore increase our chances of encountering microorganisms previously unknown to humans. Of course, the role of food animals, as well as wildlife (one of the subjects of Peter Daszak's contribution to this volume), has come very much to fore in recent years.

There are a number of examples associated with activities like agriculture, food-handling practices, and, for the vector biologists, of course, changes in water ecosystems. Table 5-1 lists just some of these cases. The basic point is that there are a number of ecological changes, many of them anthropogenic, which provide new opportunities for pathogens to emerge and gain access to human populations. Think of these as a sort of microbial explorers, discovering new niches—us—and exploring new territory.

It is important not to overlook the very important role of evolution as well. One role is obviously what evolution has already been doing for a long time, leading to the biodiversity of pathogens that we see existing in nature. It is remarkable, when you think about how great that biodiversity is. We don't even know how many viruses human beings are subject to, even how many inhabit us at this very moment. But when I think about just the herpesviruses, which are pretty well studied, that number could be very large indeed. There are eight known human herpesviruses, and at least six of them—you might argue, even seven of them, except for Human herpesvirus 8, the one that causes Kaposi's sarcoma—are ubiq-

TABLE 5-1 New Opportunities for Pathogens: Ecological Changes

Agriculture	Hantaan, Argentine hemorrhagic fever, Nipah, West Nile (Israel), possibly pandemic influenza
Food-handling practices	SARS, H5N1 influenza, HIV?, enteropathogenic *E. coli*
Dams, changes in water ecosystems	Rift Valley fever, other vectorborne diseases, Schistosomiasis
Deforestation, reforestation	Kyasanur Forest, Lyme disease
Climate changes	Hantavirus pulmonary syndrome (HPS), vectorborne diseases

uitous in the human population. They can be found all over the world. Several of them are present at very high prevalence in the human population.

That just gives you an idea of some of that great biodiversity. As it happens, these herpesviruses are all specialized for humans. There are, of course, herpesviruses of other species. So a lot of coevolution between host and pathogen goes on as well.

Of course, there is adaptation to new hosts and environments through natural selection. We see this with influenza most notably, but with many other examples—the coronaviruses, like SARS—as well. Of course, antimicrobial resistance has been mentioned so many times. If anyone needs to be convinced about the role of evolution in the world, I think this is a pretty good demonstration—one of the rare examples in which you can do *in vitro* exactly the same thing as what happens in the real world, just on a different scale.

There are many case studies. I'll briefly discuss a few, just to illustrate some key points.

Hantavirus pulmonary syndrome was ironically one of the first things to happen suddenly in the United States after the original Institute of Medicine *Emerging Infections* report came out in October 1992. Hantavirus pulmonary syndrome suddenly appeared in the southwestern United States in the following spring and summer.

My friend Richard Preston wrote a book called *The Hot Zone*. He has a very philosophical chapter at the end where he talks about the "revenge of the rainforest." I think it is a good thought, in that we should be kinder to our environment, for many good reasons. The rainforests are great sources of biodiversity and, to a great extent, that biodiversity was largely unexplored.

But an emerging infection can occur anywhere. Even the southwestern United States, which looks so dry, arid, and inhospitable to life, has its share, different from the rainforest, but just as significant.

Jim Hughes, who is a Forum member and was the director of the National Center for Infectious Diseases (NCID) at the Centers for Disease Control and Prevention (CDC) at the time of the outbreak, knows this story firsthand. Starting in the late spring and then going through the summer of 1993, people started appearing at emergency departments and clinics with respiratory distress. Many of them were hospitalized. I believe the case fatality rate at that time was about 60 percent, even with treatment. It is a little lower now, but it is still hovering near 40 to 50 percent.

The health departments did the usual investigations: There is a pocket of plague in that area, so the local health departments tested for that. Another possibility could be influenza out of season. These, and other likely possibilities, were ruled out. The state health departments then called in CDC, which did a number of tests and identified, perhaps surprisingly, a hantavirus as the most likely culprit. This was tested both by serology and, later, shedding of virus was tested by polymerase chain reaction (PCR). Of course, when you think of hantavirus, you usually think of rodents, with a few minor exceptions. So a number of rodent species trapped near patients' homes were tested. The most frequent rodent was apparently also the most frequently infected: *Peromyscus maniculatus*, the deer mouse. This is a very successful and prolific rodent that is essentially the major wild rodent in this entire area. Ruth Berkelman likes to refer to this as your typical hardworking single mom, as shown in the illustration (Figure 5-3).

Of course, once a test was developed and people started looking for the virus, they were able to find it in a great number of other places, including serum and tissue samples that had been saved earlier because the etiology was unknown, but

FIGURE 5-3 A deer mouse (*Peromyscus maniculatus*), natural host for the Sin Nombre (hantavirus pulmonary syndrome) virus, with her young.
SOURCE: Image courtesy of Bet Zimmerman, www.sialis.org.

FIGURE 5-4 Hantaviruses of the Americas. Viruses associated with human disease are shown in bold.

SOURCE: Adapted from Peters (1998) with permission from ASM Press and Jim Mills.

odd—cases of acute respiratory distress. There were even some cases outside the geographic range of *Peromyscus maniculatus*, which turned out to be hantaviruses that were natural infections of other rodent species.

This point is illustrated in Figure 5-4 (I thank C. J. Peters, then at CDC, for the illustration). Before 1993, the United States had one known hantavirus, not associated with human disease (Prospect Hill virus) and another hantavirus of rats, Seoul virus, and related variants that could be found in port cities; neither was associated with serious acute disease in the United States. After 1993, we had to add another: the virus that causes hantavirus pulmonary syndrome. Then, when people started looking for hantaviruses, there was no shortage of previously unrecognized cases. In Figure 5-4, the virus names in bold have been associated with human disease, while many others have not. So throughout North and South America, suddenly there was a whole rash of hantaviruses that nobody knew existed.

That is evolution at work. We do not know how long ago this diversification occurred. It could have been as long as 2 million years ago, when some of the rodent species separated, but I would defer to the mammalologists on that issue.

As with HIV, at first we think it is an orphan, but, of course, it has its relatives; we just hadn't found them yet.

What about the respiratory viruses? We have been thinking about that question a great deal lately. Some of our most serious historical examples—influenza, measles, smallpox, and many others—have been respiratory viruses. Right now pandemic influenza and H5N1 avian flu are very much on our minds.

Figure WO-11 was one of Josh Lederberg's favorite slides. It shows U.S. mortality rates. You can see an enormous peak in 1918, coinciding with the 1918 flu pandemic. It was a big event, and even in the United States it killed at least a half-million people, most of them young and previously healthy.

Several pandemics have been documented. The 1918 pandemic was by far the worst. The Asian flu of 1957—as it happened, I lived through the next two twentieth-century influenza pandemics—was not much fun, to put it mildly, but nothing was ever like 1918. I wasn't there to experience that one, thankfully. Later, the pandemic Hong Kong flu of 1968 appeared but was relatively mild compared with 1918 and even 1957.

There have been some other events along the way: the reappearance in 1977 of H1N1, and the famous swine flu scare in 1976, which, in fact, Harvey Fineberg, now president of the IOM, wrote about when he was at Harvard ("the epidemic that never was," as he and his coauthor Richard Neustadt dubbed it [Neustadt and Fineberg, 1983]).

Figure 5-5 shows a ward filled with patients suffering from influenza during the 1918 pandemic. These are soldiers who were about to go overseas to fight in World War I. The photo shows graphically the impact that a disease like the 1918 flu had. CDC has since recalculated the case fatality rates adjusted to today's population, just extrapolating what the expected deaths would be. With today's population, a 1918-like pandemic would be expected to cause almost 2 million deaths in the United States alone. If it were like the 1957 or 1968 pandemic, a much milder pandemic, it might be fewer than 100,000 deaths. In any case, it is not something to take lightly.

In pandemic influenza viruses, the novel or new genes tend to come from avian influenza viruses that then reassort, often with mammalian influenza genes (or at times the virus may possibly go directly from avian to human, although that seems to be a relatively rare event).

We hear a great deal recently about the H5N1 avian flu in humans, and about the next pandemic. These two terms, "pandemic" and "avian flu," are really not synonymous, although nonscientists sometimes mistakenly use them that way. Rob Webster and Virginia Hinshaw discovered some years ago that the waterfowl of the world appear to be the natural reservoirs for influenza viruses. Therefore,

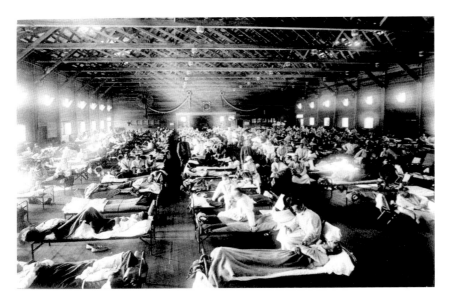

FIGURE 5-5 Influenza pandemic 1918 at Camp Funston, Kansas.
SOURCE: Image NCP 1603 courtesy of the National Museum of Health and Medicine, Armed Forces Institute of Pathology, Washington, DC.

there is certainly open territory for influenza virus dissemination along any of the Old World flyways for bird migration.

As a result of all of those movements of birds, both migratory fowl and domestic poultry, we have seen a number of outbreaks of H5N1 avian influenza, starting in Asia, but extending into Europe and Africa as well. There have been some human cases, mostly (although not all) occupational, and with a high case fatality rate. Fortunately, there have been only a few instances of human-to-human transmission so far, all apparently quite limited. Obviously, everyone is watching this closely, just in case there is a change in the ability of the virus to transmit from person to person. If this virus were able to infect people readily and transmit itself, let's say, as well as ordinary seasonal influenza does, then it could well be the next pandemic.

I am not putting money on H5N1, however. The next pandemic is going to happen, but so far nobody in this field can predict exactly when and where, and which influenza strain will be responsible. The only people who claim they can, at least as of now, are either charlatans or great risk takers. It is safer to bet on horse races.

Let me now move on briefly to that second step in emergence—establishment and dissemination. Luckily for us, this is much harder for a newly-minted pathogen. So many infections that can get into human beings from time to time may not have a good way of transmitting or propagating themselves. We have given them some help in this regard—think about HIV, for example, spreading in the blood supply or through contaminated injection equipment—and provided some highways for what I like to call "microbial traffic": pathogens moving into new areas or new populations. Of course, environmental changes can be important here as well.

It used to take a long time to get around the world, but now you can do it, if you make all your connections, in 24 to 48 hours. If you do not make all your connections, as happens to most of us, then you spend time in a usually crowded airport, where you have even more opportunity to infect others.

Consider SARS, for example. By the way, ironically, Hong Kong decided to embark on a new promotional campaign just before SARS started. The slogan was "Hong Kong will take your breath away." I do not know what inspired them to come up with it just then. Maybe they are better prognosticators than we are when it comes to the flu and other respiratory diseases. They certainly have had much more direct experience.

The consequences of SARS on global travel were enormous. The usually bustling Hong Kong airport was deserted. At least the few who did arrive there did not have to worry about waiting for their luggage. And the hotel rooms were cheaper, especially at the Hotel Metropole, which, we now know, was site of the "Big Bang" of SARS.

The spread of SARS was a remarkable event, when you think about it. One infected individual—a physician, in fact—from south China treated a patient who had an unusual pneumonia. Clinicians usually assume community-acquired pneumonia is not very transmissible—a major mistake here, as this turned out to be, unfortunately, an exception. He then went to Hong Kong, where he stayed at the Hotel Metropole, a popular business hotel, and became sick. He believed he had the same disease that had killed the patient he had treated earlier. He went to the hospital, told his healthcare providers about his odd patient, and warned them to be careful. Apparently they did not pay much attention. There were 99 healthcare workers infected in Hong Kong alone.

At the same time, another dozen people were infected in the Hotel Metropole by this index patient. This is what was responsible for the dissemination of SARS essentially worldwide. Of course, everyone likes to say that it was an interesting coincidence that he stayed in Room 911. There no longer is a Room 911 at the Hotel Metropole, by the way. This is a little bit like the first Legionnaires' out-break and what it did to that hotel's image, but that is another story.

We had a few near-misses with SARS. The man who went to Vietnam was actually a New York businessperson who did not go back to New York. One doc-tor from Singapore did go to New York, but did not get sick until he was on his way home and was put into isolation in Germany.

Just to put in a small plug for one of my favorite causes (of course, this is completely biased): ProMED-mail, the listserv for reporting and discussion of emerging infections. There was a little item that appeared there in February 2003, just questioning whether something odd was going on in China, with reports of deaths. The next day China admitted to having 305 cases of SARS.

Yi Guan, as he first reported at an IOM Forum meeting on SARS in October 2003, actually was able to find earlier cases, going back at least to November 2002. There were several different cases in perhaps five cities in southern China, but they were not reported or recognized at the time. He did a survey and found that animal slaughterers and wild animal handlers had a much greater chance of becoming seropositive. Why? Because the ultimate link to humans was another cute little animal, *Paguma larvata*, the palm civet, which is actually a prized food animal in south China, particularly during the winter. It is very expensive. The civets became infected, it would appear, in the live animal markets, probably from contact with bats (according to work by Peter Daszak and colleagues). Wild-caught and farmed civets—yes, they do farm them—that were tested were all negative for the SARS coronavirus.

Then, of course, SARS came to Canada, as we well know, and wreaked havoc. Those of you who know Don Low, as many do, know that he was right at the front line there; I remember that when I saw him at one of our Forum meetings just after the crisis was over, he was exhausted.

By the end of all this, there were about 8,000 cases, most of them in the original area, but a few in other widely scattered places, with over 700 deaths, or about a 10 percent case fatality rate. Not a trivial disease.

This also was the first time the World Health Organization (WHO) had really acted aggressively, which got the Canadians very annoyed, since WHO issued a travel advisory recommending that travelers avoid Toronto. But WHO acted very effectively and was able, I think, to solve some of the scientific and disease-control problems rather quickly.

There is probably a parallel story with HIV origins. We do not know how it entered the human population. It may very well have been through a similar mechanism as SARS. It came from chimpanzees, most likely, and humans may have become infected by preparing or handling infected nonhuman primates for the "bushmeat" trade.

Hospitals also provide opportunities for emerging infections. Transmission of infections by contaminated injection equipment is well known. Most of the Ebola cases arose this way.

In summary, there are some recognizable factors responsible for precipitating or enabling emergence, such as ecological factors or globalized travel and trade. This was the framework, which I had originally developed, that we used in the *Emerging Infections* (IOM, 1992) report. These factors have since been augmented and embellished in the new version of the IOM *Emerging Infections* report, titled *Microbial Threats to Health*, published in 2003 (Box WO-3; IOM, 2003). So there are even more of them now, but I think they are recognizable. We know what is responsible for emerging infections and should be able to prevent them.

What are we going to do about this? One thing we can do is improve disease surveillance. I will put in another plug for ProMED here. There is a sort of backhanded compliment, I guess, from a recent popular book about John Snow and cholera, *The Ghost Map*, by Steven Johnson (2006). On page 219, he states: "The popular ProMED-mail e-list offers a daily update on all the known disease outbreaks flaring up around the world, which surely makes it the most terrifying news source known to man."

The reality is that we need better early-warning systems and more effective disease control, implemented without delay. If we had let SARS go the way we had let AIDS go, probably very few of us would be here to talk about it, especially the physicians.

To summarize, these are my central themes:

- There are factors responsible for the emergence of infectious disease.
- Often, interspecies transfer is responsible or facilitates emergence.
- Things we do (anthropogenic changes) often increase the risk of transmission by altering the environment and interposing ourselves into an environment containing pathogens unfamiliar to humans.
- We can manage those risks in some ways using our wits.
- You might ask, what should we be doing to make the world safer? Effective global surveillance is one, as are better diagnostics, political will to respond to these events, and research to help understand the ecology and pathogenesis of these "new" infections and to help develop effective preventive or therapeutic measures.

I am sure the other contributors to this chapter will have additional suggestions and insights into the problem and about how we might begin to make the world safer. We must get serious about this. Our future as a species may well depend on it someday.

ECOLOGICAL ORIGINS OF NOVEL HUMAN PATHOGENS[3]

Mark Woolhouse, Ph.D.[4]
University of Edinburgh

Eleanor Gaunt, B.Sc.[4]
University of Edinburgh

A systematic literature survey suggests that there are 1399 species of human pathogen. Of these, 87 were first reported in humans in the years since 1980.

---

[3]This article is reprinted with permission from *Critical Reviews in Microbiology* 33:231-242 (2007).

[4]Centre for Infectious Diseases, University of Edinburgh, Edinburgh, United Kingdom.

The new species are disproportionately viruses, have a global distribution, and are mostly associated with animal reservoirs. Their emergence is often driven by ecological changes, especially with how human populations interact with animal reservoirs. Here, we review the process of pathogen emergence over both ecological and evolutionary time scales by reference to the "pathogen pyramid." We also consider the public health implications of the continuing emergence of new pathogens, focusing on the importance of international surveillance.

Introduction

In this review, we will be particularly concerned with species of pathogen that have recently been reported to be associated with an infectious disease in humans for the first time. As discussed more fully below, not all such pathogens (possibly very few of them) will be truly "new," at least in the sense that the pathogen has only recently discovered us rather than we have only recently discovered the pathogen. This focus on novel pathogens differs somewhat from the more general topic of "emerging infectious diseases," which is often taken to include previously rare disease which are now on the increase, and sometimes diseases once considered to be in decline but which are now resurgent—the so-called "re-emerging" diseases. However, our focus does fairly reflect one of the major public health concerns of the early 21st century, the possible emergence of new pathogens species and novel variants (OSI 2006).

At first glance, a pre-occupation with yet-to-emerge disease problems may seem extravagant, given the massive and all too immediate health burdens imposed by malaria, tuberculosis, measles, and other familiar examples. An obvious counterargument is the relatively recent advent of HIV-1, unrecognized less than a generation ago and yet now one of the world's biggest killers. As we shall discuss, the great majority of novel pathogens have not caused public problems on anything like this scale. However, AIDS (reinforced by knowledge of other plagues occurring throughout human history—see Diamond 2002) reminds us that the possibility that they could do so is real. In the early stages of the emergence of a new disease, it is a possibility that all too often cannot easily be dismissed as current concerns about H5N1 influenza A virus attest. A second reason for concern is that outbreaks of new diseases, and the public reaction to them, can cause economic and political shocks far greater than might be anticipated. The 2003 SARS epidemic, for example, resulted in fewer than 1000 deaths but cost the global economy many billions of dollars (King et al. 2006). Variant CJD, which has caused just over 100 deaths mostly confined to the UK, has had a global economic impact of a similar magnitude. Moreover a better understanding of the natural history of the emergence of new infectious diseases should inform our ability to combat them and, as the 2003 SARS epidemic illustrated, rapid, coordinated intervention can be highly effective.

Pathogen Diversity

Surveys of Pathogen Species

Although the existence of pathogens has been recognized for centuries, the first comprehensive list of human pathogen species was not published until 2001 (Taylor, Latham, and Woolhouse 2001). This list was generated from a comprehensive review of the secondary literature available at the time (see Taylor, Latham, and Woolhouse 2001 for full details). Each entry was a distinct species known to be infectious to and capable of causing disease in humans under natural transmission conditions. Species only known to cause infection through deliberate laboratory exposure were excluded. Species only known to cause disease in immuno-compromised patients and species only associated with a single human case of infection (e.g., Zika virus) were included. Ectoparasites such as ticks and leeches were not included. The 2001 list included species names that appeared in either (1) a text book published within the previous 10 years, or (2) standard web-based taxonomy browsers (see below), or (3) an ISI Web of Science citation index search covering the preceding 10 years. In subsequent work (e.g., Woolhouse and Gowtage-Sequeira 2005) NCBI taxonomies were used throughout (www.ncbi.nlm.nih.gov.library.vu.edu.au/Taxonomy/).

This methodology has the advantage that it is (or, at least, aspires to be) systematic, transparent and reproducible by other researchers. However, it does have its limitations and two of these in particular are worth highlighting. First, the criterion "capable of causing disease" has been variously interpreted and not all text book reports of disease-causing organisms can be confirmed from the primary literature. Second, some taxonomies have been revised since 2001, altering which pathogen variants are regarded as "species." Further revisions can reasonably be anticipated. More fundamentally, using the species as the unit of analysis ignores a wealth of important and interesting variation that occurs within species in traits such as virulence factors, antigenicity, host specificity or antibiotic resistance. Moreover, what is meant by "species" may differ from one group to another; some pathogens have complex subspecific taxonomies (e.g., *Salmonella enterica*, *Listeria monocytogenes*, human rhinoviruses, Candiru virus complex, *Trypanosoma brucei* complex), making direct comparisons of different "species" potentially problematic. With these caveats noted however, a survey of recognized species represents a natural starting point for investigations of the diversity of human pathogens.

Surveys of New Pathogen Species

A subset of human pathogen species of special interest here is those that have only recently been discovered. In this context, "recently" is taken (arbitrarily) as meaning from 1980 onwards and "discovered" means recognized as causing infection and disease in humans. Thus there are several possible reasons for a pathogen to appear in the list of "new" species.

1. Both the pathogen and the disease it causes did not occur before 1980.
2. The disease was already recognized but the pathogen was not identified as the etiological agent before 1980.
3. The pathogen was already recognized but had not been associated with human disease before 1980.
4. Neither the pathogen nor the disease it causes were recognized or reported before 1980, but they did occur.
5. What was considered to be a single pathogen before 1980 was subsequently recognized as comprising two or more species.

Strictly speaking, only the first of these possibilities constitutes an "emerging" infectious disease as defined earlier. In practice, however, most post-1980 pathogens probably fall into categories (2) to (5). For example, phylogenetic evidence has demonstrated clearly that the evolutionary origins of the human immunodeficiency viruses pre-date their discoveries in the 1980s by at least several decades (van Heuverswyn et al. 2006).

To provide a more complete picture of new pathogens the list of species described above was supplemented in early 2007 by searching the WHO, CDC, and ProMed web sites and the primary literature.

Results of Pathogen Surveys

Based on the above methodologies an updated version of the previously reported surveys generates a list of 1399 species of human pathogen. The most diverse group is the bacteria (over 500 species) with fungi, helminths and viruses making up most of the remainder (Table 5-2).

Of these 1399 species of human pathogen, 87 have been discovered from 1980 onwards (Table 5-3). The composition of the subset of new species is very different from the full list. New species are dominated by viruses, and there are relatively few bacteria, fungi or helminths (Table 5-2). Within these broad categories certain taxa stand out: human retroviruses were not reported until 1980; most of the new fungi are microsporidia; and almost half the new bacteria are rickettsia. Although the over-representation of viruses is highly statistically significant (odds ratio (OR) = 18.0, $P < 0.001$), it is not clear that (excluding retroviruses) particular *kinds* of viruses have special status. Single-stranded RNA viruses make up the largest subset of new species (45 species) but are only marginally over-represented. Similarly, bunyaviruses are the largest single family but are also only marginally over-represented in the list of new viruses.

In summary, since 1980 new human pathogen species have been discovered at an average rate of over 3 per year. Almost 75% of these have been virus spe-

TABLE 5-2 Numbers of Pathogen Species by Taxonomic Category

	No. spp.	No. spp. since 1980
TOTAL	1399	87
Bacteria	541	11
Fungi	325	13
Helminths	285	1
Prions	2	1
Protozoa	57	3
Viruses	189	58
DNA viruses	36	9
RNA viruses	153	49

cies even though viruses still represent a small fraction (less than 14%) of all recognized human pathogen species.

Geographic Origins of Novel Pathogens

For those pathogen species discovered in the post-1980 period, the geographic location of the first reported human case(s) can often be determined from the primary literature, at least to within specific countries and often to specific regions or municipalities. However, this is not possible for all new pathogen species. For example, although the early history of HIV-1 has been exhaustively investigated the exact origin of the first reported human case remains unclear (Barre-Sinoussi et al. 1983). Similarly, the only reported human case of European bat lyssavirus 2 in a human could have resulted from exposure in Finland, Switzerland or Malaysia (Lumio et al. 1986). Moreover, some new human pathogens were already endemic or ubiquitous in the human population when they were first discovered; examples include human metapneumovirus and human bocavirus. For those pathogens which were discovered previously, but were only recently associated with human disease (such as commensals which have become pathogenic in patients immunosuppressed due to infection with HIV) the geographic origin is taken as the location in which the patient became sick (if the patient was not reported as having recent travel history).

Figure 5-6 shows a map of the points of origin of the first human cases of disease caused by 51 of the 87 pathogen species discovered since 1980. Data of this kind must be interpreted cautiously, not least because of likely ascertainment bias (variable likelihood of detection and identification of novel pathogens) in dif-

TABLE 5-3 Dates of First Reports of Human Infection with Novel Pathogen Species

Human bocavirus	2005	Dobrava-Belgrade virus	1992
Human coronavirus HKU1	2005	*Ehrlichia chaffeensis*	1991
Human T-lymphotropic Virus 3	2005	*Encephalitozoon hellem*	1991
Human T-lymphotropic Virus 4	2005	Guanarito virus	1991
Human coronavirus NL63	2004	*Nosema ocularum*	1991
SARS coronavirus	2003	Banna virus	1990
Cryptosporidium hominis	2002	Gan gan virus	1990
Baboon cytomegalovirus	2001	Reston Ebola virus	1990
Human metapneumovirus	2001	Semliki Forest virus	1990
Cryptosporidium felis	2001	Trubanaman virus	1990
Whitewater Arroyo virus	2000	*Vittaforma corneae*	1990
Brachiola algerae	1999	*Corynebacterium amycolatum*	1989
Ehrlichia ewingii	1999	European bat lyssavirus 1	1989
Nipah virus	1999	Hepatitis C virus	1989
TT virus	1999	Barmah Forest virus	1988
Brachiola vesicularum	1998	Picobirnavirus	1988
Menangle virus	1998	Dhori virus	1987
Trachipleistophora anthropophthera	1998	Sealpox virus	1987
		Suid herpesvirus 1	1987
Bartonella clarridgeiae	1997	*Cyclospora cayetanensis*	1986
Laguna Negra virus	1997	European bat lyssavirus 2	1986
Andes virus	1996	Human herpesvirus 6	1986
Australian bat lyssavirus	1996	Human immuno-deficiency virus 2	1986
BSE/CJD agent	1996	Kasokero virus	1986
Ehrlichia canis	1996	Kokobera virus	1986
Juquitiba virus	1996	Rotavirus C	1986
Metorchis conjunctus	1996	Borna disease virus	1985
Trachipleistophora hominis	1996	*Enterocytozoon bieneusi*	1985
Usutu virus	1996	*Pleistophora ronneafiei*	1985
Bayou virus	1995	Human torovirus	1984
Black Creek Canal virus	1995	Rotavirus B	1984
Cote d'Ivoire Ebola virus	1995	*Scedosporium prolificans*	1984
Hepatitis G virus	1995	Candiru virus	1983
New York virus	1995	*Capnocytophaga canimorsus*	1983
Anaplasma phagocytophila	1994	*Helicobacter pylori*	1983
Hendra virus	1994	Hepatitis E virus	1983
Human herpesvirus 7	1994	Human adenovirus F	1983
Human herpesvirus 8	1994	Human immuno-deficiency virus 1	1983
Sabia virus	1994	*Borrelia burgdorferi*	1982
Bartonella elizabethae	1993	Human T-lymphotropic Virus 2	1982
Encephalitozoon intestinalis	1993	Seoul virus	1982
Gymnophalloides seoi	1993	*Microsporidian africanum*	1981
Sin Nombre virus	1993	Human T-lymphotropic Virus 1	1980
Bartonella henselae	1992	Puumala virus	1980

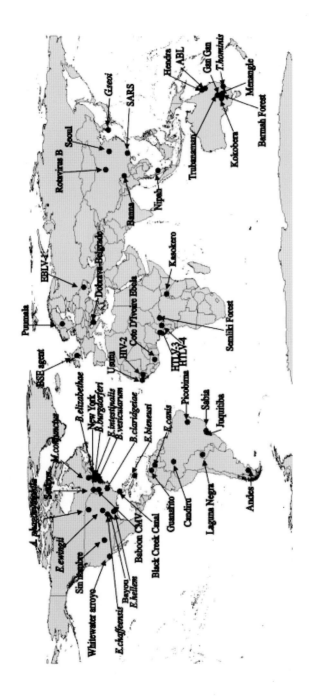

FIGURE 5-6 World map indicating points of origin of the first reported human cases of disease caused by 51 novel pathogen species since 1980. Locations are identified to municipality or region (occasionally country), jiggled as necessary to avoid overlap.

ferent parts of the world. Nonetheless, Figure 5-6 does make the important point that the emergence of new pathogens shows a truly global pattern, with multiple incidents being reported from every continent except Antarctica (with other gaps apparent in, for example, the Middle East and central Asia). There is no striking tendency for new pathogens to be more likely to be reported from tropical rather than temperate regions, or from less developed regions, or from more densely populated regions.

Process of Pathogen Emergence

Reservoirs of Infection

Relatively few human pathogens are known solely as human pathogens. The remainder also occur in other contexts: as commensals; or free-living in the wider environment; or as infections of hosts other than humans.

Overall, probably no more than 50 to 100 species are specialist human pathogens. These range from major killers such as *Plasmodium falciparum*, mumps virus, *Treponema pallidum*, smallpox and HIV-1 to those causing more minor problems such as the human adenoviruses and rhinoviruses.

Hundreds of species which can cause human disease occur naturally as "commensals" found on the skin, on mucosal surfaces, or in the gut. They are normally benign but are sometimes pathogenic, for example if introduced into the blood system via a wound or in association with AIDS or other immunosuppressive conditions. Examples include the streptococci and *Candida* spp.

Several hundred human pathogen species have environmental reservoirs; these are referred to as "sapronoses." Examples include *Bacillus anthracis*, *Legionella pneumophila*, and *Cryptococcus neoformans*. Here, we do not take sapronotic to include pathogens which are transmitted via the fecal-oral route or via a free-living stage of a complex parasite life cycle. Most sapronoses are bacteria or fungi, plus some protozoa, and cause sporadic infections of humans. Few are highly transmissible (directly or indirectly) between humans, an important exception being *Vibrio cholerae*. Some human pathogens (e.g., *Listeria* spp.) are both sapronotic and zoonotic.

Many more pathogens—over 800 species—are capable of infecting animal hosts other than humans. These range from species where humans are largely incidental hosts—such as rabies or *Bartonella henselae*—to species in which the main reservoir (sensu Haydon et al. 2002) is the human population and animals may be largely incidental hosts, that is, the so-called "reverse zoonoses" such as *Schistosoma haematobium*, rubella virus, *Mycobacterium tuberculosis*, or *Necator americanus*. We refer to all of these as "zoonotic," following the World Health Organization's definition of zoonoses as "diseases or infections which are naturally transmitted between vertebrate animals and humans." In contrast to some other authors (e.g., Hubalek 2003) we do not consider pathogens with invertebrate reservoirs, and especially pathogens which are transmitted by arthropod vectors, as zoonotic. Note that the WHO definition does not include human pathogen species which recently evolved from animal pathogens, such as HIV-1. Nor does it include pathogens with complex life cycles where vertebrate animals are involved only as intermediate hosts with humans as the sole definitive host. It does, however, include reverse zoonoses.

Few of the 87 new human pathogen species in Table 5-3 are commensals or sapronoses. The great majority—around 80%—are associated with nonhuman vertebrate reservoirs (e.g., SARS coronavirus, vCJD agent and *Borrelia burgdorferi*) and most of the remainder appear to be long-standing human pathogens which have only recently been identified (e.g., Hepatitis G virus). Even some of the nonzoonotic pathogens, notably HIV-1 and HIV-2, are recently evolved from pathogens of nonhuman vertebrates (Keele et al. 2006). Compared with human pathogen species reported before 1980 the new species are statistically significantly more likely to be associated (or, at least, are more likely to be *known to be* associated) with a nonhuman animal reservoir (OR = 2.75, P < 0.001).

The reservoirs of the new, zoonotic human pathogens are mainly mammals, although a small number are associated with birds (Figure 5-7). However, the reservoirs include a wide range of mammal groups with ungulates, carnivores, and rodents most frequently involved, but also bats, primates, marsupials and occasionally other taxa (Figure 5-7). These observations must be interpreted with some caution because our knowledge of the host range of many pathogens is still incomplete. Nevertheless, the data available give the impression that taxonomic relatedness is less important than ecological opportunity as a determinant of the reservoirs of novel human pathogens. *Homo sapiens* as a species is classified within primates and, beyond that, the most closely related major groups are the rodents and lagomorphs. Ungulates, carnivores, and bats are more distant relatives. One related observation is that emerging human pathogens are especially likely to have a broad host range which includes more than one of these groups (Woolhouse and Gowtage-Sequeira 2005).

Drivers of Pathogen Emergence

As discussed earlier, not all the pathogens in the list of new species should be regarded as truly emerging; some have only recently been identified as the causative agents of established infectious diseases. However, for 30 or more of new species the literature suggests various drivers deemed to be associated with their emergence at the present time. These drivers can be considered within a framework originally suggested by the Institute of Medicine (IOM 2003), noting that this framework was devised with reference to all emerging and re-emerging infectious diseases, not just newly discovered pathogen species.

The most commonly cited drivers fall within the following IOM categories: economic development and land use; human demographics and behavior; inter-

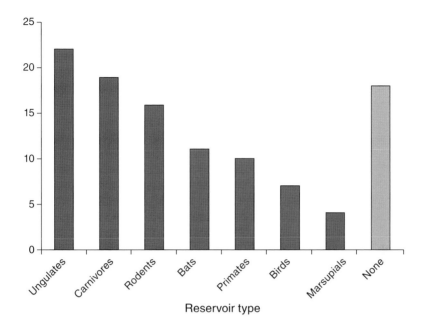

FIGURE 5-7 Counts of recently discovered human pathogens species (see Table 5-3) associated with various categories of non-human animal reservoirs. Some pathogens species are associated with more than one category of reservoir. These data should be regarded as no more than indicative since for some species the reservoir(s) is suggested in the literature but has not yet been confirmed and for others a reservoir is suspected but has not yet been even tentatively identified.

national travel and commerce; changing ecosystems; human susceptibility; and hospitals. Economic development and land use, and especially *changes* in economic development and land use, are associated with the emergence of pathogens such as Nipah virus and *Borrelia burgdorferi* through activities such as intensification of farming and forest encroachment respectively. Human demographics and behavior, and especially *changes* in human demographics and behavior, are associated with the emergence of pathogens such as HIV-1 and Hepatitis C virus through activities such as sexual activity and intravenous drug use. International travel and trade are increasing as part of the process of globalization and are associated with the emergence of pathogens such as SARS coronavirus. Changing ecosystems covers unintended consequences of human activities such as desertification, pollution, and climate change and is associated with the emergence of pathogens such as the hantaviruses. Broadly speaking, the set of drivers listed so far are all "ecological" in nature; they are to do with the ways that humans interact with their wider environment (especially with other vertebrate animals both domestic and wild), providing opportunities for pathogens to infect humans, and with the ways that humans interact with each other, providing opportunities for pathogens to spread within human populations. A particular concern—implicit but not highlighted in the IOM's list—is increasing use of "exotic" animal species, whether as food, farm animals or pets, and the trade that accompanies this.

The other most commonly cited drivers are to do with human population health. Human susceptibility is particularly important in the context of coinfections associated with AIDS (e.g., several species of microsporidia) but also covers the effects of malnutrition and other immunosuppressive conditions. The hospitals category covers iatrogenic transmission (e.g., vCJD), and xenotransplantation (e.g., baboon cytomegalovirus), as well as nosocomial infections (e.g., Ebola viruses and Rotavirus C).

Other categories listed by the IOM—such as "intent to harm"—have not been or are not commonly cited as associated with the emergence of novel human pathogen species. Among these is the category "microbial adaptation and change," an observation that we expand on below.

Transmission and Disease

The 87 new species of human pathogen are associated with public health problems of hugely variable magnitudes. At one extreme is HIV-1 which has killed an estimated 25 million people since it was first reported in 1983, with 40 million more currently infected (UNAIDS 2007). HIV-1 has a high transmission potential within many human populations (combining transmission mainly by sexual contact or by needle-sharing associated with intravenous drug use with an infectious period of several years) and is highly pathogenic (with a case fatality rate close to 100% in the absence of treatment). At the other extreme, Menangle virus is known to have infected only 2 farm workers in which it may have caused a mild febrile illness (Chant et al. 1998). Menangle virus does not appear to be highly infectious to or transmissible between humans and has not so far been associated with severe disease. In the following section we consider the kinds of epidemiological and biological differences that underlie the vast difference in public health impacts between pathogens such as HIV-1 and pathogens such as Menangle virus.

Pathogen Pyramid

A useful aid to conceptualizing the process of pathogen emergence is the pathogen pyramid. The concept of the pathogen pyramid was first put forward by Wolfe et al. (2004) and developed further in Wolfe, Dunavan, and Diamond (2007). A very similar framework but with a more formal mathematical under-pinning was adopted by Woolhouse, Haydon, and Antia (2005). The pyramid we use here has four levels corresponding to exposure, infection, transmission, and epidemic spread (Figure 5-8). Wolfe, Dunavan, and Diamond (2007) subdivided epidemic spread into (in their terminology): Stages 4a, b, and c, infectious diseases that exist in animals but with different balances of animal-to-human and human-to-human spread (where Stage 4c corresponds to reverse zoonoses as defined above); and Stage 5, pathogens exclusive to humans (corresponding to specialist human pathogens as defined above).

Level 1: Exposure The first stage of the emergence of a new pathogen is the exposure of humans to that pathogen. Exposure requires "contact" between humans and the pathogen reservoir (which may be animal or environmental; exposure to commensals is implicit). The nature of "contact" is determined by the mode of transmission of the pathogen, e.g., animal bite, contamination of food with fecal material, blood-feeding by arthropod vectors or exposure to aerosols. The only barrier to exposure is insufficient overlap between habitats occupied by

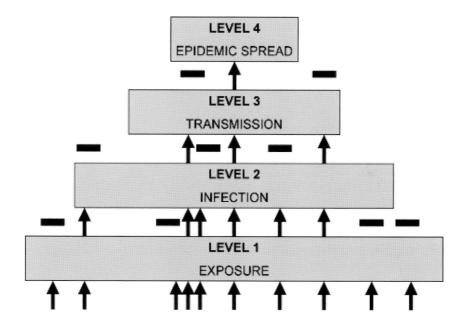

FIGURE 5-8 The pathogen pyramid (adapted from Wolfe, Dunavan, and Diamond 2007). Each level represents a different degree of interaction between pathogens and humans, ranging from exposure through to epidemic spread. Some pathogens are able to progress from one level to the next (arrows); others are prevented from doing so by biological or ecological barriers (bars)—see main text.

humans and habitats occupied by the pathogen. Changes in human ecology, particularly patterns of land use and interactions with animal reservoirs, are likely to change our exposure to potential new pathogens, as are changes in the ecology of the pathogens, their reservoirs or their vectors, e.g., as a result of climate change or other kinds of environmental change.

We do not know how many potential human pathogen species there are which we have not yet been exposed to, but we do know that human pathogens make up only a fraction of the known biodiversity of viruses, bacteria, fungi, protozoa and helminths, which in turn probably makes up only a fraction of the biodiversity which exists (Dykhuizen 1998).

Level 2: Infection The second stage of pathogen emergence is reached if the pathogen proves capable of infecting humans, possibly causing disease. As reviewed above, we know of 1399 species that have reached this stage. Others may have done so but have yet to be identified. Others may do so in the future but, to date, we have had no or insufficient exposure to them. Clearly, there will often be significant biological barriers—referred to as species barriers—preventing organisms infecting other kinds of host from infecting humans. We do not, for example, share any pathogens with plants, very few with invertebrates, and only a small number with cold-blooded vertebrates (e.g., *Salmonella* spp. in reptiles and amphibians—Mermin et al. 2004; helminth infections from fish—Chai et al. 2005). In contrast, we share many more of our pathogens with birds, and we share more than half with other species of mammal.

Indeed, the species barrier (at least between humans and other mammals) may not be as profound as is sometimes implied. According to Cleaveland et al. (2001) over 500 different species of pathogen are known to occur in domestic livestock and as many as 40% of these are zoonotic. The same authors report for domestic carnivores (dogs and cats) that almost 400 pathogen species are known, of which almost 70% are zoonotic. These data imply that, given the opportunities for exposure to pathogens that proximity to domestic animals must surely provide, many pathogens, perhaps even a majority, are capable of crossing the species barrier and infecting humans.

As suggested by the IOM (2003) report, an important contributor to the ability of a new pathogen to infect humans is variation in human susceptibility. In some cases this variation might have a genetic basis; for example, apparently pre-existing genetic variation in human susceptibility to HIV (Arien, Vanham, and Arts 2007). More commonly, phenotypic variation in the human population will be important, particularly factors which compromise the human immune system. The most striking examples come from the wide range of opportunistic infections associated with the immunosuppressive effects of HIV infection; these include several pathogen species, such as the microsporidia *Brachiola algerae* and *Enterocytozoon bieneusi* which were first recognized in AIDS patients.

Level 3: Transmission The third stage of pathogen emergence is reached if a pathogen that can infect humans also proves capable of transmission from one human to another. Transmission in this context need not be direct (e.g., by aerosol spread or sexual contact); it might be indirect (e.g., via contamination of food) or via an arthropod vector. The requirement is simply that an infection of one human leads ultimately to an infection of another.

In most cases the barriers preventing transmission will be biological, often reflecting tissue tropisms within the human host since pathogens normally need to access the gut, upper respiratory tract, urogenital tract or (especially for vector-borne infections) blood in order to be able to exit the body. However, sometimes such barriers can be overcome by changes in human behavior. The two best examples concern prion diseases. Kuru is only transmitted through cannibalism, which is extremely rare in most human societies. vCJD is not transmissible between humans except iatrogenically as a result of surgical procedures or blood transfusions.

Again, these barriers to human-to-human transmission are far from insuperable. Although information is lacking for many pathogen species (Taylor, Latham, and Woolhouse 2001), the literature suggests that a substantial minority—at least 500 species, over one third of the total, and possibly many more—are transmissible between humans.

Level 4: Epidemic Spread The fourth and, in our version, final level of the pathogen pyramid is reached if a pathogen is sufficiently transmissible within the human population to cause major epidemics or pandemics and/or to become endemic, without the involvement of the original reservoir. This represents a quantitative rather than qualitative distinction and it can be made more formally precise by reference to the concept of the basic reproduction number, R_0. R_0 can be defined as the average number of secondary cases of infection produced when a primary case is introduced into a large population of previously unexposed hosts (adapted from Anderson and May 1991). The distinction between Level 3 and Level 4 pathogens can be expressed in terms of R_0. If R_0 is less then one then, on average, a single primary case will fail to replace itself and although there may be chains of transmission these will be self-limiting—this corresponds to Level 3. On the other hand, if R_0 is greater than one then, on average, a single primary case will produce more than one secondary case and, at least initially, there will be an exponential increase in the number of cases and ultimately a major epidemic is possible—this corresponds to Level 4. (A proviso is that, even if R_0 >1, stochastic extinction of the infection chain is quite possible, especially in the early stages of the epidemic when numbers of cases are low—see May, Gupta, and McLean 2001.)

The barriers between Level 3 and Level 4 are both biological and epidemiological. The biological barriers are to do with pathogen infectivity, host susceptibility, the infectiousness of the infected host and for how long the host is infectious (whether this is terminated by recovery or death). The epidemiological barriers are to do with the rate and pattern of contacts between infectious and susceptible hosts. Here again, the nature of a "contact" reflects the mode of transmission of the pathogen (see above). The rate and pattern of contacts can increase, and hence R_0 can increase, independently of the pathogen, as a result of shifts in host demography or behavior. In the context of human hosts such shifts could constitute changes in factors such as population density (e.g., urbanization), living conditions, water supply and sanitation, patterns of travel and migration, or sexual behavior and intravenous drug use, depending on the specific pathogen involved. These might be augmented by changes in host susceptibility due to the kinds of factors listed earlier. Clearly, for the same pathogen R_0 can vary considerably from one human population to another. Similarly, different strains of the same pathogen species may have very different R_0 values in humans, e.g., different subtypes of influenza A virus.

In principle, this barrier might seem quite fragile; the kinds of changes in host demography and behavior alluded to above are certainly occurring. In practice, it is not clear how many species of human pathogen have reached Level 4 since we have estimates of R_0 values within human populations for only a handful of them. Based on earlier studies (Taylor, Latham and Woolhouse 2001; Woolhouse and Gowtage-Sequeira 2005) a plausible estimate is that 100 to 150 pathogen species are capable of causing major outbreaks within human populations, with half to two-thirds of these being specialist human pathogens and the remainder also occurring in animal reservoirs or the wider environment. This implies considerable attrition between levels 3 and 4 of the pathogen pyramid.

Status of New Pathogens

We can now consider where the 87 new human pathogen species fit within the pathogen pyramid. It is immediately clear that the majority of them are at Level 2; they can infect humans but are rarely if at all transmitted between humans. Examples include *Borrelia burgdorferi*, vCJD agent, most of the hantaviruses and *Ehrlichia* spp. At the other extreme, although there are a number that appear to be at Level 4, most of these are pathogens which are probably long established in human populations but have only recently been recognized, such as human metapneumovirus or hepatitis C virus. Only a very small number are likely to be recent additions to the repertoire of Level 4 human pathogens, namely HIV-1, HIV-2 and, arguably before its spread was contained, SARS coronavirus. In between, at Level 3, there is a significant minority of new pathogens that are somewhat transmissible between humans but which have so far been restricted to relatively minor outbreaks. These include Andes virus, human torovirus and some *Encephalitozoon* spp. For these species the value of the basic reproduction number R_0 is of particular interest, especially if it lies close to one, the threshold for potential epidemic spread. R_0 can be estimated from data on the distribution of outbreak sizes as follows.

The quantitative analysis of outbreak data used to estimate R_0 is based upon a methodology developed by Jansen et al. (2003) for measles case data from the UK (to monitor the effect of changes in childhood vaccination coverage). Here, we apply the technique (see also Matthews and Woolhouse 2005) to data on human outbreaks of Andes virus (see Figure 5-9 for details). Andes virus is an emerging South American hantavirus and there are concerns that, unusually for hantaviruses, it can be transmitted directly between humans (Wells et al. 1997). Most reports of Andes virus represent sporadic cases (i.e., outbreaks of size 1) but clusters of cases also occur, ranging in size from 2 to 20 (Figure 5-9). This pattern—many small outbreaks and a few larger ones—is typical of a wide range of infectious diseases (Woolhouse, Taylor, and Haydon 2001). The best estimate of R_0 based on these data lies in the range 0.22 to 0.37. This is well below one and in reality is likely to be an over-estimate since at least some of the clusters

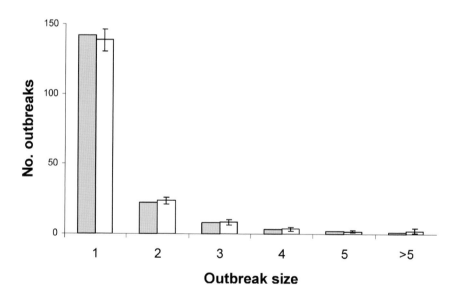

FIGURE 5-9 Analysis of Andes virus outbreaks. Frequencies of outbreaks of different sizes (grey bars) are compared with the fit of a statistical model to the data (open bars). Outbreak data are taken from Wells et al. (1997) and Lazaro et al. (2007). The model is described in full in Matthews and Woolhouse (2005) and is fitted using maximum likelihood methods, giving an estimate of a single model parameter, the basic reproduction number or R_0. Vertical lines show the variation in frequencies associated with the 95% confidence intervals for the estimate of R_0. In this case the best estimate of R_0 is 0.28 with 95% confidence intervals 0.22 to 0.37.

of cases may reflect exposure to a common source rather than, as is assumed in the analysis, person-to-person spread. However, the analysis does suggest that occasional larger outbreaks will occur (the R_0 estimates are consistent with up to 1 in 200 outbreaks being of size 10 or more) without necessarily implying that there has been a major change in Andes virus epidemiology. This same approach can be applied to other "Level 3" pathogens to determine how close they are to reaching Level 4 of the pyramid (cf. Jansen et al. 2003).

Evolution and Emergence

So far we have examined the emergence of new species of human pathogens over time scales of a few decades. However, the origins of many human pathogens are considerably more ancient, extending back over time scales of thousands to millions of years. This process has been reviewed by, among others, Weiss (2001), Diamond (2002), and Wolfe, Dunavan, and Diamond (2007). Of particular interest here are examples of pathogens which have emerged in human populations as a result of successfully crossing the species barrier from an animal reservoir and reaching Level 4 status. Any analysis must be prefaced by the observation that we have good evidence for the origins of only a small minority of pathogens, plausible hypotheses (usually based on the epidemiologies of related species) for some of the remainder, and no information at all for the majority. Wolfe, Dunavan, and Diamond (2007) have proposed that this lack is addressed by a research program they term an "origins initiative". That said, 16 examples of putative species jumps are listed in Table 5-4. Inspection of this list suggests two tentative observations. First, although a variety of different kinds of pathogen are listed including several species of bacteria and protozoa, the majority are viruses. Second, a variety of different animal reservoirs are involved: primates, ungulates, rodents and birds. Wolfe, Dunavan, and Diamond (2007) point out that primates are much better represented in this list than might be expected given their much more modest role as reservoirs of modern zoonoses. This may reflect both the much greater ecological overlap between humans and other primates in the distant past and the notion that pathogens of our closest relatives are more likely to be epidemiologically successful in humans. The latter idea is supported by the observation that two of the most recent examples of successful species jumps—HIV-1 and HIV-2—have primate origins (Keele et al. 2006). Similarly, several human pathogens with much deeper evolutionary origins, perhaps even pre-dating *Homo sapiens* as a distinct species, are also most closely related to modern primate pathogens. Examples include the hepatitis B and G viruses (Simmonds 2001). It is worth noting that species jumps can occur in both directions. For example, it is thought that *Mycobacterium bovis*—predominantly a cattle pathogen—evolved from the human pathogen *M. tuberculosis* (Brosch et al. 2002).

HIV-1 and HIV-2 illustrate that the evolution of new species of pathogen is an ongoing process. Both are sufficiently divergent from their closest relatives—

TABLE 5-4 List of Human Pathogens Which Have Successfully Crossed the Species Barrier and Proved Capable of Epidemic Spread and, in Some Cases, Endemic Persistence in Human Populations. The Original Hosts Have Been Identified with Varying Degrees of Certainty.

Pathogen	Disease	Original Hosts
Corynebacterium diphtheriae	Diphtheria	Domestic herbivores?
Dengue fever virus	Dengue fever	Old World primates
Hepatitis B virus	Hepatitis B	Apes
HIV-1	AIDS	Chimpanzee
HIV-2	AIDS	Sooty mangabey
Influenza A virus	Influenza	Wildfowl
Measles virus	Measles	Sheep/goat
Mumps virus	Mumps	Mammals (pigs?)
Plasmodium falciparum	Malaria	Birds?
Plasmodium vivax	Malaria	Asian macaques
Rickettsia prowazeckii	Typhus	Rodents
SARS coronavirus	SARS	Bats/palm civets
Trypanosoma brucei subspp.	Sleeping sickness	Wild ruminants
Variola virus	Smallpox	Ruminants (camels?)
Yellow fever virus	Yellow fever	African primates
Yersinia pestis	Plague	Rodents

SOURCE: Modified from Weiss (2001) and Wolfe, Dunavan, and Diamond (2007).

SIV_{cpz} and SIV_{smg} respectively—in terms of both their genome sequences and their biologies to be regarded as distinct species. This has probably occurred within the last 100 years. In a nonhuman context, over even shorter time scales we have seen the evolution of another new species of pathogen, canine parvovirus (CPV), associated with a cat virus, feline panleukopenia virus (FPV), jumping into dogs (Parrish and Kawaoka 2005). CPV has spread to dog populations around the world in only a few years.

All of these examples concern RNA viruses, and RNA viruses differ from pathogens with DNA genomes in having far higher nucleotide substitution rates and so the potential for rapid adaptation to new host species (Holmes and Rambaut 2004). The importance of this kind of genetic lability has been explored by Antia et al. (2003) using simple mathematical models. These authors suggested that the potential for successful adaptation (which they defined as becoming sufficiently transmissible that R_0 in humans became greater than one) is sensitive both to the size of initial outbreaks (determined mainly by the initial R_0 value) and, especially, to the rate of genetic change and the genetic distance to be traveled. As discussed earlier, the initial R_0 value is a function not only of pathogen biology but also of features of human demography and behavior which promote transmission and thus the kinds of changes in these mentioned above have the potential to increase the likelihood of the evolution of new human pathogens.

The successful adaptation of a nonhuman pathogen to humans is itself a highly stochastic process. This is illustrated by the early evolution of the human immunodeficiency viruses (see Van Heuverswyn et al. 2006). There is phylogenetic evidence for numerous introductions of SIVs into human populations; most of these failed to become established (Arien et al. 2007) and only HIV-1 M subtype C has become truly pandemic.

This pattern raises the question of where, in practice, the relevant genetic changes that allow a pathogen to successfully invade a human population occur. Antia et al.'s analysis focuses on the process of adaptation within the human population. However, it may be that genetic change within the original reservoir (whether animal or environmental) is also critical for producing variants which are capable of infecting humans in the first place. With a handful of exceptions, such as the simian immunodeficiency viruses, we typically have very little information on the genetic and functional diversity of human pathogens or their immediate ancestors in nonhuman reservoirs.

This is a potentially important topic for future research but a reasonable working hypothesis, supported by our knowledge of the origins of HIV, is that genetic variation in nonhuman pathogen populations does occasionally and incidentally produce human infective variants, and this explains why so many novel human pathogens are RNA viruses (Woolhouse, Taylor, and Haydon 2001). This idea is further supported by the observation that RNA viruses tend to have broader host ranges than DNA viruses (Cleaveland, Laurenson, and Taylor 2001; Woolhouse, Taylor, and Haydon 2001), implying that they can more easily adapt to new host species.

The implication of the preceding discussion is that pathogen evolution is not only an important driver of progression up the pathogen pyramid over long time scales but that, especially for RNA viruses, this process may be relevant over much shorter time scales as well. In addition, we note that evolution is clearly a key driver of the emergence of new variants of existing human pathogen species, with potentially significant epidemiological consequences. This is evident in the generation of antibiotic resistant bacteria and chloroquine resistant malaria, as well as variants expressing novel virulence factors (e.g., *E. coli* O157) or with distinct pathogenicities (e.g., H5N1 influenza A).

Finally, we note that an important feature of new pathogens is that they have not been previously subject to evolutionary constraints on their virulence (i.e., the degree of harm they do to the host) in the new host (Ebert 1998). Moreover, the new host may make only a small contribution to the epidemiology of the pathogen (Level 3 of the pathogen pyramid), or even none at all [if] it is an epidemiological "dead end" in the sense that although infection can occur there is no onward transmission of infection (Level 2). In such cases evolutionary constraints on pathogen virulence may be weakened or absent (Woolhouse, Taylor and Haydon 2001). Putting these observations together it is unsurprising that many new human pathogens (e.g., Nipah and Ebola viruses, some hantaviruses, SARS coronavirus, and HIV-1) are very virulent, as indicated by their high case-fatality rates.

Public Health Implications

Future Emergence Events

It seems likely that the kinds of ecological changes that have been associated with pathogen emergence in the recent past (see IOM 2003) will continue to occur in the immediate future, e.g., continued deforestation for agriculture, intensification of livestock production, globalization, bush meat trade, urbanization, and so on. In that case, we can reasonably anticipate the reporting of yet more new species of human pathogen (currently happening at a rate of over 3 per year—Table 5-3) in the immediate future as well.

The survey of new pathogen species reported since 1980 suggests the kinds of pathogens that are most likely to emerge in the future. Four characteristics are expected to be particularly important:

1. RNA virus (most new pathogens are RNA viruses);
2. Nonhuman animal reservoir (most new human pathogens are associated with or originate from other kinds of host, usually other species of mammal);
3. Broad host range (pathogens that are already capable of exploiting a range of different hosts species are more likely to have the potential to infect us);
4. Some, perhaps initially limited, potential for transmission between humans (in which case, evolution of the pathogen and/or changes in the human population that increase the pathogen's transmission potential could lead to a marked increase in the size of outbreaks).

The above criteria are certainly not intended as absolute predictors of pathogen emergence; a good historical counterexample is syphilis (new to the Old World in the late 15th century, its origins remain disputed but it is a bacterium not associated with nonhuman reservoirs—Weiss, 2001). Even so, it is helpful to have some indication of what kinds of new pathogen we are most likely to encounter.

Surveillance

The first line of defense against any emerging pathogen is its rapid detection and identification. Recent practical experience with BSE and SARS demonstrates that rapid detection and identification leading to the rapid introduction of preventive measures can prove highly effective in combating outbreaks of novel diseases

(Wilesmith 1994; Stohr 2003). Moreover, computer simulation studies motivated by concerns about the possible emergence of pandemic influenza suggest that only if a new strain is detected in the very earliest stages and interventions are put in place extremely promptly is their any realistic prospect of curtailing an epidemic (Ferguson et al. 2006).

Surveillance for novel pathogens, however, does present some particular challenges. Initially, this is likely to depend on clinical observation, such as the reporting of clusters of cases of disease with unusual symptoms. Internet surveillance for reports of unusual disease outbreaks is also possible and, in the longer term, generic diagnostic tools—for example, lab-on-a-chip tests for all known human viruses—should become available (OSI 2006).

The map of reports of new pathogen species (Figure 5-6) argues strongly that surveillance needs to be global, especially considering the unprecedented rates of international travel and trade that can allow new infectious diseases, such as SARS, to spread around the world over time scales of days or weeks. Pathogen emergence is an international problem.

Multi-Disciplinarity

Another key lesson from surveying novel pathogens is the importance of animal reservoirs in the emergence of new infectious diseases. One implication of this is that surveillance in reservoir populations likely to be an effective tool for monitoring risks to humans (Cleaveland, Meslin, and Breiman 2007). On top of this, it may often be the case that most scientific knowledge of the basic biology of an unusual human pathogen lies, at least initially, with the veterinary community rather than the medical community. Palmarini (2007) lists a number of examples of this: infectious cancers, retroviruses, lentiviruses, transmissible spongiform encephalopathies, rotaviruses, and papilloma viruses. To this list could be added coronaviruses and ehrlichiosis. More generally, it is now widely recognized that humans share the majority of their pathogens with other animals (Taylor, Latham, and Woolhouse 2001).

Together, these observations underline the importance of close linkages between medical and veterinary researchers, resonating with the "one medicine" concept originally put forward by Schwabe (1969) and seeming especially appropriate in the context of emerging infectious diseases.

However, understanding the process of emergence requires much more than an understanding of the basic biology of the host-pathogen interaction, important though this undoubtedly is. A theme of this review has been the importance of ecological factors for the emergence of new pathogens. But we have used "ecological" to cover a very wide range of environmental, agricultural, entomological, demographic, behavioral, cultural, economic, and sociological drivers of pathogen emergence. In specific contexts these could include the bush meat trade (associated with the emergence of HIV and SARS), livestock feed production (associated with BSE/vCJD) or changes in pig farming practices (associated with Nipah virus). These examples emphasize that disease emergence is a multidisciplinary problem and needs to be understood at a number of scientific levels. Collaborations need to be developed not just between the human and animal health branches of the biomedical research community but also with researchers covering a much wider range of disciplines.

Conclusions

The pathogen pyramid provides a useful conceptual framework for thinking about the process of the emergence of a new species of human pathogen. However, it is immediately clear that at each level of the pyramid there are some important gaps in our knowledge.

First, we still have very little idea of the diversity of pathogens to which humans are being or could be exposed. Systematic surveys across a range of possible sources of new pathogens (notably other mammal species) using techniques such as shotgun sequencing are possible in principle, and would provide this information.

Establishing a priori which pathogens are capable of infecting humans is even more challenging. A first step would be to identify the cell receptors used by the 189 recognized species of human virus. At present, we have this information for only around half of the virus species.

Estimating the transmission potential of a new pathogen within the human population can only be achieved by closely monitoring initial outbreaks. Analysis of such data can provide some early warning of crucial epidemiological changes (as illustrated by the analysis of measles data mentioned above). Real time analysis of epidemic data can also provide timely estimates of the transmission potential (see Lipsitch et al. 2003 for application to the SARS epidemic) which can help inform control efforts. On the other hand, for many of the rarer human pathogens we do not currently know whether or not they are transmissible between humans (Woolhouse 2002).

It is extremely likely that we will encounter new species of human pathogen in the near future. We urgently need the scientific and logistic capacity to rapidly detect and evaluate the threat that new pathogens present and to intervene quickly and effectively wherever necessary. Experience of SARS provides some encouragement that, given adequate resources, efforts to combat emerging pathogens can be successful, but further challenges lie ahead.

GENOMIC EVOLVABILITY AND THE ORIGIN OF NOVELTY: STUDYING THE PAST, INTERPRETING THE PRESENT, AND PREDICTING THE FUTURE

Jonathan A. Eisen, Ph.D.[5]
University of California, Davis

Introduction

One of the unifying goals of this workshop, as well as of the Forum on Microbial Threats, has been to promote the study of microbes, not only to enhance our understanding of their present roles in the world but also, we hope, to predict their future changes (e.g., the emergence of new infectious diseases). This was, of course, one of the life missions of Joshua Lederberg, who helped create the Forum and who this workshop is honoring.

Studies of evolution are central to these goals because "nothing in biology makes sense except in the light of evolution" (Dobzhansky, 1973). Evolutionary studies help us understand the past and interpret the present, and from a combination of those two we have some possibility of being able to predict the future. Since Lederberg was also keen on evolutionary studies (Lederberg, 1997, 1998), it is appropriate for a workshop in his honor to focus on *Microbial Evolution and Co-Adaptation*.

I would like to note that I feel a personal connection to Joshua Lederberg, as I received much of my microbiology training from Ann Ganesan who had been a Ph.D. student in his lab. However, anyone, with or without a specific connection to Lederberg, can learn a great deal about him and his work through a wonderful website made available by the National Library of Medicine.[6] There you can find many of his scientific papers, as well as his letters, scientific articles he wrote, articles he read, columns he wrote for *The Washington Post* on science policy, and more. In addition, I would like to point out that Lederberg was an ardent supporter of "open access" to scientific publications. He was on the board of PubMed when PubMed Central was created. PubMed Central is a centralized archive of freely available, full-text versions of scientific publications. Although not all of his papers are in PubMed Central,[7] most are available at the National Library of Medicine site.

In this paper, I am focusing on one key aspect of evolution: the *origin of novelty*, or how new forms, functions, processes, and properties originate. In addition, I consider some of the factors that influence the likelihood that novelty

---

[5]University of California, Davis Genome Center; Department of Medical Microbiology and Immunology and Section of Evolution and Ecology, Davis, CA 95616; E-mail: jaeisen@ucdavis.edu; Website: http://phylogenomics.blogspot.com.
[6]See http://profiles.nlm.nih.gov/BB.
[7]See http://www.pubmedcentral.nih.gov.

will originate—something generally referred to as *evolvability*. Why do some organisms invent new functions readily while others are "novelty challenged"? I note that I focus here on work from my lab and am not attempting to review the entire field.

I have been interested in the origin and novelty and evolvability, particularly as they occur in microbes, since I was introduced to microbes as an undergraduate through studies of hydrothermal vent ecosystems. Actually, I had written a paper on this back in high school, but it was not until college that I truly focused on the topic. A bit later, in 1995, my career—and that of most other microbiologists and evolutionary biologists—was changed forever with the publication of the first complete genome of any free-living organism (Fleischmann et al., 1995). It was then that I shifted my research to the integration of evolutionary analyses with studies of genome sequences. For better or for worse, I coined the term for this field: *phylogenomics* (Eisen et al., 1997). Note that the way I use this term is a bit different than some others in the community. Many people use the term phylogenomics to refer to the use of genome-scale data (e.g., genome sequences) for phylogenetic studies. With that introduction, I will now relate some "phylogenomic tales" as examples of how phylogenomic analysis can help us understand the origin of novelty. This will also demonstrate the usefulness of this approach for understanding the past, interpreting the present, and—maybe—predicting the future.

Phylogenomics and Novelty I:
Predicting Gene Functions Using Evolutionary Trees

Throughout this workshop, we have seen many examples of genome sequencing leading to wonderful insights about the microbial world. Indeed, it can be said that genome sequence data have sparked a renaissance in microbiology. It is important to realize, however, that much of this renaissance rests on one particular step in the analysis—the prediction of gene function based on gene sequence. This step is critical because typically one generates the genome sequence of a particular organism, most of whose genes will not have been studied experimentally. Prediction of gene function adds value to the genome sequence data because such predictions can guide further computational and experimental studies of the organism. My first phylogenomic tale illustrates how, in the course of a genome-sequencing project, the evolutionary analysis of a particular gene can enable us to make more accurate predictions about the function of that gene in a particular organism and, in some cases, can also provide insight into the evolutionary processes in that organism, as well.

This is the story of one such organism, *Helicobacter pylori*, a bacterium that dwells in the stomachs of humans and some other mammals. For many years, these stomach dwellers were generally ignored. However, thanks in a large part to the work of people like Barry Marshall, it is now known that *H. pylori* is a causative agent of stomach ulcers as well as gastric cancers (Marshall, 2002). Due to its medical importance as well as its novel ability to tolerate very high acidity, this species was one of the first targeted for genome sequencing. In 1997, the genome of one strain was published (Tomb et al., 1997).

At that time, as a Ph.D. student at Stanford, I was relentlessly badgering everyone I knew, attempting to convince them that evolutionary analysis could help in the prediction of gene function. I had become convinced of this myself through analysis of the trickle of genome sequence data for humans, yeast, and other organisms that had already begun to flow before the first complete genome was published. Back in 1995, I had even published a paper showing the benefits of evolutionary reconstructions in studies of one family of proteins, the SNF2 family (Eisen et al., 1995). Although the benefits were clear to me, others were not so sure. Fortunately, at the time I was teaching a class with Rick Myers, a professor in the genetics department and the head of the Stanford Human Genome Center. He had been asked to write a "News and Views" piece for *Nature Medicine* commenting on the recent papers reporting the sequencing and analysis of the genomes of *H. pylori* (Tomb et al., 1997) and *Escherichia coli* K12 (Blattner et al., 1997). Also, since he was one of the people I had been badgering, he suggested I try to come up with an example of where the inclusion of evolutionary analysis could have benefited their work.

Luckily for me, there was a claim in the *H. pylori* paper that was a perfect candidate for evolutionary analysis. The authors reported (Tomb et al., 1997):

> The ability of *H. pylori* to perform mismatch repair is suggested by the presence of methyl transferases, mutS and uvrD. However, orthologues of MutH and MutL were not identified.

This was right up my alley because I was working on the evolution of DNA mismatch repair at the time.

A DNA mismatch can be created when the wrong base is put into a newly synthesized strand by the enzyme carrying out DNA replication (i.e., DNA polymerase). Thus, a mismatch indicates a replication error. Mismatch repair is a process whereby, immediately following DNA replication, repair enzymes scan for mismatches between the template and newly synthesized DNA strands. When the mismatch repair machinery finds one, it removes a section of DNA containing the mismatch from the newly synthesized strand. That section is then resynthesized using the original (and presumably accurate) template strand as a guide. Mismatch repair is vital. It greatly reduces the mutation rate by correcting many of the replication errors made by DNA polymerase.

It was because of my knowledge of the evolution of mismatch repair that the report in the *H. pylori* paper caught my attention. I knew that every time a mismatch repair system had been found in an organism, regardless of whether that organism was from the bacteria, mammals, plants, yeast, or a variety of other groups, and regardless of whether it was found by genetics, by biochemistry, or even by targeted cloning, the pattern was the same. The system always required at least one member of the MutS family of proteins and one member of the MutL family. Yet, according to the paper, *H. pylori* did not encode a MutL homolog. So I decided to look at this in more detail.

My first step was to recheck the genome sequence analysis. I did this by first using the Basic Local Alignment Search Tool (BLAST), which compares a given DNA, RNA, or protein sequence with corresponding sequences in a database and determines if there are similar sequences therein and, if so, generates a list of the closest matches. First, I took all known MutL-like proteins and searched them against the *H. pylori* genome data and found, as Tomb et al. did, that there were no close matches. Given that they had determined the complete genome of this strain, the absence of BLAST match suggested there was indeed no MutL encoded in the genome. I note that this "determining the absence" of something from a genome is one of the key benefits of determining *complete* genome sequences (Fraser et al., 2002).

Then I did some BLAST searches with MutS-like proteins as "queries" and found, as the authors had reported, that there was one, and only one, protein encoded in the genome that was similar to proteins in the MutS family. So I took this protein and then used it to search against all known sequences from other organisms, to see to what it was most similar. This in essence was mimicking the searches done in the analysis of the genome, and the result seemed quite convincing (Table 5-5). All of the proteins that were most similar to the *H. pylori* protein were described in the database as "Mismatch repair protein MutS" or something similar. This description of the related proteins, also known as their annotation, was clearly what led the authors to conclude that this protein was involved in mismatch repair. This left me with a conundrum. There was no MutL protein encoded in the genome, yet there was, apparently, a MutS protein. Many possible explanations came to mind, all of which were interesting. *H. pylori* might have been the first species to be found with a mismatch repair system that did not require a MutL homolog. Or it might have recently lost its MutL, as had been seen in many strains of *E. coli* and *Salmonella* (LeClerc et al., 1996). Alternatively, perhaps the MutS-like protein was not a normal MutS involved in mismatch repair, but rather was used for a different function in this organism.

Although these, along with yet other explanations, seemed plausible, one observation suggested to me that the latter explanation—that the MutS-like protein was doing something else—might be the correct one. In the list of BLAST matches, I had noticed that members of the MutS family that I knew to have documented roles in mismatch repair were not high on the list, indicating the *H. pylori* protein was not as similar to these as it was to some other MutS-like proteins that might have a novel function (Table 5-5). In addition, I knew from my prior work (Eisen et al., 1995), and from the work of others (e.g., Tatusov et al., 2000), that BLAST scores were not a reliable indicator of evolutionary relat-

TABLE 5-5 BLAST Search Results as They Were Seen in 1997 Using the MutS-like Protein from *Helicobacter pylori* as a Query

Sequences producing significant alignments	Score (bits)	E Value
sp|P73625|MUTS_SYNY3 DNA MISMATCH REPAIR PROTEIN	117	3e-25
sp|P74926|MUTS_THEMA DNA MISMATCH REPAIR PROTEIN	69	1e-10
sp|P44834|MUTS_HAEIN DNA MISMATCH REPAIR PROTEIN	64	3e-09
sp|P10339|MUTS_SALTY DNA MISMATCH REPAIR PROTEIN	62	2e-08
sp|O66652|MUTS_AQUAE DNA MISMATCH REPAIR PROTEIN	57	4e-07
sp|P23909|MUTS_ECOLI DNA MISMATCH REPAIR PROTEIN	57	4e-07

edness. So my next step was to investigate the evolutionary history of the MutS proteins, including the new one from *H. pylori*. I did this by generating a multiple sequence alignment of all available MutS sequences and inferring an evolutionary tree from that alignment. The evolutionary tree revealed that there were two sub-families of MutS homologs in bacteria, one containing the "normal" MutS-like proteins known to be involved in mismatch repair and the other containing the *H. pylori* protein along with a few others. None of the proteins in this second sub-family had ever been studied experimentally and all were only distantly related to the "normal" MutS subfamily. Given this finding along with the observation that *H. pylori* lacked a MutL homolog, we wrote in our *Nature Medicine* article (Eisen et al., 1997) that it was premature to predict that mismatch repair would be found in this species. I followed this up with a more comprehensive evolutionary study (Eisen, 1998b) that came to the same conclusion.

I would like to point out that this was not simply an esoteric exercise. Mismatch repair has great significance due to its role in modifying the mutation rate. Without mismatch repair, an organism's mutation rate usually goes way up (and in addition the rate of acquisition of DNA from other organisms also tends to go up). This has important implications for the evolution of virulence, pathogenicity, and drug resistance. Many papers published since this initial analysis have confirmed that *H. pylori* actually does have a high baseline mutation rate (Kang et al., 2006). In fact, the entire group of epsilon proteobacteria (of which *H. pylori* is a member) does not have a normal MutS homolog. Thus, the question arises: Do all of these organisms have high mutation rates? Or have they evolved some compensatory process that reduces mutation rate even without mismatch repair? At least from current data, it seems that many members of this group do have somewhat elevated mutation rates. For example, when the Sanger Center was sequencing the genome of a close relative of *H. pylori*, *Campylobacter jejuni* (which also does not encode a normal MutS homolog), even in the few generations required to grow up a sample of this strain for sequencing, many mutations were acquired (Parkhill et al., 2000). This suggests that the mutation rate for this strain is quite high. Awareness of this dynamic is vitally important when designing therapeutics to target organisms that lack mismatch repair. This example illustrates how evolutionary analysis of a gene found in a genome can not only tell us something about the biology of that organism, but can also help us to predict its evolvability.

This *H. pylori* story is but one of many that demonstrate the value of including evolutionary analysis when predicting gene function. In this regard, I must point out that I am far from unique in holding this view. For example, while I was working on the use of phylogenetic trees, multiple groups were showing how classifying proteins into families and subfamilies was critical for predicting function (Sonnhammer et al., 1997; Tatusov et al., 2000). My approach to this functional prediction was somewhat different from these subfamily- or ortholog-focused approaches in that I have argued that one needs to actively use the tree itself by using an approach known as character state reconstruction (Figure 5-10). Character state reconstruction is a commonly used method in phylogenetics whereby one can infer for particular traits (also known as characters) the history of change between different forms of those traits (also known as states). Normally, character state reconstruction is used to infer information about ancestral nodes in a tree (e.g., the common ancestor of two extant organisms), but it can also be used to infer the likely state of modern organisms. It is relatively straightforward to use these methods to infer information about protein function by treating each protein much as you would treat different organisms. Importantly, not only can one infer likely functions for proteins using this approach, but this has a benefit over subfamily classification approaches in that it is less likely to make incorrect predictions of function (such as, when function changes rapidly; Eisen, 1998a). It is worth noting that this adaptation of character state reconstruction methods for predicting the functions of uncharacterized genes is analogous to predicting the biology of a species based on the position of that organism in the tree of life. Such predictions tend to work better for gene function, in a large part because organism-level biology can change much more rapidly than the function of specific genes.

Regardless of whether one uses my character state-based approach or one of the subfamily-based approaches, it is clear that adding information about the evolutionary history of a gene can help predict its functions. Last, and perhaps most important, methods that make use of evolutionary information have been automated (Brown et al., 2007; Haft et al., 2001; Tatusov et al., 2000; Zmasek and Eddy, 2002) and thus can be employed more readily and on larger genomic data sets.

FIGURE 5-10 Phylogenetic prediction of gene function. Outline of a phylogenomic methodology. In this method, information about the evolutionary relationships among genes is used to predict the functions of uncharacterized genes (see text for details). Two hypothetical scenarios are presented and the path of trying to infer the function of two uncharacterized genes in each case is traced. (A) A gene family has undergone a gene duplication that was accompanied by functional divergence. (B) Gene function has changed in one lineage. The true tree (which is assumed to be unknown) is shown at the bottom. The genes are referred to by numbers (which represent the species from which these genes come) and letters (which in A represent different genes within a species). The thin branches in the evolutionary trees correspond to the gene phylogeny and the thick gray branches in A (bottom) correspond to the phylogeny of the species in which the duplicate genes evolve in parallel (as paralogs). Different colors (and symbols) represent different gene functions; gray (with hatching) represents either unknown or unpredictable functions.
SOURCE: Reprinted from Eisen (1998a) with permission from Cold Spring Harbor Laboratory Press.

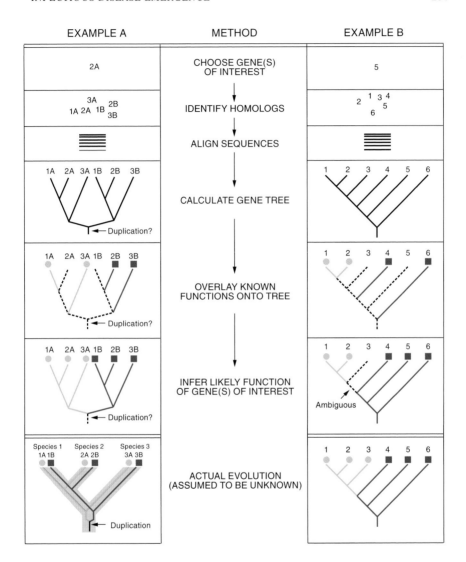

Phylogenomics and Novelty II: Recent Evolution

The methods for predicting function outlined above focus on making use of known information about some genes to predict the functions of uncharacterized genes. These methods do not work well, or even at all, if completely novel functions have arisen in an organism over short evolutionary time scales. Fortunately, over the last few years researchers have developed suites of methods to scan through genomic data for evidence of recent evolutionary diversification. Thus, my second phylogenomic tale relates to how knowledge about the origin of novelty helps us both carry out and interpret these scans.

The key to leveraging information about recent evolutionary events is to first get an understanding of how new functions arise on short time scales. Fortunately, we know a decent amount about this and have heard a great deal of recent new insights at this meeting. Examples include clustered regularly interspaced short palindromic repeats (CRISPRs) loci, which appear to be immune-system analog in bacteria and archaea that provides for immunity from phage, the rapid loss of genes that are not under strong positive selection, the use of contingency loci to rapidly change the sequence of a protein, and so forth. In fact, many of these phenomena have been either discovered or characterized in detail through comparative genomic analysis of closely related organisms.

Based upon this we can design a relatively simple process for taking a genome and identifying recent events in its history: sequence the genome and the genomes of some close relatives; compare the genomes to each other (including documentation of gene order conservation, gene gain and loss, gene duplication, and generation of simple polymorphisms); and then catalog the variation into different classes that correspond to different mechanisms of novelty generation. For example, polymorphisms in protein coding regions can be classified into synonymous (do not change protein sequence) and non-synonymous (change amino acid sequence) and then the pattern of synonymous versus non-synonymous substitutions can be used to screen a genome for the selective pressure different genes are under. Similarly, one can build evolutionary trees of all genes in a genome and look for those with longer branches in one lineage over another as evidence for an acceleration of evolutionary rate (Pollard et al., 2006). This sort of logic can be applied to just about any type of recent evolutionary event in genomes. Here I go into a bit more detail about how one can use this approach focusing on recent gene duplication events.

We know from the classic work of Ohta (2000) and others that gene duplication followed by subsequent divergence of the duplicates is a very important mechanism for the generation of novelty in virtually all organisms. Thus, to identify those genes within a lineage that are most likely to have recently diversified functions, we can turn this around and look for recent duplications. We did this by scanning complete genomes, looking for gene families that are expanded in one lineage compared to related lineages. As far as I know, we were the first to use this method when we applied it to the *Deinococcus radiodurans* genome (White et al., 1999). Subsequently, this general approach has been used in the analysis of many genomes and developed into a robust tool for characterizing them (Jordan et al., 2001).

The work I am going to describe here involves analysis of the genome of *Vibrio cholerae*. John Heidelberg had led a project to sequence this genome at The Institute for Genomic Research (TIGR) and asked me for help in carrying out some analyses (Heidelberg et al., 2000). One thing I did was to scan the genome for gene families that had undergone lineage-specific duplications (i.e., duplications that occurred since the organism last shared a common ancestor with any other organism for which we also had the complete genome sequence available). This was done "function blind"—meaning we simply analyzed the raw sequence data and not the known or predicted functions of genes. We found something very striking. In one gene family, the number of genes in this species was much greater than that in other related species. More importantly, the "extra" genes in *V. cholerae* were apparently the result of multiple rounds of gene duplication that occurred in the evolutionary branch leading up to this species (i.e., since it diverged from other lineages for which genomes were available). This family encoded the methyl-accepting chemotaxis proteins (MCPs) which were predicted to be involved in sensing and responding to chemical gradients in the environment (Figure 5-11).

Given the known biology of *V. cholerae* as an aquatic microbe, it seemed even more likely that this protein family might indeed have experienced recent evolutionary adaptations. Of course, not all duplications are related to evolutionary diversification, but with a genome encoding more than 4,000 proteins, identifying a candidate subset to pursue with more careful informatics and with experimental studies was definitely helpful.

Phylogenomics III: Uncharacterized Genes

Both of the approaches described above predict the function of particular genes by making use of experimental information about homologs of those genes. Unfortunately, this does not always work well, for many reasons. For instance, much of the time a gene of interest will have homologs in other species but none of those homologs have been studied experimentally. Such genes, known as "conserved hypothetical" genes, pose a significant challenge for function prediction. Fortunately, over the last 10 years, many new methods have been developed that are particularly useful for characterizing their functions (see Marcotte, 2000, for review). Since these methods make use of other types of experimental information (such as coexpression patterns, protein-protein interaction networks) or computational analysis (including chromosomal location, shared promoter sequences, protein domain patterns), they are generally known as "nonhomology" methods.

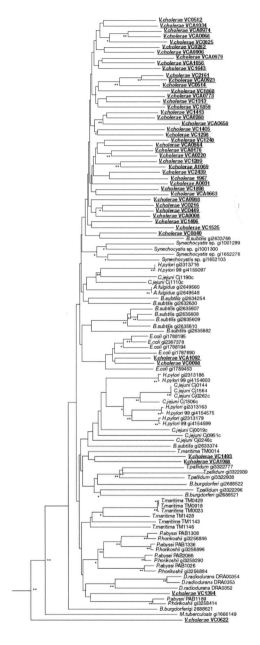

FIGURE 5-11 Phylogenetic tree of methyl-accepting chemotactic protein (MCP) homologs in completed genomes. Homologs of MCPs were identified by FASTA3 searches of all available complete genomes. Amino acid sequences of the proteins were aligned using CLUSTALW, and a neighbor-joining phylogenetic tree was generated from the alignment using the PAUP* program (using a PAM-based distance calculation). Hypervariable regions of the alignment and positions with gaps in many of the sequences were excluded from the analysis. Nodes with significant bootstrap values are indicated: two asterisks, .70 percent; asterisk, 40 ± 70 percent.

SOURCE: Adapted from Heidelberg et al. (2002), with permission from Macmillan Publishers Ltd. Copyright 2002.

I'm going to introduce you to one of them, my favorite: phylogenetic profiling (Pellegrini et al., 1999).

In phylogenetic profiling, we first determine the distribution of genes of interest across many species. Genes with similar patterns of distribution are then grouped together. The underlying idea here is that often several genes interact in some way, for example, all being subunits of a complex protein or being involved in carrying out a particular process such as methanogenesis. For one gene to be functional, all must be present in an organism. Such genes would thus tend to be found in groupings that have similar patterns of distribution across species. It is important to point out that when interpreting these profiles, one must take into account two key processes in the evolution of microbial genomes. First, unless genes are used or are under strong selection to be maintained, they tend to disappear. Second, microbes don't just inherit genes vertically within a lineage; they also acquire genes from other organisms by horizontal gene transfer. Significantly, when genes that work together are acquired horizontally, they tend to all get added or deleted simultaneously, or nearly simultaneously, with the result that when we compare genomes, we see that all members of such a group are either present or absent.

Here's how one actually carries out phylogenetic profiling. You start with a set of genes in which you are interested, perhaps all the genes in the complete genome sequence of "your" organism. You then compare them against each complete genome sequence in a genome database by asking a simple yes-or-no question: For each gene in your organism, is there a homolog in the other genome? After you have done this for every gene in your genome, you create a profile for each gene by plotting its presence or absence across all the species. With such profiles in hand, one can then identify genes with similar profiles. One way to do this is to simply cluster genes by their profiles and look for tight clusters of genes with highly similar distribution patterns. An example of this is shown in Figure 5-12 in which each row corresponds to a gene and each column represents one species. Conceptually, this is analogous to microarray clustering of gene expression patterns. In fact, microarray clustering software is often used for analyzing phylogenetic profiles.

Once you have such groupings of genes with similar cross-species distribution patterns, you can then use them to aid in predicting gene functions. For example, we used phylogenetic profiling to analyze the genome of the bacterium *Carboxydothermus hydrogenoformans* (Wu et al., 2005). There we found a very tight cluster of genes shared among many sporulating species (e.g., *Bacillus subtilis*) but absent from species that did not sporulate—even if closely related. Many of the gene families in this cluster were known to be involved in sporulation in other species. Based on this information, we predicted that *C. hydrogenoformans* had the ability to sporulate, and indeed we subsequently confirmed this experimentally. Moreover, our analysis revealed that there were also many other gene families of unknown function that were shared by sporulating species

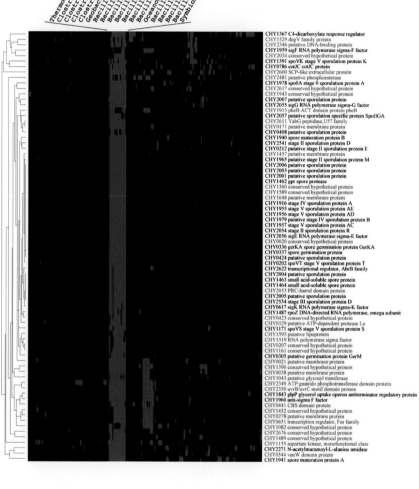

and absent from nonsporulating ones. Such genes were likely candidates for carrying out novel sporulation-associated activities. A bit of confirmation came just as we were finishing our paper. Richard Losick at Harvard published a set of studies on sporulation in *B. subtilis* that identified a few new sporulation genes (Eichenberger et al., 2004; Silvaggi et al., 2004), and many of our candidates were in their list of novel sporulation genes. Perhaps most interestingly, many of our candidates were still not identified as likely sporulation genes and likely represent novel sporulation-associated functions yet to be characterized.

I note that the approach of phylogenetic profiling can be strengthened by modifying the basic yes-or-no question. Instead of asking if there is a *homolog* of your gene present in another species, you want to ask if there is an *ortholog*, thus using some evolutionary information to improve your clustering (Eisen and Wu, 2002). With either method, phylogenetic profiling is a powerful tool for finding sets of genes that function in related processes or in a pathway. Although it does not characterize their biochemical activity well, it can provide insight into the process in which they participate (e.g., sporulation) and thus guide experimental studies. As we sequence more and more genomes, this method will become more and more informative.

Phylogenomics IV: Acquisition of Function from Others

There are two basic strategies by which organisms evolve new functions. One option is through modification of their own genome (e.g., mutation, gene

FIGURE 5-12 Phylogenetic profile analysis of sporulation in *Carboxydothermus hydrogenoformans*. For each protein encoded by the *C. hydrogenoformans* genome, a profile was created of the presence or absence of orthologs of that protein in the predicted proteomes of all other complete genome sequences. Proteins were then clustered by the similarity of their profiles, thus allowing the grouping of proteins by their distribution patterns across species. Examination of the groupings showed one cluster consisting mostly of homologs of sporulation proteins. This cluster is shown with *C. hydrogenoformans* proteins in rows (and the predicted function and protein ID indicated on the right) and other species in columns with the presence of an ortholog indicated in red and its absence in black. The tree to the left represents the portion of the cluster diagram for these proteins. Note that most of these proteins are found only in a few species represented in red columns near the center of the diagram. The species corresponding to these columns are indicated. We also note that though most of the proteins in this cluster, for which functions can be predicted, are predicted to be involved in sporulation and some have no predictable functions (highlighted in blue). This indicates that functions of these proteins' homologs have not been characterized in other species. Since these proteins show similar distribution patterns to so many proteins with roles in sporulation, we predict that they represent novel sporulation functions.
SOURCE: Wu et al. (2005).

duplication, domain swapping, invention of new genes), but these processes can sometimes be quite slow. In many cases, it is much easier instead to acquire the function from another organism that already has it. How is this done? By acquisition or affiliation. In other words, they can acquire the requisite genes via sex or lateral gene transfer, or they can gain access to the products of those genes through some type of affiliation with organisms that have those functions. Such affiliations include long-term symbioses. Symbioses are categorized as being parasitic (when one partner obviously benefits and the other is harmed), commensal (where one benefits and the other is unaffected), or mutualistic (where both benefit), but often we do not actually know the full extent of mutual impacts.

I am going to give an example of function acquisition by symbiosis, and demonstrate how genomic studies, combined with an understanding of the biology and evolution of the symbiosis, can aid in functional predictions. One partner in this symbiosis is the glassy-winged sharpshooter. This insect, like other sharpshooters, is an obligate xylem feeder that makes its living by feeding on the fluids in the xylem portion of the circulatory system of the host plant. This particular species has received special attention because it is a vector for Pierce's disease, a nasty problem in grape vineyards. The disease agent is a bacterium, *Xylella fastidiosa*, that infects the xylem and can be transmitted between plants by the sharpshoorters, much like bloodborne pathogens are transmitted between animal hosts (see Chatterjee et al., 2008, for a review).

Obligate sap-feeding insects face a serious challenge. As part of their defenses, many plants make their sap less useful to sap-feeding insects by removing some nutrients that are essential for animals. For example, the essential amino acids (that all animals cannot synthesize and thus require in their diet) tend to be present in very low concentration in phloem sap. To counter this, many obligate phloem-feeding insects have bacterial symbionts living inside specialized cells in their gut. The insect provides the bacteria with sugars from sap, and the bacteria, in turn, make amino acids for their hosts. Xylem sap, which moves from the roots to the rest of the plant, tends to be even more nutrient-poor than phloem sap, and obligate xylem feeders also have bacterial symbionts living inside specialized cells in the gut (see Moran et al., 2008, for a review on heritable symbionts).

When we started our project, obligate phloem-feeding insects, such as aphids, had already been studied extensively, but much less was known about the obligate xylem-feeders. We (especially our collaborator, Nancy Moran) thought there might be a different twist to the story for the bacterial symbionts living in xylem-feeding hosts. At that time, all the species of sharpshooters examined had been found to host *Baumannia cicadellinicola*, a close relative of the symbionts that make amino acids for the aphids (Moran et al., 2003). Our first step was to apply shotgun genome sequencing methods to the DNA obtained from endosymbiont-containing tissue dissected from this sharpshooter. Using this approach we were able to determine the complete genome of *B. cicadellinicola*. Examination of the genome revealed many very interesting things (Wu et al., 2006).

First, we found that this organism had many of the hallmarks typical of intracellular symbionts: a small genome, low G+C content, and high evolutionary rates. As an aside, the high evolutionary rates often seen in intracellular symbionts are thought to be due, in large part, to the small effective population sizes for intracellular organisms. However, we found significant variation in the rate of evolution among endosymbionts, with the highest rates tending to be found in those that lack homologs of mismatch repair genes (Figure 5-13). Adaptation to an intracellular existence is typically accompanied by marked reduction in genome size, probably due to random forces. When important DNA repair genes are lost, mutation rates may go up—an evolutionarily significant consequence of their small genome size. Furthermore, whereas free-living species would have opportunities to reacquire the repair genes from other organisms, this is very unlikely for intracellular ones, isolated as they are. Thus another evolutionary consequence of an intracellular existence is reduced evolvability by means of lateral gene transfer.

Secondly, further examination of the genome and prediction of gene functions revealed pathways for synthesizing diverse vitamins and cofactors, suggesting that this symbiont was also helping its xylem-feeding host to deal with a very nutrient-poor diet. Based on our prior knowledge of these types of symbioses, we expected to find pathways for the synthesis of the essential amino acids required by the sharpshooter—but we could not find any. In thinking about the mechanisms for the evolution of novelty, it seemed unlikely to us that this host, the

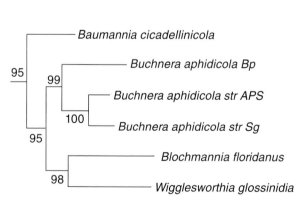

FIGURE 5-13 There is significant variation in the rate of evolution among endosymbionts, with the highest rates tending to be found in those that lack homologs of mismatch repair genes.
SOURCE: Adapted from Wu et al. (2006).

glassy-winged sharpshooter, would have evolved the ability to synthesize essential amino acids, given that this capability has never been found, as far as I know, in any animal species. Nevertheless, the observations were that the sharpshooter eats only xylem sap, xylem does not contain the essential amino acids, and the genes for essential amino acid synthesis pathways were not present in the genome of either the sharpshooter or its *Baumannia* endosymbiont. We were vexed.

There were three possibilities we considered that could reasonably explain this conundrum. One was that the sharpshooter was acquiring amino acids from other food sources. This seemed unlikely as sharpshooters are generally considered to be obligate xylem sap feeders. A second possibility was that the glassy-winged sharpshooter was getting the essential amino acids from the xylem sap. Though we could not rule this out, it seemed unlikely because there should be strong selection on the plants to keep essential amino acids out of the xylem sap and because xylem generally was not known to have such amino acids. A third possibility was that another organism in the sharpshooter system was making amino acids. This seemed to be the most likely possibility especially since our collaborator Nancy Moran had just recently shown that there was a second type of bacterial symbiont living inside the guts of all sharpshooters (Moran et al., 2005). We had not paid much attention to this second type of symbiont since the *Baumannia* symbionts were so closely related to the *Buchnera* symbionts of aphids that provided all the nutritional supplements needed by their host to feed on phloem sap (and since these new symbionts were from a completely different phylum of bacteria).

Fortunately, we had a quick, though somewhat dirty, way to test for the possibility that another organism in the system was making essential amino acids. To sequence the *Baumannia* genome we did not use a pure culture since these symbionts had never been grown in the lab. Instead, we had done a "metagenomics" project in which Nancy Moran's lab had dissected hundreds of sharpshooters and removed as carefully as possible the tissue that was known to contain the *Baumannia* symbionts. We then extracted DNA from this material and used it for whole-genome shotgun sequencing during which we sheared the DNA into moderately small pieces, cloned these pieces into a plasmid library, and then sequenced the ends of these plasmid clones. From these data we were able to generate a good assembly of the *Baumannia* genome, which we then finished with PCR and primer walking methods. The key for us was that not all of the sequence reads that we obtained were found to map to the *Baumannia* genome. Some came from other organisms in the sample. So the first thing we did was to look in these other data for genes that might be involved in synthesizing essential amino acids—and we immediately found a few.

So the next question was: From what organism did these genes originate? We knew there should be host DNA in the sample (although we thought it was unlikely that the host would be synthesizing essential amino acids since no animals are known to do so) and that there might also be DNA from the second symbiont as well as from other resident microbes. So what we needed to do was to sort the DNA sequence reads into which came from which organism. This sorting is commonly known as "binning" in metagenomic studies. We tried every binning method in use at the time including genome assembly, analysis of DNA base composition and word frequencies, examination of depth of coverage, and others. Unfortunately none of them worked well, most likely because we had very little coverage of the genomes from these other organisms. This is where phylogenomic approaches came in handy.

We decided to try to sort the sequence reads by phylogenetic analyses. So we took all the reads, identified all possible proteins or protein fragments that they could encode, then for these identified which had apparent homologs in sequence databases, and for those built phylogenetic trees. We then sorted the phylogenetic trees by which organism's genes showed up in the tree as the nearest neighbor of the protein or fragment.

Overall the trees showed only a few major patterns. In some (Figure 5-14A) the nearest neighbor was something from an animal. Thus, we concluded that the sequence reads in this "bin" likely corresponded to fragments of the host genome. In other trees, the nearest neighbor was a *Wolbachia* or some close relative (Figure 5-14B). Since *Wolbachia* (a type of bacteria related to *Rickettsia*) are common intracellular parasites of insects, we concluded that these reads came from *Wolbachia* that infected at least some of the insects that Nancy had dissected. Then there was a large collection of reads for which the trees showed a grouping with species in the Bacteroidetes phylum (Figure 5-14C). Because *Sulcia* was in this phylum, we concluded these were likely from the second symbiont.

We then asked: Of the potential essential amino acid synthesizing genes we had identified in some of the reads, to which of the bins did they belong? The answer was clear as day—all belonged to the *Sulcia* bin. We thus concluded that it was likely that the second symbiont was the provider of essential amino acids for the host and so we spent another year or so trying to finish the genome of this symbiont. Though we did not quite finish the genome, from the 130 or so kilobase pairs (kbp) of DNA we mapped to this organism, we found that it encoded in essence all the essential amino acid synthesis pathways (Wu et al., 2006); this was later confirmed by the complete genome (McCutcheon and Moran, 2007). What we had discovered was a dual symbioses where one symbiont (*Baumannia*) makes vitamins and cofactors and the other (*Sulcia*) makes essential amino acids, and together they supplement the nutrient-poor diet of the glassy-winged sharpshooter contributing to this organism's annoying ability to spread Pierce's disease. Most importantly for this article here, we would not have been able to sort out the data from the different organisms (and thus would not have discovered the dual symbioses) without phylogenetic analysis of the metagenomic data.

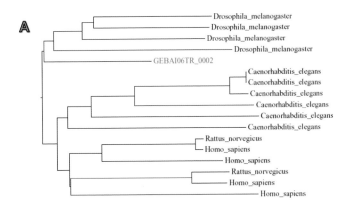

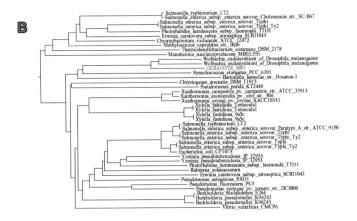

Phylogenomics V: Knowing What We Do Not Know

As has often been heard at this workshop, Lederberg was very fond of emphasizing that we need to know what we do not know. In that spirit, I want to discuss how knowing what we do not know can help with functional predictions. One aspect of what we do not know that influences our ability to make useful functional predictions is that genome-sequencing projects are highly biased in terms of what types of organisms have been sequenced. For example, I and many others noticed a few years ago (Eisen, 2000; Hugenholtz, 2002) that most of the genomes of bacteria were coming from just three of the 40+ phyla of bacteria (Figure 5-15). The same trend was seen in Archaea and microbial eukaryotes. So based on this we applied for, and in 2002 received, a grant from the "Assembling the Tree of Life" program at the National Science Foundation (NSF) to sequence

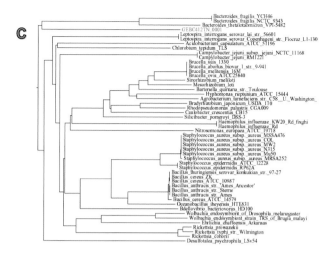

FIGURE 5-14 Phylogenetic trees of putative proteins encoded by single sequence reads of DNA isolated from symbiont-containing tissue of the glassy-winged sharpshooter. Trees were constructed by aligning putative proteins encoded by the reads to homologs from complete genome sequences. (A) Phylogenetic tree for read GEBA106TR, translated in the second reading frame (thus the label *002*). Note how the encoded protein groups in the tree with proteins from *Drosophila*. This read is likely from a piece of DNA from the host genome. (B) Phylogenetic tree for read GEBA193TR_001. Note how the encoded protein groups with sequences from *Wolbachia*; thus, this read likely corresponds to DNA from a *Wolbachia* infecting one of the dissected sharpshooters. (C) Phylogenetic tree of read GEBA412TN_001. Note how the encoded protein groups in trees with *Bacteroidetes* species which are relatives of the second (*Sulcia*) symbiont.
SOURCE: Based on data in Wu et al. (2006).

the first genomes from representatives of 8 phyla of bacteria. We have now finished this project and are in the process of writing up a series of papers on our findings. Yet even from the initial analyses, what was abundantly clear was that a single genome from these phyla was simply not enough. Each phylum represents something on the order of 1 billion to 2 billion years of evolution, and a lot happens in that time in bacteria. So a single genome cannot do justice to the diversity of genes and features of each phylum.

Based on this, in collaboration with the Joint Genome Institute (a Department of Energy [DOE]-funded genome center), we have started a new initiative to really fill in the genomes from across the tree of life. This Genomic Encyclopedia of Bacteria and Archaea (GEBA)[8] is just getting started, with 100 genomes

---

[8]See http://www.jgi.doe.gov/programs/GEBA/ for more detail.

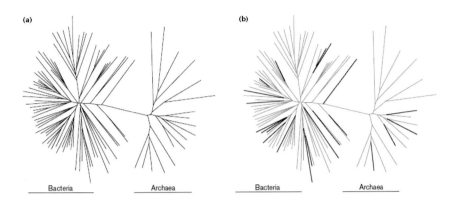

FIGURE 5-15 The diversity of Bacterial and Archaeal species for which complete genomes are available is still poor. (a) Phylogenetic tree, based on rRNA sequences, of representatives of many major Bacterial and Archaeal lineages. (b) Lineages for which complete genomes are available are highlighted. The rRNA tree for representative species was downloaded from the Ribosome Database Project (available at http://www.cme.msu. edu/RDP/html/index.html).
SOURCE: Reprinted from Eisen (2000) with permission from Elsevier. Copyright 2000.

being sequenced from across the tree in the first year. Already the results are quite convincing that sampling from across the tree leads to enormous benefits. For example, the phylogenetic profiling method outlined above works best when you have sampling of diverse genomes from different phylogenetic groups. Adding these GEBA genomes to the mix makes phylogenetic profiling work much better.

Sampling from across the tree will take some effort because there are many, many, major groups of Bacteria, Archaea, and microbial Eukaryotes, and many of these do not have any cultured representatives. However, the benefits will likely be enormous.[9]

It is important to point out however that just having genome sequences from across the tree is not sufficient. Functional information from diverse organisms is also critical. I give one example here of why. When I first went to TIGR, Owen White was in charge of a project to sequence the genome of the bacterium *Deinococcus radiodurans*, the most radiation-resistant organism known. This was very exciting to me since I did my Ph.D. research in part on the evolution of radiation resistance. So I volunteered to help Owen analyze the genome. Since there was some experimental evidence that active DNA repair processes contributed to the

---

[9]I note it will be good to sample from across viral diversity, although since there is no phylogenetic tree linking all viruses it is unclear exactly how to do this sampling.

resistance of this organism, I spent some time looking for likely DNA repair genes in the genome, making use of the phylogenomic approaches I had been advocating. Indeed we were able to find many genes that appeared likely to be involved in DNA repair processes.

The problem was that the list we came up with was very similar to the list that we could make for nonradiation-resistant organisms such as *E. coli* and *B. subtilis*. However, a little thinking about what we did not know helps explain this. Imagine if sometime in the recent history of *D. radiodurans* a novel DNA repair gene evolved. The method we were using to look for DNA repair genes would not have found this since we were looking primarily for homologs of genes that were shown in other species to be involved in DNA repair processes. Even using novel methods such as phylogenetic profiling would not necessarily help if the new genes in this species were not connected in any way to known DNA repair pathways. The problem here is that most experimental studies of repair genes in bacteria were done in two phyla and we would have a hard time identifying novel repair genes if they had been invented anywhere else in the bacterial tree of life.

This is a general lesson for all functional predictions. Such predictions rely upon some functional characterizations done in some organism. The more these functional studies are done across the tree of life, the better will our functional predictions become. Similarly, functional predictions rely in part upon comparative genome analysis; thus, the more genomes we have from across the tree, the better functional predictions we will get. Thus, knowing what we do not know is critical in guiding experimental and sequencing studies to get the most out of the diversity of organisms.

Summary: A Call for a Field Guide to the Microbes

Overall, what I have tried to do here is present examples of how evolutionary and genome analysis can be integrated into "phylogenomic" studies. I have focused on predicting functions of genes, but the benefits of phylogenomics extend to all aspects of the biology of microbes.

I should emphasize that this is not some radical or overly novel concept as the integration of phylogeny and function is well known to be critical for understanding the diversity of life. What I have tried to show here is that this is as true for genomics and micreobes as it is for physiology, behavior, genes, ecosystems and other arenas in which evolution has been shown to be a powerful tool. I should note that there is a third piece of information that is useful in addition to integrating phylogeny and function—the biogeographical patterns of the distributions of organisms. For microbes, figuring out the distribution patterns of organisms and the rules determining these patterns is one of the final frontiers. If we are able to integrate phylogeny, function and genomes, and biogeography, we will have something for microbes that is known to be useful in many other organisms—a field guide. A field guide to microbes would no doubt be useful in many arenas, and I am certain it would be a book that Josh Lederberg would have carried with him wherever he went.

Acknowledgments

I thank Merry Youle for assistance in the editing of this manuscript and Martin Wu and Dongying Wu for help in many of the genome analyses reported here. In addition, I thank the many people at TIGR and UC Davis who provided help in generating and analyzing the genome and metagenomes under discussion. This work was supported by the Defense Advanced Research Projects Agency under grant HR0011-05-1-0057.

CAN WE PREDICT FUTURE TRENDS IN DISEASE EMERGENCE?

Peter Daszak, Ph.D.[10]
Consortium for Conservation Medicine, Wildlife Trust

In this essay, I examine the process of zoonotic disease emergence and pose the question: Can we use past trends to predict future patterns and better control this public health threat? Stephen Morse, in this chapter, wrote about the ecological and demographic factors that drive disease emergence. This process is essentially evolutionary, with these "drivers" forcing selection of novel, or already present, pathogen strains that are better suited to transmission within our changed populations or environment. Here, I will provide examples of how these drivers can be analyzed and their influence on pathogen transmission measured. I will then demonstrate how our group has used these analyses to make predictions of different steps in the process of disease emergence. First, however, I will briefly review what history has to teach us about disease emergence.

Historical Trends in Disease Emergence

As we look into history, we can see that the process of disease emergence has occurred in a series of earlier phases. For example, many pathogens now endemic in humans appear to have evolved from an ancestor that moved into our populations at around the time we first domesticated animals—between 10,000 to 15,000 B.C.E. Measles virus, for example, is phylogenetically most closely related to the ungulate pathogens rinderpest and peste des petits ruminant viruses.

Epidemiological studies of host-parasite dynamics demonstrate that infectious diseases cannot become endemic until host populations reach a certain threshold density. In the case of measles, repeated introductions of the virus to

---

[10]Executive director.

the South Pacific Islands did not result in establishment of this disease except in regions where human population densities approached 500,000 (Black, 1966). The first zoonotic pathogens apparently emerged in humans between 200 B.C.E. and 300 C.E., as networks of communities in the Fertile Crescent,[11] reached threshold densities for diseases such as measles and smallpox (Dobson and Carper, 1996). This early phase of human infectious disease emergence may have been preceded by another such period during the Pleistocene Era,[12] when the human population contracted during glaciation, then expanded and migrated afterward. Later, global transportation networks, exploration, conquest, and trade made possible the emergence and spread of the Black Death (plague) in Europe in the fourteenth century, and brought the Conquistadors and smallpox to the Aztecs in the fifteenth and sixteenth centuries. With the Industrial Revolution came pandemic influenza and with air travel and other features of globalization, HIV/AIDS and our current phase of emerging diseases.

As these grand episodes of disease emergence suggest, one simple predictor of infectious disease emergence is globalization: As we move into new regions, we expand endemic pathogen geographic ranges and provide a new source of susceptible hosts for others. As Figure 5-16 shows, that rate has accelerated to the point where we are connected as never before via globalized travel and trade networks (Hufnagel et al., 2004).

In these cases, infectious agents are exploiting increased human population density and increased contact rates between people. As globalization progresses and human populations become increasingly dense and interconnected, we can already make one simple prediction: The rate of disease emergence will increase correspondingly. Furthermore, as our global patterns of trade and economy rely on this connectivity, we can also predict that the economic cost of these outbreaks will increase, as illustrated in Figure 5-17.

But there is another important pattern in these historical trends: The first emergence of many now-common endemic diseases (e.g., measles, smallpox) is a zoonotic event explained by ecology—increasing contact with recently-domesticated wildlife led to spill-over of their pathogens and evolution into human-adapted strains. This process has been repeated throughout recent history (e.g., plague), culminating in the recent phase of new zoonoses, HIV-1 and -2, Ebola, severe acute respiratory syndrome (SARS), and others.

We can break down this process of zoonotic disease emergence into three discrete phases, each with different dynamics and drivers. In the first stage ("preemergence"), human changes to the environment drive wildlife into new

---

[11]Fertile Crescent historic region of the Middle East. A well-watered and fertile area, it arcs across the northern part of the Syrian desert. It is flanked on the west by the Mediterranean and on the east by the Euphrates and Tigris rivers, and includes all or parts of Israel, the West Bank, Jordan, Lebanon, Syria, and Iraq. From antiquity this region was the site of settlements and the scene of bloody raids and invasions (see http://www.encyclopedia.com/doc/1E1-FertileC.html).

[12]Geological era dating from 1.8 million to 10,000 years ago.

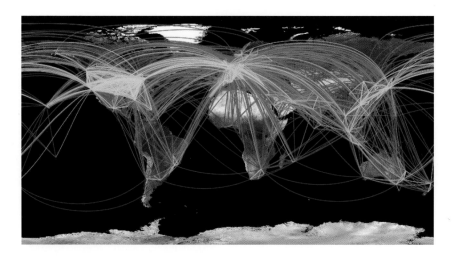

FIGURE 5-16 The rate of globalization has accelerated to the point where we are connected as never before via globalized travel and trade networks.
SOURCE: Hufnagel et al. (2004).

regions, or introduce livestock, leading to spill-over among animal populations and outbreaks of novel wildlife diseases. In the second stage, wildlife pathogens "spill over" to human populations causing single cases, small clusters of cases, or localized outbreaks, as has occurred with Nipah virus or Ebola virus. In the final stage, "pandemic spread," pathogens either adapt to, or are already adapted to, human-to-human transmission, and move rapidly across nations. This pandemic stage is likely rarely reached, but can lead to severe impact due to high mortality (e.g., HIV/AIDS) or economic impact (e.g., SARS).

As we examine these three phases, we see that they are driven by different causal factors, or drivers, but that in almost all cases these are anthropogenic[13] and are measurable. For example, the rise in bushmeat hunting (the original cause of HIV emergence [Peeters et al., 2002]) has been analyzed and measured in Southeast Asia and Africa, and the landscape changes underlying Lyme disease emergence in the United States have also been measured and analyzed. Below, I discuss how analyses of the ecological or demographic drivers of disease can be used to measure and predict the future risk of disease emergence across these different phases.

---

[13]Caused by humans.

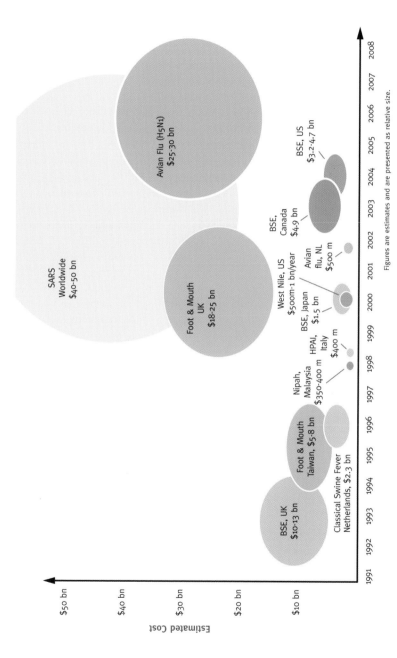

FIGURE 5-17 Economic impacts of selected emerging infectious diseases.
SOURCE: Figure courtesy of BioEra.

Predicting International (Pandemic) Spread—West Nile Virus, Avian Influenza, and Nipah Virus

Of the three steps in disease emergence, the most simple to develop predictions around is the process of international or pandemic spread. This is because the drivers responsible for this are simple and well-defined: human travel and trade patterns. Our group has used this approach to predict future patterns of West Nile virus and avian influenza spread. For West Nile virus (WNV), which has moved across the United States and into South America, we examined the most likely pathways of introduction to Hawaii, the Galapagos, and Barbados.

We examined the only plausible pathways for WNV spread: when carried by people, wind-transported mosquitoes, mosquitoes hitching a ride on boats or planes, animals (pets or poultry), or by migratory birds. We conducted simple calculations of the likely number of "bird days" of risk via these pathways, incorporating data on the average number of mosquitoes per plane; the numbers of migratory birds that pass through Hawaii, the Galapagos, and Barbados from West Nile-infected locations and their reservoir competence; and the number of infected people who travel to these locations. We determined that West Nile is likeliest to be transmitted to all three locations by mosquitoes on airplanes (Figure 5-18; Kilpatrick et al., 2004, 2006b)—a risk two to three orders of magnitude higher than any of the others. Because most mosquitoes on planes are found in the cargo hold, simple measures to eradicate them (i.e., residual insecticide use in cargo holds) would likely reduce the risk of WNV introduction significantly.

We used a similar approach to examine whether the most likely pathways of H5N1 avian influenza spread into and across Europe was due to migratory birds or poultry. We used data from the Food and Agriculture Organization (FAO) on country-to-country poultry trade, and from the Royal Society for the Protection of Birds (RSPB) and Smithsonian on migratory birds. We considered the numbers

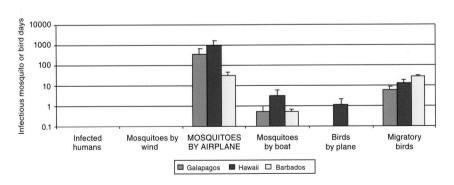

FIGURE 5-18 WNV is most likely to be spread by airplane.
SOURCE: Adapted from Kilpatrick et al. (2004, 2006b).

of birds that migrate along the major flyways, the dimensions of their summer breeding areas, the timing of their migration, and whether their migrations routes pass over H5N1-infected locations (Kilpatrick et al., 2006a). We examined each spreading event of H5N1 from 1998 until the present to determine whether it was more likely to have been caused by the poultry trade or by migratory birds. Our analyses suggest that the virus spread initially within Southeast Asia due to the poultry trade, but once it moved out of the region, it was rapidly spread via migratory pathways to and within Europe.

For countries where H5N1 had not yet been reported, we then proceeded to calculate the risk of introduction associated with trade in poultry and wild birds, and with bird migration (expressed as infectious bird days, as shown in Figure 5-19; Kilpatrick et al., 2006a).

Although the United States does not trade in poultry with any country that has reported H5N1, its neighbors do; thus, it is not hard to imagine that the virus could travel easily into the United States across its borders. Our analysis also showed much lower (two to three orders of magnitude) risk of H5N1 introduction to the United States through migratory birds. These results contrasted with the publicly stated position of the U.S. Department of Agriculture (USDA) and other U.S. government entities that migratory birds moving through the Siberia-Alaskan flyway would be the most likely cause of H5N1 introduction to the United States.

A large outbreak of encephalitis in Malaysia in 1998-1999 was traced to a novel bat-borne paramyxovirus, Nipah virus (NiV). There was no human-to-human transmission of NiV in Malaysia, but the virus caused the death of over 100 people, with a fatality rate of around 40 percent. The virus is carried by fruit bats (*Pteropus spp.*), infected pigs, and people in close contact with infected pigs in Malaysia and Singapore (Figure 5-20).

To analyze what caused this outbreak, we investigated the first cases of NiV infection in pigs and humans at the "index" farm—a large (30,000-head) intensively managed export farm in Ipoh, Malaysia. The human NiV infections reflect the transmission dynamics of NiV in the pig population and showed an interesting pattern—a long period (>18 months) of slow, smoldering infection followed by a large epidemic spike. After five years of field studies, virology, experimental work, and mathematical modeling, we have been able to piece together the factors that led to this pattern (Daszak et al., 2006). First, we have shown that NiV is able to survive for prolonged periods on fruit juice at room temperature (Fogarty et al., 2008), supporting the notion that bats visiting the index farm to feed on fruit trees adjacent to pigsties were the origin of infection.

In Malaysia, we captured wild bats, tested them for NiV antibodies, and isolated the virus. We identified evidence of NiV in every colony we examined in Malaysia and reported changes in seroprevalence in young bats, suggesting that the virus circulates endemically in the Malaysian bat population. We found one colony around a kilometer from the index farm and obtained anecdotal evidence

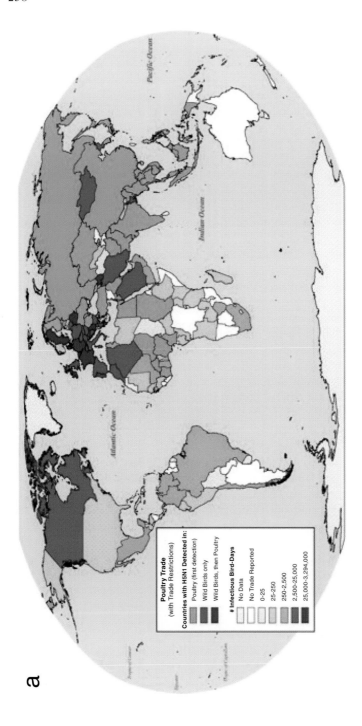

FIGURE 5-19 Predicted risk of H5N1 avian influenza introduction from countries that have had H5N1 outbreaks (in blue). Risk was estimated as the number of infectious bird days (number of infected birds × days shedding virus) caused by trade (presented as yearly totals per 12 months) in: (a) live poultry with no exports from countries reporting H5N1 in poultry (France, Denmark, Sweden, and Germany are considered H5N1-free). (b) Estimated number of ducks, geese, and swans migrating between mainland continents, number of infectious bird

b

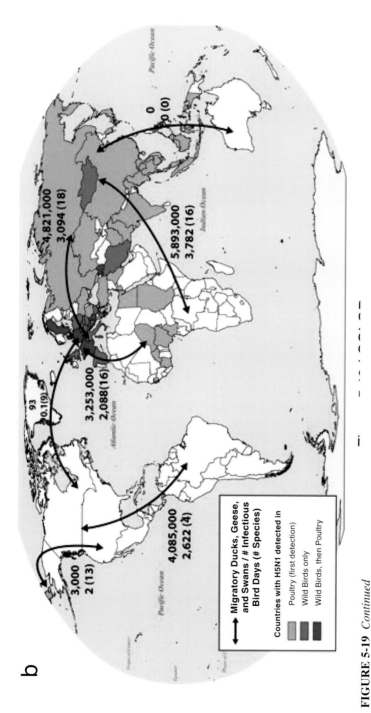

FIGURE 5-19 *Continued*

days, and number of species (in parentheses). Numbers given between Asia and North America include only those that breed on mainland Asia and winter in North America south of Alaska; an additional 200,000-400,000 ducks breed in Siberia and molt or winter in or off the coast of Alaska. In addition, about 20,000 geese migrate between Ireland and North America.

SOURCE: Kilpatrick et al. (2006a).

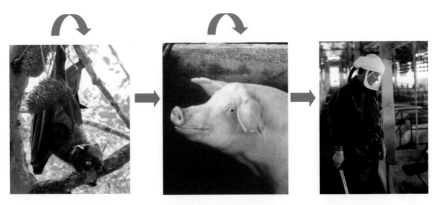

FIGURE 5-20 Species chain for Nipah virus in Malaysia. This emerging virus is carried by fruit bats across the Old World tropics. It emerged in people in Malaysia after spilling over from bats to pigs, which act as amplifier hosts.
SOURCE: Wildlife Trust Inc. (bat); M. L. Bunning (pig and human).

of bats visiting the site. These findings suggest that the virus could have been introduced repeatedly to the pigs at the index farm. We also tracked the bats and found that they fly large distances and frequently cross between Malaysia and Indonesia, migrating to follow the fruiting and flowering of orchard crops. Our data from these field studies were able to refute a previously published hypothesis that El Niño/Southern Oscillation (ENSO)-driven forest fires in Sumatra caused bats to move into Malaysia during the 1990s and introduce the virus there for the first time.

We were able to obtain excellent demographic data on Malaysian pig farms, thanks to the legal requirements for export farms in Malaysia. As a result, birth and death rates among the pigs were carefully tracked. We used these data to develop a mathematical model of infection dynamics that simulates precisely the structured system of the pig farm where NiV first emerged. This model allowed us to "recreate" the original outbreak as a computer simulation and analyze why it persisted for almost two years prior to the large-scale outbreak. Our analyses showed that the emergence of NiV in humans appears to be closely linked to the intensification of pig farming in Malaysia. It is unusual that a wildlife pathogen is enhanced by intensive livestock production and then becomes a human problem. This system, however, involves both pigs and mangoes, and the production of both commodities in Malaysia rose sharply in the two decades preceding the 1999 outbreak, as farmers used pig manure to fertilize fruit trees. When pig populations declined as a result of the 1999 Nipah outbreak, mango production declined as well. We, therefore, used our modeling approach to predict the likelihood of future NiV emergence and spread should the Malaysian government decide to reconstitute these large export farms in this or other regions.

Can We Predict Spillover of the Next Emerging Zoonosis?

Challenges

Here, I believe, is the "Holy Grail" for emerging disease research—to develop a valid strategy to predict the next emerging zoonosis—the next HIV or SARS. To do this will require a fusion of evolution, ecology, virology, and microbiology. This is clearly a challenge, and it has been proposed that it will be difficult if not impossible to overcome. For example, Murphy (1998) commented, "In general, there is no way to predict when or where the next important new zoonotic pathogen will emerge or what its ultimate importance might be."

Before I address the challenge, I first need to acknowledge that Murphy raises an important point in his article—that we have a very poor knowledge of the true diversity of microbes able to emerge as new zoonoses in the future—described by Morse (1993) as the "zoonotic pool." We can assemble crude estimates of this diversity. Consider that there are approximately 50,000 vertebrates (see also paper by Morse in this chapter) and that zoonotic viruses emerging in humans tend to be vertebrate viruses. If we estimate that each vertebrate species carries 20 endemic, unknown viruses (almost certainly a gross underestimate), then there is a global diversity of 1 million viruses (bats alone would carry about 20,000 unknown viruses). With only approximately 2,000 different species of viruses identified, we can crudely say that we underestimate the zoonotic pool by at least 99.8 percent!

This is one of two obstacles we must overcome if we are going to predict the next zoonosis: We do not know the real size or global distribution of the zoonotic pathogen pool. In addition, we must account for surveillance bias, because people are looking harder for emergent diseases in some places than in others, and that determines where previous zoonoses have been identified.

Emerging Disease Database

With the above challenges in mind, we constructed a database (based on an earlier published list of emerging infectious diseases [EIDs]; Taylor et al., 2001) of EID "events," defined as the original case or case cluster representing a given infectious disease (including drug-resistant strains) emerging in human populations for the first time, that occurred between 1940 and 2004 (Jones et al., 2008). These EID events include newly evolved strains of pathogens (e.g., drug-resistant and multiply drug-resistant strains), pathogens that have recently entered human populations for the first time, and pathogens that have probably been present in humans historically, but which have recently increased in incidence.

The geographic origin of emergence for each of these diseases is shown in Figure 5-21. Surveillance bias is evident in this figure: Europe and the United States, which support the most comprehensive disease surveillance efforts, detect

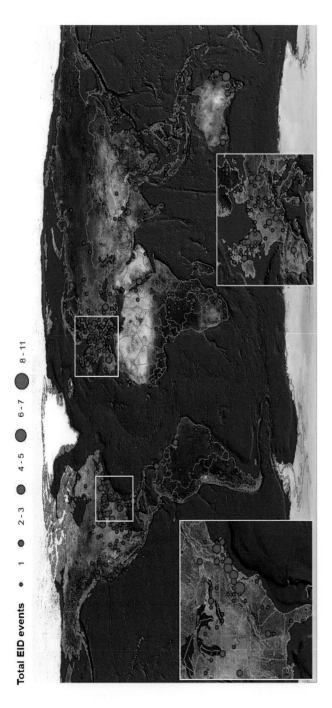

FIGURE 5-21 Global richness map of the geographic origins of EID events from 1940 to 2004. The map is derived for EID events caused by all pathogen types. Circles represent one degree grid cells, and the area of the circle is proportional to the number of events in the cell. SOURCE: Jones et al. (2008).

the greatest numbers of EID events. We corrected for that by geographically plotting the coordinates of every author—about 17,000 of them—of every paper published in the *Journal of Infectious Diseases* (*JID*) for the last 20 years, and used this information in our analyses.

We were able to use this database to address some key questions in emerging disease biology. First, whether EIDs are really on the rise (Jones et al., 2008). Decade by decade, from the 1940s to the 1990s, the number of EID events has increased significantly, even after accounting for the increasing numbers of scientists over this period. This has another implication: It is reasonable to expect that this trend will continue in the future. We also found that a majority of EID events were associated with drug-resistant microbes. Second, we were able to examine whether zoonoses such a HIV/AIDS, which are the most high-profile EIDs, are truly the most significant threat. We found that zoonoses emerging from wildlife (i.e., HIV, SARS, Ebola and Nipah viruses) are indeed significantly rising over time and during the 1990s, represented the dominant type of emerging disease.

Testing Hypotheses

We used our database approach to examine two simple questions: (1) Is disease emergence an "anthropogenic" process (i.e., are human changes to demography, the environment, and other factors the key drivers of EIDs)? (2) Can we obtain a more accurate map of the emerging disease "hotspots"—the regions most likely to cause the next new emerging disease?

To test these theories, we first found a way around the dilemma of not knowing where the diversity of pathogens resides by assuming that each mammalian species harbors a similar number of host-specific pathogens. If this is true, then the global distribution of wildlife diversity approximates the potential zoonotic pathogen diversity. In our analysis, we used a global dataset on mammalian host richness.

We then used a simple multiple logistic regression to assess the correlation between the risk of an EID historically and some key factors thought responsible for disease emergence, correcting for reporter bias with the dataset on *JID* authors. We addressed the first hypothesis by testing global human population density against EID risk and showed that this is a significant predictor of risk for each group of pathogen. This specifically shows that the risk of a disease *emerging* (not spreading) is dependent on human population density (i.e., those regions with dense human populations and presumably lots of human-driven changes are most likely to lead to a new EID).

By plotting out our risk measures globally, we were able to produce the first ever global distribution maps of emerging disease risk, corrected for reporter bias, and based on correlated trends in EIDs. These predictive maps of EID "hotspots" show different global distribution patterns when we sorted EID events according to their origins (e.g., zoonotic diseases from wildlife; vector-borne pathogens; drug-resistant pathogens; Jones et al., 2008). For EIDs of wildlife origin (the high-profile zoonoses), these hotspots are primarily tropical areas where wildlife diversity is highest, and particularly where human density is also high, as occurs in southern Brazil, northern India and Bangladesh, and Southeast Asia (Figure 5-22). However, Europe and the United States also have significant potential for zoonotic disease emergence, due to continued, high-level environmental changes.

However, perhaps one of the key findings of our analysis is that if we plot out the geographic distribution of all 17,000 *JID* authors, we find that the global effort for infectious disease research has largely focused on regions from where the next EID is least likely to emerge. Indeed, few EID hotspots—located primarily in developing countries—are under thorough surveillance for infectious pathogens (Jones et al., 2008). We therefore concluded that global efforts to detect emerging infections should be slightly refocused to the Tropics if we are to rapidly intervene with this process of emergence.

Using Predictive Approaches: "Smart Surveillance"

Can we use this hotspot approach to increase our capacity for preventing the next EID? If we return to Nipah virus, we see that this emerging pathogen fits into the high-profile group of zoonoses that are lethal to humans and have emerged from wildlife in tropical regions. During the last decade, antibodies to this pathogen have been reported in bats across Southeast Asia, South Asia, Madagascar, China, and even continental Africa. But this knowledge has been gleaned through different groups working independently and often serendipitously. There has been no focused, global surveillance for viruses related to NiV in bats.

If we examine the wildlife zoonotic disease hotspot map (Figure 5-22) in one of the highest risk regions, Bangladesh, the human population has been subject to a series of repeated outbreaks of NiV with higher case-fatality rates than in Malaysia (average around 70 percent), evidence of foodborne infection, and evidence of up to five chains of human-to-human transmission. Bangladesh has the densest population of any country on Earth that is not an urban city-state: 2,595 people per square mile, as compared with a global average of 128 persons per square mile (the United States has 80 people per square mile; http://www.worldatlas.com, 2006). The country also has surprisingly high wildlife diversity, given its population. Thus, it appears that in Bangladesh, Nipah virus is closer to stage three, or pandemic emergence. This raises important questions: Why were there no programs to identify NiV in Bangladesh once the virus was discovered in Malaysia? What other regions globally might harbor spillover of NiV or related viruses? What other zoonotic pathogens might be lurking in the South Asia hotspot within bats or other wildlife hosts?

I propose that a more efficient strategy to address future emerging diseases

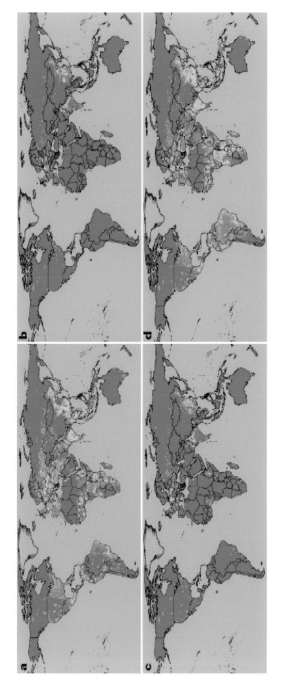

FIGURE 5-22 Global distribution of the relative risk of an EID event. Maps are derived for EID events caused by (a) zoonotic pathogens from wildlife; (b) zoonotic pathogens from nonwildlife; (c) drug-resistant pathogens; and (d) vector-borne pathogens. Green corresponds to lower values; red to higher values.

SOURCE: Jones et al. (2008).

is to combine rigorous analyses of the fine-scale ecological and demographic changes within hotspot regions (the risk factors) with state-of-the-art molecular approaches to viral discovery. This will give us a more accurate predictive model for future disease emergence, and better definition of the size and diversity of the zoonotic pool. Techniques such as pyrosequencing and mass tag polymerase chain reaction (PCR) will rapidly decrease the expense and logistical challenges involved in identifying new viral groups, and if applied to key groups of wildlife species (those most often responsible for disease emergence in the past) within hotspot regions, will provide the most cost-effective way to proactively address the EID challenge. This model for virus-hunting in the future is, of course, still somewhat crude. It is impossible, for example, to determine the future ability of a novel virus to jump hosts successfully to humans, and its likely pathogenicity. However, by focusing first on viral groups known to be pathogenic, and by targeting viral discovery within these clades, significant progress can be made toward dealing with the EID threat.

Acknowledgments

The work described in this chapter was carried out by a large number of collaborators, including members of the Henipavirus Ecology Research Group (HERG),[14] especially Jon Epstein (Consortium for Conservation Medicine) and Juliet Pulliam (Fogarty International Center). The work on West Nile virus and avian influenza was led by A. Marm Kilpatrick (Consortium for Conservation Medicine, University of California Santa Cruz) and the hotspots analyses were conducted in collaboration with Kate Jones (Institute of Zoology) and Marc A. Levy (Center for International Earth Science Information Network, Columbia). This work was supported in part by a National Institutes of Health/National Science Foundation "Ecology of Infectious Diseases" award from the John E. Fogarty International Center R01-TW00824, by core funding to the Consortium for Conservation Medicine from the V. Kann Rasmussen Foundation and is published in collaboration with the Australian Biosecurity Cooperative Research Center for Emerging Infectious Diseases (AB-CRC).

REFERENCES

Overview References

IOM (Institute of Medicine). 1992. *Emerging infections: microbial threats to health in the United States*. Washington, DC: National Academy Press.
———. 2003. *Microbial threats to health: emergence, detection, and response*. Washington, DC: The National Academies Press.

---

[14]See http://www.henipavirus.org.

Morse References

CDC (Centers for Disease Control and Prevention). 2008. *HIV mortality (through 2005)*, http://www.cdc.gov/hiv/topics/surveillance/resources/slides/mortality/index.htm (accessed February 10, 2009).

Crichton, M. 1969. *The andromeda strain*. New York: Knopf.

IOM (Institute of Medicine). 1992. *Emerging infections: microbial threats to health in the United States*. Washington, DC: National Academy Press.

———. 2003. *Microbial threats to health: emergence, detection, and response*. Washington, DC: The National Academies Press.

Johnson, S. 2006. *The ghost map*. New York: Riverhead Books/Penguin.

McNeill, W. H. 1976. *Plagues and peoples*. New York: Bantam Doubleday Dell Publishing Group, Inc.

Morens, D., G. K. Folkers, and A. S. Fauci. 2004. The challenge of emerging and re-emerging infectious diseases. *Nature* 430(6996):242-249.

Neustadt, R. E., and H. V. Fineberg. 1983. *The epidemic that never was: policy-making and the swine flu scare*. New York: Vintage Books.

Peters, C. J. 1998. Hantavirus pulmonary syndrome in the Americas. Chapter 2 in *Emerging infections 2*, edited by W. M. Scheld, W. A. Craig, and J. M. Hughes. Washington, DC: ASM Press.

Woolhouse and Gaunt References

Anderson, R.M., and May, R.M. 1991. Infectious Disease of Humans Dynamics and Control. Oxford Scientific Press, Oxford, U.K.

Antia, R., Regoes, R.R., Koella, J.C., and Bergstrom, C.T. 2003. The role of evolution in the emergence of infectious diseases. *Nature* 426, 658–661.

Arien, K.K.,Vanham, G., and Arts, E.J. 2007. Is HIV-1 evolving to a less virulent form in humans? *Nat. Rev. Microbiol.* 5, 141–151.

Barre-Sinoussi, F., Chermann, J.C., Rey, F., Nugeyre, M.T., Chamaret, S., Gruest, J., Dauguet, C., Axler-Blin, C.,Vezinet-Brun, F., Rouzioux, C., Rozenbaum, W., and Montagnier, L. 1983. Isolation of a T-lymphotropic retrovirus from a patient at risk for Acquired Immune Deficiency Syndrome (AIDS). *Science* 220, 868–871.

Brosch, R., Gordon, S.V., Marmiesse, M., Brodin, P., Buchrieser, C., Eiglmeier, K., Garnier, T., Gutierrez, C., Hewinson, G., Kremer, K., Parsons, L.M., Pym, A.S., Samper, S., van Soolingen, D., and Cole, S.T. 2002. A new evolutionary scenario for the *Mycobacterium tuberculosis* complex. *Proc. Natl. Acad. Sci. USA* 99(6), 3684–3689.

Chai, J.Y., Murrell, K.D., and Lymbery, A.J. 2005. Fish-borne parasitic zoonoses: status and issues. *Int. J. Parasitol.* 35, 1233–1254.

Chant, K., Chan, R., Dwyer, D.E., and Kirkland, P. 1998. Probable human infection with a newly described virus in the family Paramyxoviridae. *Emerg. Inf. Dis.* 4, 273–275.

Cleaveland, S., Laurenson, M.K., and Taylor, L.H. 2001. Diseases of humans and their domestic mammals: pathogen characteristics, host range and the risk of emergence. *Philos. Trans. R. Soc. Lond. B. Biol. Sci.* 356, 991–999.

Cleaveland, S., Meslin, F.X., and Breiman, R. 2006. Dogs can play useful roles as sentinel hosts for disease. *Nature* 440, 605.

Diamond, J. 2002. Evolution, consequences and future of plant and animal domestication. *Nature* 418, 700–707.

Dykhuizen, D.E. 1998. Santa Rosalia revisited: why are there so many species of bacteria. *Ant. v. Leeuwenhoek* 73, 25–33.

Ebert, D. 1998. Experimental evolution of parasites. *Science* 282, 1432–1436.

Ferguson, N.M., Cummings, D.A., Fraser, C., Cajka, J.C., Cooley, P.C., and Burke, D.S. 2006. Strategies for mitigating an influenza pandemic. *Nature* 442, 448–452.

Haydon, D.T., Cleaveland, S., Taylor, L.H., and Laurenson, M.K. 2002. Identifying reservoirs of infection: a conceptual and practical challenge. *Emerg. Infect. Dis.* 8, 1468–1473.

Holmes, E.C., and Rambaut, A. 2004. Virus evolution and the emergence of SARS coronavirus. *Phil. Trans. R. Soc. B. Biol. Sci.* 359, 1059–1065.

Hubalek, Z. 2003. Emerging human infectious diseases: anthroponoses, zoonoses, and sapronoses. *Emerg. Inf. Dis.* 9, 403–404.

IOM (Institute of Medicine). 2003. *Microbial threats to health: emergence, detection, and response.* National Academy Press, Washington, DC, USA.

Jansen, V.A.A., Stollenwerk, N., Jensen, H.J., Ramsay, M.E., Edmunds, W.J., and Rhodes, C.J. 2003. Measles outbreaks in a population with declining vaccine uptake. *Science* 301, 804.

Keele, B.F., Van Heuverswyn, F., Li, Y., Bailes, E., Takehisa, J., Santiago, M.L., Bibollet-Ruche, F., Chen,Y.,Wain, L.V., Liegeois, F., Loul, S., Mpoudi Ngole, E., Bienvenue, Y., Delaporte, E., Brookfield, J.F.Y., Sharp, P.M., Shaw, G.M., Peeters, M., and Hahn, B.H. 2006. Chimpanzee reservoirs of pandemic and nonpandemic HIV-1. *Science* 313, 523–526.

King, D.A., Peckham, C., Waage, J.K., Brownlie, J., and Woolhouse, M.E.J. 2006. Infectious diseases: preparing for the future. *Science* 313, 1392–1393.

Lázaro, M.E., Cantoni, G.E., Calanni, L.M., Resa, A.J., Herrero, E.R., Iacono, M.A., Enria, D.A., and González Cappa, S.M. 2007. Clusters of hantavirus infection, southern Argentina. *Emerg. Inf. Dis.* 13, 104–110.

Lipsitch, M., Cohen, T., Cooper, B., Robins, J.M., Ma, S., James, L., Gopalakrishna, G., Chew, S.K., Tan, C.C., Samore, M.H., Fisman, D., and Murray, M. 2003. Transmission dynamics and control of severe acute respiratory syndrome. *Science* 300, 1966–1970.

Lumio, J., Hillbom, M., Roine, R., Ketonen, L., Haltia, M., Valle, M., Neuvonen, E., and Lahdevirta, J. 1986. Human rabies of bat origin in Europe. *Lancet* 1, 378.

Matthews, L., and Woolhouse, M.E.J. 2005. New approaches to quantifying the spread of infection. *Nat. Rev. Microbiol.* 7, 529–536.

May, R.M., Gupta, S., and McLean, A.R. 2001. Infectious disease dynamics: What characterizes a successful invader? *Phil. Trans. R. Soc. B. Biol. Sci.* 356, 901–910.

Mermin, J., Hutwagner, L., Vugia, D., Shallow, S., Daily, P., Bender, J., Koehler, J., Marcus, R., Angulo, F.J. Emerging Infections Program FoodNet Working Group. 2004. Reptiles, amphibians and human *Salmonella* infection: a population-based, case-control study. *Clin. Inf. Dis.* 38, S253–261.

OSI (Office of Science and Innovation). 2006. *Foresight. Infectious Diseases: Preparing for the Future.* Office of Science and Innovation, London, UK.

Palmarini, M. 2007. A veterinary twist on pathogen biology. *PLoS Path.* 3, e12.

Parrish, C.R., and Kawaoka, Y. 2005. The origins of new pandemic viruses: the acquisition of new host ranges by canine parvovirus and influenza A viruses. *Ann. Rev. Microbiol.* 59, 553–586.

Simmonds, P. 2001. Reconstructing the origins of human hepatitis viruses. *Phil. Trans. R. Soc. B. Biol. Sci.* 356, 1013–1026.

Stohr, K. 2003. A multicentre collaboration to investigate the cause of severe acute respiratory syndrome. *Lancet* 361, 1730–1733.

Taylor, L.H., Latham, S.M., and Woolhouse, M.E.J. 2001. Risk factors for human disease emergence. *Philos. Trans. R. Soc. Lond. B. Biol. Sci.* 356, 983–989.

UNAIDS. 2007. *AIDS Epidemic Update December 2006.* UNAIDS/WHO, Geneva, Switzerland.

Van Heuverswyn, F., Li, Y., Neel, C., Bailes, E., Keele, B.F., Liu, W., Loul, S., Butel, C., Liegeois, F., Bienvenue, Y., Mpoudi Ngolle, E., Sharp, P.M., Shaw, G.M., Delaporte, E., Hahn, B.H., and Peeters, M. 2006. Human immunodeficiency viruses: SIV infection in wild gorillas. *Nature* 444, 164.

Weiss, R.A. 2001. Animal origins of human infectious disease. *Philos. Trans. R. Soc. Lond. B. Biol. Sci.* 356, 957–977.

Wells, R.M., Sosa Estani, S., Yadon, Z.E., Enria, D., Padula, P., Pini, N., Mills, J.N., Peters, C.J., Segura, E.L., and the Hantavirus Pulmonary Syndrome Study Group for Patagonia. 1997. An unusual hantavirus outbreak in southern Argentina: person-to-person transmission. *Emerg. Infect. Dis.* 3, 171–174.

Wilesmith, J.W. 1994. An epidemiologist's view of bovine spongiform encephalopathy. *Philos. Trans. R. Soc. Lond. B. Biol. Sci.* 343, 357–361.

Wolfe, N.D., Dunavan, C.P., and Diamond, J. 2007. Origins of major human infectious diseases. *Nature* 447, 279–283.

Wolfe, N.D., Switzer, W.M., Carr, J.K., Bhullar, V.B., Shanmugam, V., Tamoufe, U., Prosser, A.T., Torimiro, J.N., Wright, A., Mpoudi-Ngole, E., McCutchan, F.E., Birx, D.L., Folks, T.M., Burke, D.S., and Heneine, W. 2004. Naturally acquired simian retrovirus infections in central African huters. *Lancet* 363, 932–937.

Woolhouse, M.E.J. 2002. Population biology of emerging and re–emerging pathogens. *Trends Microbiol.* 10, S3–7.

Woolhouse, M.E.J., and Gowtage-Sequeria, S. 2005. Host range and emerging and reemerging pathogens. *Emerg. Infect. Dis.* 11, 1842–1847.

Woolhouse, M.E.J., Haydon, D.T., and Antia, R. 2005. Emerging pathogens: the epidemiology and evolution of species jumps. *Trends in Ecology and Evolution* 20, 238–244.

Woolhouse, M.E.J., Taylor, L.H., and Haydon, D.T. 2001b. Population biology of multihost pathogens. *Science* 292, 1109–1112.

Eisen References

Blattner, F. R., G. Plunkett 3rd, C. A. Bloch, N. T. Perna, V. Burland, M. Riley, J. Collado-Vides, J. D. Glasner, C. K. Rode, G. F. Mayhew, J. Gregor, N. W. Davis, H. A. Kirkpatrick, M. A. Goeden, D. J. Rose, B. Mau, and Y. Shao. 1997. The complete genome sequence of *Escherichia coli* K-12. *Science* 277(5331):1453-1474.

Brown, D. P., N. Krishnamurthy, and K. Sjölander. 2007. Automated protein subfamily identification and classification. *PLoS Computational Biology* 3(8):e160.

Chatterjee, S., R. P. Almeida, and S. Lindow. 2008. Living in two worlds: the plant and insect lifestyles of *Xylella fastidiosa*. *Annual Review of Phytopathology* 46:243-271.

Dobzhansky, T. 1973. Nothing in biology makes sense except in the light of evolution. *The American Biology Teacher* 35(March):125-129.

Eichenberger, P., M. Fujita, S. T. Jensen, E. M. Conlon, D. Z. Rudner, S. T. Wang, C. Ferguson, K. Haga, T. Sato, J. S. Liu, and R. Losick. 2004. The program of gene transcription for a single differentiating cell type during sporulation in *Bacillus subtilis*. *PLoS Biology* 2(10):e328.

Eisen, J. A. 1998a. Phylogenomics: improving functional predictions for uncharacterized genes by evolutionary analysis. *Genome Research* 8(3):163-167.

———. 1998b. A phylogenomic study of the MutS family of proteins. *Nucleic Acids Research* 26(18):4291-4300.

———. 2000. Assessing evolutionary relationships among microbes from whole-genome analysis. *Current Opinion in Microbiology* 3(5):475-480.

Eisen, J. A., and P. C. Hanawalt. 1999. A phylogenomic study of DNA repair genes, proteins, and processes. *Mutation Research* 435(3):171-213.

Eisen, J. A., and M. Wu. 2002. Phylogenetic analysis and gene functional predictions: phylogenomics in action. *Theoretical Population Biology* 61(4):481-487.

Eisen, J. A., K. S. Sweder, and P. C. Hanawalt. 1995. Evolution of the SNF2 family of proteins: subfamilies with distinct sequences and functions. *Nucleic Acids Research* 23(14):2715-2723.

Eisen, J. A., D. Kaiser, and R. M. Myers. 1997. Gastrogenomic delights: a movable feast. *Nature Medicine* 3(10):1076-1078.

Fleischmann, R. D., M. D. Adams, O. White, R. A. Clayton, E. F. Kirkness, A. R. Kerlavage, C. J. Bult, J. F. Tomb, B. A. Dougherty, J. M. Merrick, K. McKenney, G. Sutton, W. FitzHugh, C. Fields, J. D. Gocayne, J. Scott, R. Shirley, L. Liu, A. Glodek, J. M. Kelley, J. F. Weidman, C. A. Phillips, T. Spriggs, E. Hedblom, M. D. Cotton, T. R. Utterback, M. C. Hanna, D. T. Nguyen, D. M. Saudek, R. C. Brandon, L. D. Fine, J. L. Fritchman, J. R. Fuhrmann, N. S. M. Geoghagen, C. L. Gnehm, L. A. McDonald, K. V. Small, C. M. Fraser, H. O. Smith, and J. C. Venter. 1995. Whole-genome random sequencing and assembly of *Haemophilus influenzae* Rd. *Science* 269(5223):496-512.

Fraser, C. M., J. A. Eisen, K. E. Nelson, I. T. Paulsen, and S. L. Salzberg. 2002. The value of complete microbial genome sequencing (you get what you pay for). *Journal of Bacteriology* 184(23):6403-6405.

Haft, D. H., B. J. Loftus, D. L. Richardson, F. Yang, J. A. Eisen, I. T. Paulsen, and O. White. 2001. TIGRFAMs: a protein family resource for the functional identification of proteins. *Nucleic Acids Research* 29(1):41-43.

Heidelberg, J. F., J. A. Eisen, W. C. Nelson, R. A. Clayton, M. L. Gwinn, R. J. Dodson, D. H. Haft, E. K. Hickey, J. D. Peterson, L. Umayam, S. R. Gill, K. E. Nelson, T. D. Read, H. Tettelin, D. Richardson, M. D. Ermolaeva, J. Vamathevan, S. Bass, H. Qin, I. Dragoi, P. Sellers, L. McDonald, T. Utterback, R. D. Fleishmann, W. C. Nierman, O. White, S. L. Salzberg, H. O. Smith, R. R. Colwell, J. J. Mekalanos, J. C. Venter, and C. M. Fraser. 2000. DNA sequence of both chromosomes of the cholera pathogen *Vibrio cholerae*. *Nature* 406(6795):477-483.

Hugenholtz, P. 2002. Exploring prokaryotic diversity in the genomic era. *Genome Biology* 3(2): REVIEWS0003.

Jordan, I. K., K. S. Makarova, J. L. Spouge, Y. I. Wolf, and E. V. Koonin. 2001. Lineage-specific gene expansions in bacterial and archaeal genomes. *Genome Research* 11(4):555-565.

Kang, J. M., N. M. Iovine, and M. J. Blaser. 2006. A paradigm for direct stress-induced mutation in prokaryotes. *FASEB Journal* 20(14):2476-2485.

LeClerc, J. E., B. Li, W. L. Payne, and T. A. Cebula. 1996. High mutation frequencies among *Escherichia coli* and *Salmonella* pathogens. *Science* 274(5290):1208-1211.

Lederberg, J. 1997. Infectious disease as an evolutionary paradigm. *Emerging Infectious Diseases* 3(4):417.

———. 1998. Emerging infections: an evolutionary perspective. *Emerging Infectious Diseases* 4(3):366.

Marcotte, E. M. 2000. Computational genetics: finding protein function by nonhomology methods. *Current Opinion in Structural Biology* 10(3):359-365.

Marshall, B. 2002. *Helicobacter pylori*: 20 years on. *Clinical Medicine* 2(2):147-152.

McCutcheon, J. P., and N. A. Moran. 2007. Parallel genomic evolution and metabolic interdependence in an ancient symbiosis. *Proceedings of the National Academy of Sciences* 104(49): 19392-19397.

Moran, N. A., C. Dale, H. Dunbar, W. A. Smith, and H. Ochman. 2003. Intracellular symbionts of sharpshooters (Insecta: Hemiptera: Cicadellinae) form a distinct clade with a small genome. *Environmental Microbiology* 5(2):116-126.

Moran, N. A., P. Tran, and N. M. Gerardo. 2005. Symbiosis and insect diversification: an ancient symbiont of sap-feeding insects from the bacterial phylum Bacteroidetes. *Applied and Environmental Microbiology* 71(12):8802-8810.

Moran, N. A., J. P. McCutcheon, A. Nakabachi. 2008. Genomics and evolution of heritable bacterial symbionts. *Annual Review of Genetics* 42:165-190.

Ohta, T. 2000. Evolution of gene families. *Gene* 259(1-2):45-52.

Parkhill, J., B. W. Wren, K. Mungall, J. M. Ketley, C. Churcher, D. Basham, T. Chillingworth, R. M. Davies, T. Feltwell, S. Holroyd, K. Jagels, A. V. Karlyshev, S. Moule, M. J. Pallen, C. W. Penn, M. A. Quail, M. A. Rajandream, K. M. Rutherford, A. H. van Vliet, S. Whitehead, and B. G. Barrell. 2000. The genome sequence of the food-borne pathogen *Campylobacter jejuni* reveals hypervariable sequences. *Nature* 403(6770):665-668.

Pellegrini, M., E. M. Marcotte, M. J. Thompson, D. Eisenberg, and T. O. Yeates. 1999. Assigning protein functions by comparative genome analysis: protein phylogenetic profiles. *Proceedings of the National Academy of Sciences* 96(8):4285-4288.

Pollard, K. S., S. R. Salama, B. King, A. D. Kern, T. Dreszer, S. Katzman, A. Siepel, J. S. Pedersen, G. Bejerano, R. Baertsch, K. R. Rosenbloom, J. Kent, and D. Haussler. 2006. Forces shaping the fastest evolving regions in the human genome. *PLoS Genetics* 2(10):e168.

Silvaggi, J. M., D. L. Popham, A. Driks, P. Eichenberger, and R. Losick. 2004. Unmasking novel sporulation genes in *Bacillus subtilis*. *Journal of Bacteriology* 86(23):8089-8095.

Sonnhammer, E. L., S. R. Eddy, and R. Durbin. 1997. Pfam: a comprehensive database of protein domain families based on seed alignments. *Proteins* 28(3):405-420.

Tatusov, R. L., M. Y. Galperin, D. A. Natale, and E. V. Koonin. 2000. The COG database: a tool for genome-scale analysis of protein functions and evolution. *Nucleic Acids Research* 28(1):33-36.

Tomb, J. F., O. White, A. R. Kerlavage, R. A. Clayton, G. G. Sutton, R. D. Fleischmann, K. A. Ketchum, H. P. Klenk, S. Gill, B. A. Dougherty, K. Nelson, J. Quackenbush, L. Zhou, E. F. Kirkness, S. Peterson, B. Loftus, D. Richardson, R. Dodson, H. G. Khalak, A. Glodek, K. McKenney, L. M. Fitzegerald, N. Lee, M. D. Adams, E. K. Hickey, D. E. Berg, J. D. Gocayne, T. R. Utterback, J. D. Peterson, J. M. Kelley, M. D. Cotton, J. M. Weidman, C. Fujii, C. Bowman, L. Watthey, E. Wallin, W. S. Hayes, M. Borodovsky, P. D. Karp, H. O. Smith, C. M. Fraser, and J. C. Venter. 1997. The complete genome sequence of the gastric pathogen *Helicobacter pylori*. *Nature* 388(6642):539-547.

White, O., J. A. Eisen, J. F. Heidelberg, E. K. Hickey, J. D. Peterson, R. J. Dodson, D. H. Haft, M. L. Gwinn, W. C. Nelson, D. L. Richardson, K. S. Moffat, H. Qin, L. Jiang, W. Pamphile, M. Crosby, M. Shen, J. J. Vamathevan, P. Lam, L. McDonald, T. Utterback, C. Zalewski, K. S. Makarova, L. Aravind, M. J. Daly, K. W. Minton, R. D. Fleischmann, K. A. Ketchum, K. E. Nelson, S. Salzberg, H. O. Smith, J. C. Venter, and C. M. Fraser. 1999. Genome sequence of the radioresistant bacterium *Deinococcus radiodurans* R1. *Science* 286(5444):1571-1577.

Wu, D., S. C. Daugherty, S. E. Van Aken, G. H. Pai, K. L. Watkins, H. Khouri, L. J. Tallon, J. M. Zaborsky, H. E. Dunbar, P. L. Tran, N. A. Moran, and J. A. Eisen. 2006. Metabolic complementarity and genomics of the dual bacterial symbiosis of sharpshooters. *PLoS Biology* 4(6):e188.

Wu, M., Q. Ren, A. S. Durkin, S. C. Daugherty, L. M. Brinkac, R. J. Dodson, R. Madupu, S. A. Sullivan, J. F. Kolonay, D. H. Haft, W. C. Nelson, L. J. Tallon, K. M. Jones, L. E. Ulrich, J. M. Gonzalez, I. B. Zhulin, F. T. Robb, and J. A. Eisen. 2005. Life in hot carbon monoxide: the complete genome sequence of *Carboxydothermus hydrogenoformans* Z-2901. *PLoS Genetics* 1(5):e65.

Zmasek, C. M., and S. R. Eddy. 2002. RIO: analyzing proteomes by automated phylogenomics using resampled inference of orthologs. *BMC Bioinformatics* 3:14.

Daszak References

Black, F. L. 1966. Measles endemicity in insular populations: critical community size and its evolutionary implication. *Journal of Theoretical Biology* 11(2):207-211.

Daszak, P., R. Plowright, J. H. Epstein, J. Pulliam, S. Abdul Rahman, H. E. Field, A. Jamaluddin, S. H. Sharifah, C. S. Smith, K. J. Olival, S. Luby, K. Halpin, A. D. Hyatt, A. A. Cunningham, and Henipavirus Ecology Research Group (HERG). 2006. The emergence of Nipah and Hendra virus: pathogen dynamics across a wildlife-livestock-human continuum. In *Disease ecology: community structure and pathogen dynamics,* edited by R. S. Collinge. Oxford, UK: Oxford University Press. Pp. 186-201.

Dobson, A. P., and E. R. Carper. 1996. Infectious diseases and human population history. *Bioscience* 46(2):115-126.

Epstein, J. H., V. Prakash, C. S. Smith, P. Daszak, A. B. McLaughlin, G. Meehan, H. E. Field, and A. A. Cunningham. 2008. Henipavirus infection in fruit bats (*Pteropus giganteus*), India. *Emerging Infectious Diseases* 14(8):1309-1311.

Fogarty, R., K. Halpin, A. D. Hyatt, P. Daszak, and B. A. Mungall. 2008. Henipavirus susceptibility to environmental variables. *Virus Research* 132(1-2):140-144.

Hufnagel, L., D. Brockmann, and T. Geisel. 2004. Forecast and control of epidemics in a globalized world. *Proceedings of the National Academy of Sciences* 101(42):15124-15129.

Jones, K. E., N. G. Patel, M. A. Levy, A. Storeygard, D. Balk, J. L. Gittleman, and P. Daszak. 2008. Global trends in emerging infectious diseases. *Nature* 451(7181):990-993.

Kilpatrick, A. M., Y. Gluzberg, J. Burgett, and P. Daszak. 2004. Quantitative risk assessment of the pathways by which West Nile virus could reach Hawaii. *Ecohealth* 1(2):205-209.

Kilpatrick, A. M., A. A. Chmura, D. W. Gibbons, R. C. Fleischer, P. P. Marra, and P. Daszak. 2006a. Predicting the global spread of H5N1 avian influenza. *Proceedings of the National Academy of Sciences* 103(51):19368-19373.

Kilpatrick, A. M., P. Daszak, S. J. Goodman, H. Rogg, L. D. Kramer, V. Cedeno, and A. A. Cunningham. 2006b. Predicting pathogen introduction: West Nile virus spread to Galapagos. *Conservation Biology* 20(4):1224-1231.

Morse, S. S. 1993. Examining the origins of emerging viruses. In *Emerging viruses*, edited by S. S. Morse. New York: Oxford University Press.

Murphy, F. A. 1998. Emerging zoonoses. *Emerging Infectious Diseases* 4(3):429-435.

Peeters, M., V. Courgnaud, B. Abela, P. Auzel, X. Pourrut, F. Bibollet-Ruche, S. Loul, F. Liegeois, C. Butel, D. Koulagna, E. Mpoudi-Ngole, G. M. Shaw, B. H. Hahn, and E. Delaporte. 2002. Risk to human health from a plethora of simian immunodeficiency viruses in primate bushmeat. *Emerging Infectious Diseases* 8(5):451-457.

Taylor, L. H., S. M. Latham, and M. E. J. Woolhouse. 2001. Risk factors for human disease emergence. *Philosophical Transactions of the Royal Society of London. Series B, Biological sciences* 356(1411):983-989.

WorldAtlas.com. 2006. *Countries of the world*, http://www.worldatlas.com/aatlas/populations/ctydensityh.htm (accessed December 18, 2008).

Appendix A

Agenda

Microbial Evolution and Co-Adaptation:
A Workshop in Honor of Joshua Lederberg

May 20-21, 2008
The National Academies
500 Fifth Street, NW—Room 100
Washington, DC

DAY 1: MAY 20, 2008

9:00-9:30:   Registration and continental breakfast

9:30-9:45:   Welcoming remarks
               Forum Leadership

9:45-10:45:   Reflections on the life and scientific legacies of Josh Lederberg

Moderator: Margaret "Peggy" Hamburg, M.D., Nuclear Threat Initiative

- David A. Hamburg, M.D.
  Carnegie Corporation
- Stephen S. Morse, Ph.D.
  Columbia University

10:45-11:15:   Questions from Forum members and audience

11:15-11:30:   Break

11:30-12:30:   Lunch and continuation of Day 1 morning discussion

Session I
The Microbiome and Co-Evolution

Moderator: Jo Handelsman, Ph.D., University of Wisconsin, Madison

12:30-1:00:   Evolutionary and ecological processes that shape natural
             microbial communities
                 Jill Banfield, Ph.D.
                 University of California, Berkeley

1:00-1:30:   Plant-microbe symbioses: the good, the bad, and the cheater!
                 Jean-Michel Ané, Ph.D.
                 University of Wisconsin, Madison

1:30-2:00:   The experimental deciphering of the complex molecular
             dialogue between host and bacterial symbiont
                 Margaret McFall-Ngai, Ph.D.
                 University of Wisconsin, Madison

2:00-2:30:   War and peace: humans and their microbiome
                 David Relman, M.D.
                 Stanford University

2:30-3:15:   Questions from Forum members and audience

3:15-3:30:   Break

Session II
Microbial Evolution and the Emergence of Virulence

Moderator: P. Frederick Sparling, Ph.D., University of North Carolina,
Chapel Hill

3:30-4:00:   A schizophrenic view of bacterial pathogenicity
                 Stanley Falkow, Ph.D.
                 Stanford University

4:00-4:30:   Recent evolution in invasive salmonellosis
                 Gordon Dougan, Ph.D.
                 The Sanger Institute

4:30-5:00: Whole genome analysis of pathogen evolution
Julian Parkhill, Ph.D.
The Sanger Institute

5:00-5:30: The evolution of virulence in bacteria and viruses: an
opinionated rant
Bruce Levin, Ph.D.
Emory University

5:30-6:00: Questions from Forum members and audience

6:15: Adjourn—Day 1

7:00-9:00: Dinner with speakers and Forum members and continuing
discussion of Day 1

DAY 2: MAY 21, 2008

8:30-9:00: Continental breakfast

9:00-9:15: Summary of Day 1
David Relman, M.D.
Chair, Forum on Microbial Threats

9:15-9:45: Further reflections on Josh Lederberg's life and legacies
Adel Mahmoud, M.D., Ph.D.
Princeton University

Session III
Mechanisms of Resistance

Moderator: Steven Brickner, Ph.D., Pfizer, Inc.

9:45-10:15: Microbial drug resistance: an old problem that requires new
solutions
Stanley N. Cohen, M.D.
Stanford University

10:15-10:45: Antibiotic resistance and the future of antibiotics
Julian Davies, Ph.D.
University of British Columbia

10:45-11:00: Break

11:00-11:30:   Expanding the resistance universe with metagenomics
               Jo Handelsman, Ph.D.
               University of Wisconsin, Madison

11:30-12:15:   Questions from Forum members and audience

12:15-1:00:   Lunch and continuation of Day 2 morning discussion

Session IV
Anticipation of Future Emerging Infectious Diseases—
Is the Past, Prologue?

Moderator: Stanley Lemon, M.D., University of Texas, Galveston

1:00-1:30:   Emerging infections: condemned to repeat?
               Stephen S. Morse, Ph.D.
               Columbia University

1:30-2:00:   Can we predict future trends in disease emergence?
               Peter Daszak, Ph.D.
               Consortium for Conservation Medicine, Wildlife Trust

2:00-2:30:   Pathogen emergence: ecology or evolution?
               Mark Woolhouse, Ph.D.
               University of Edinburgh

2:30-3:00:   Genomic evolvability and the origin of novelty
               Jonathan Eisen, Ph.D.
               University of California, Davis

3:00-3:30:   Questions from Forum members and audience

3:30-4:00:   Open discussion and wrap-up

4:00:   Adjourn

Appendix B

Acronyms

AI	autoinducer
AMD	acid mine drainage
BLAST	Basic Local Alignment Search Tool
BSE	bovine spongiform encephalopathy
Bt	*Bacillus thuringiensis*
CDC	Centers for Disease Control and Prevention
CGEPs	cellular genes exploited by pathogens
CISAC	Committee on International Security and Arms Control
CPV	canine parvovirus
CRISPR	clustered regularly interspaced short palindromic repeat
DARPA	Defense Advanced Research Projects Agency
DOE	Department of Energy
DTRA	Defense Threat Reduction Agency
EID	emerging infectious disease
ESCRT	endosomal sorting complexes required for transport
EST	expressed sequence tag
FAO	Food and Agriculture Organization
FASEB	Federation of American Societies for Experimental Biology
FEMS	Federation of European Microbiological Societies
FISH	fluorescence *in situ* hybridization

FPV	feline panleukopenia virus

GEBA	Genomic Encyclopedia of Bacteria and Archaea
GI	gastrointestinal
GSV	gene search vector

HIV	human immunodeficiency virus
HPS	Hantavirus pulmonary syndrome
HPSD-18	Homeland Security Presidential Directive-18
HTLV	human T-cell lymphoma virus

IDSA	Infectious Diseases Society of America
IHR	International Health Regulations
IOM	Institute of Medicine

JAMA	*Journal of the American Medical Association*
JID	*Journal of Infectious Diseases*

LPS	lipopolysaccharide

MAMP	microbe-associated molecular pattern
MCP	methyl-accepting chemotaxis protein
MDR	multidrug resistant
MRSA	methicillin-resistant *Staphylococcus aureus*

NAS	National Academy of Sciences
NASA	National Aeronautics and Space Administration
NCID	National Center for Infectious Diseases
NIAID	National Institute of Allergy and Infectious Diseases
NIH	National Institutes of Health
NiV	Nipah virus
NLM	National Library of Medicine
NSF	National Science Foundation
NTS	non-typhoidal salmonellosis

OR	odds ratio
ORF	open reading frame
OSTP	Office of Science and Technology Policy

PCR	polymerase chain reaction
PGN	peptidoglycan
PIGT	proteomic-inferred genome typing

ProMED   Program for Monitoring Emerging Diseases

QS   quorum sensing

RHKO   random homozygous knockout
RSV   Rous sarcoma virus

SARS   severe acute respiratory syndrome
SNP   single nucleotide polymorphism

TB   tuberculosis
TIGR   The Institute for Genomic Research

VISA   vancomycin-intermediate *Staphylococcus aureus*
VRSA   vancomycin-resistant *Staphylococcus aureus*

WHO   World Health Organization
WNV   West Nile virus
WWTP   wastewater treatment plant

XDR   extensively drug resistant

Appendix C

Glossary

Adaptive immune response: Response of the vertebrate immune system to a specific antigen that typically generates immunological memory (Alberts, B., A. Johnson, J. Lewis, M. Raff, K. Roberts, and P. Walter. 2002. *Molecular biology of the cell, fourth edition.* New York: Garland Science).

Affymetrix chip: A type of DNA microarray gene chip that contains up to 500,000 unique probes corresponding to tens of thousands of gene expression measurements (http://cnx.org/content/m12387/latest/).

Anthropogenic: Caused or produced by humans.

Archaea: Unique group of microorganisms classified as bacteria (Archaeobacteria) but genetically and metabolically different from all other known bacteria. They appear to be living fossils, the survivors of an ancient group of organisms that bridged the gap in evolution between bacteria and the eukaryotes (multicellular organisms) (http://www.medterms.com/script/main/art.asp?articlekey=2322).

Asymptomatic carrier: An individual who serves as host for an infectious agent but who does not show any apparent signs of the illness; may serve as a source of infection for others (http://medical-dictionary.thefreedictionary. com/Asymptomatic+carrier).

Bacteriocins: Bacterially produced, small, heat-stable peptides that are active against other bacteria and to which the producer has a specific immunity mechanism. Bacteriocins can have a narrow or broad target spectrum (http://www.nature.com/nrmicro/journal/v3/n10/glossary/nrmicro1273_glossary.html).

Binary associations: A population of a single microbial phylotype associates with a host.

Biofilm: A slime layer that develops naturally when bacteria congregate on surfaces (http://publications.nigms.nih.gov/thenewgenetics/glossary.html).

Bioluminescence: Light produced by a chemical reaction that originates in an organism (primarily a marine phenomenon). The causal organisms are almost always dinoflagellates, or single-cell algae, often numbering many hundreds per liter (http://www.lifesci.ucsb.edu/~biolum/).

Clustered regularly interspaced short palindromic repeats (CRISPRs): Direct repeats found in the DNA of many bacteria and archaea. These repeats range in size from 23 to 47 base pairs and are separated by spacers of similar length. Spacers are usually unique in a genome. Some spacer sequences match sequences in phage genomes; it is proposed that these spacers derive from phage and subsequently help protect the cell from infection (http://crispr.u-psud.fr/crispr/CRISPRHomePage.php?page=about).

Commensals: Organisms in a mutually symbiotic relationship where both live peacefully together while not being completely dependent on one another (example: the gut microbiome) (http://www.bacteriamuseum.org/niches/evolution/commensals.shtml).

Consortial association: Populations of more than one phylotype of microbe living in association with a host animal.

Cytoskeleton: The internal components of animal cells that give them structural strength and motility. The major components of cytoskeleton are the microfilaments (of actin), microtubules (of tubulin) and intermediate filament systems in cells (http://www.mblab.gla.ac.uk/~julian/Dict.html).

Emerging infections: Infections that are rapidly increasing in incidence or geographic range.

Endosymbiont: An organism that lives inside another organism, most often for the benefit of the two (example: rhizobia [nitrogen-fixing soil bacterial] that live within root nodules—rhizobia cannot independently fix nitrogen but need the plant as an energy source; in turn, rhizobia supply the plant host with ammonia and amino acids).

Extensively drug resistant tuberculosis (XDR TB): A relatively rare type of MDR TB. XDR TB is defined as tuberculosis that is resistant to isoniazid and rifampicin, plus resistant to any fluoroquinolone and at least one of three injectable second-line drugs (i.e., amikacin, kanamycin, capreomycin) (http://www.cdc.gov/tb/pubs/tbfactsheets/mdrtb.htm).

Generalized transduction: The ability of certain phages to transfer any gene in the donor bacterial chromosome to a recipient bacterium (http://www.ncbi.nlm.nih.gov/books/bv.fcgi?rid=iga.section.1363).

Genus: A group of species with similar characteristics that are closely related (http://www.sedgwickmuseum.org/education/glossary.html#g).

Gnotobiotic vertebrates: Organisms or environmental conditions that have been rendered free of bacteria or contaminants or into which a known microorganism or contaminant has been introduced for research purposes (http://www.answers.com/topic/gnotobiotics).

Homolog: One of two or more genes that are similar in sequence as a result of derivation from the same ancestral gene. The term covers both orthologs and paralogs.

Hormesis: Showing different effects at low doses compared to high.

Hydrothermal vent ecosystems: The community of organisms that thrives in the extreme heat and mineral-rich waters surrounding hydrothermal vents (geysers on the sea floor). Such organisms include tubeworms, vent crabs, heat worms, and ancient bacteria (http://www.ocean.udel.edu/deepsea/level-2/geology/vents.html).

Innate immune response: Immune response (of both vertebrates and invertebrates) to a pathogen that involves the preexisting defenses of the body, such as barriers formed by skin and mucosa, antimicrobial molecules and phagocytes. Such a response is not specific for the pathogen (http://www.ncbi.nlm.nih.gov/books/bv.fcgi?rid=mboc4.glossary.4754).

Interferon: One of a group of small antiviral proteins synthesized and secreted by cells following viral infection (http://www.portlandpress.com/pp/books/online/glick/searchresdet.cfm?Term=interferon).

International units: The quantity of a biologically active substance, such as a hormone or vitamin, required to produce a specific response (http://www.answers.com/topic/international-unit).

Lamarckism: The theory that evolution is caused by inheritance of character changes acquired during the life of an individual, due to its behavior or to environmental influences (http://evolution.unibe.ch/teaching/GlossarE.htm).

Leukocyte extravasation: The movement of leukocytes out of the circulatory system, via specific cell-cell contacts with endothelial cells lining the blood vessel, toward site of tissue damage or infection (http://en.wikipedia.org/wiki/Leukocyte_extravasation; http://www.whfreeman.com/kuby/content/anm/kb01an01.htm).

Lymph nodes: Small bean-shaped organs scattered along the vessels of the lymphatic system. The lymph nodes produce white blood cells and filter bacteria and cancer cells that may travel through the system (http://www.everythingbio. com/glos/definition.php?word=Lymph+nodes).

Lymphoid follicles: Specialized regions of secondary lymphoid organs critical for B-cell development (a type of lymphocyte that makes antibodies) and, thus, are critical for the functioning of the adaptive immune system (Sompayrac, L. M. 1991. *How the immune system works, second edition.* New York: Wiley-Blackwell).

Macrophage: Phagocytic cell derived from blood monocytes, typically resident in most tissues. It has both scavenger and antigen-presenting functions in immune responses (http://www.ncbi.nlm.nih.gov/books/bv.fcgi?rid=mboc4. glossary.4754).

Malthusian fitness: A measure of reproductive potential based on the Malthusian Population Theory, which asserts that in biology, the viability of virtually any organism or species greatly exceeds the Earth's capacity to support all its possible offspring. Consequently, species diversity is preserved through mechanisms that keep population sizes in check, such as predation (http://www.economyprofessor. com/economictheories/malthusian-population-theory.php).

Metagenomics: A culture-independent analysis method that involves obtaining DNA from communities of microorganisms, sequencing it in a "shotgun" fashion, and characterizing genes and genomes comparisons with known gene sequences. With this information, researchers can gain insights into how members of the microbial community may interact, evolve, and perform complex functions in their habitats (Jurkowski, A., A. H. Reid, and J. B. Labov. 2007. Metagenomics: a call for bringing a new science into the classroom while it's still new. *CBE Life Sciences Education* 6(4):260-265; National Research Council. 2007. *The new science of metagenomics: revealing the secrets of our microbial planet.* Washington, DC: The National Academies Press).

Methicillin-resistant *Staphylococcus aureus* (MRSA): A type of staph that is resistant to antibiotics called β-lactams. β-lactam antibiotics include methicillin and other more common antibiotics such as oxacillin, penicillin and amoxicillin (http://www.cdc.gov/ncidod/dhqp/ar_MRSA_ca_public.html#2).

Microbiome: Term used to describe the collective genome of our indigenous microbes (microflora) (Hooper, L. V., and J. I. Gordon. 2001. Commensal host-bacterial relationships in the gut. *Science* 292(5519):1115-1118).

Microcins: Bacteriocins that contain a smaller number of amino acids.

Multidrug-resistant tuberculosis (MDR TB): Tuberculosis that is resistant to at least two of the best anti-TB drugs, isoniazid and rifampicin. These drugs are considered first-line drugs and are used to treat all persons with TB disease (http://www.cdc.gov/tb/pubs/tbfactsheets/mdrtb.htm).

Neutrophil: Most common blood leukocyte; a short-lived phagocytic cell of the myeloid series, which is responsible for the primary cellular response to an acute inflammatory episode, and for general tissue homeostasis by removal of damaged material (http://www.mblab.gla.ac.uk/~julian/Dict.html).

Pandemic: Disease outbreak occurring over a wide geographic area and affecting an exceptionally high proportion of the population (http://www2.merriam-webster.com/cgi-bin/mwmednlm?book=Medical&va=pandemic).

Pathogen: A specific causative agent (such as a bacterium or virus) of disease (http://www2.merriam-webster.com/cgi-bin/mwmednlm?book=Medical&va=pathogen).

Pathogenicity: Pathogenicity reflects the ongoing evolution between a parasite and host, and disease is the product of a microbial adaptive strategy for survival.

Pathogenicity islands: Large genomic regions encoding for virulence factors of pathogenic bacteria, present on the genomes of pathogenic strains but absent from the genomes of nonpathogenic members of the same or related species. (Hacker, J., and J. Kaper. 2000. Pathogenicity islands and the evolution of microbes. *Annual Review of Microbiology* 54:641-679).

Peyer's patches: Large aggregates of lymphoid tissue (similar to lymph nodes) located in the mucosa and extending into the submucosa of the small intestine. Peyer's patches facilitate the generation of an immune response within the mucosa (http://www.microbiologybytes.com/iandi/2b.html).

Phage(s): A virus that infects bacteria. Many phages have proved useful in the study of molecular biology and as vectors for the transfer of genetic information between cells. Lytic phage (e.g., the T series phage that infect *Escherichia coli* [coliphages]), invariably lyse a cell following infection; temperate phage (e.g., lambda bacteriophage) can also undergo a lytic cycle or can enter a lysogenic cycle, in which the phage DNA is incorporated into that of the host, awaiting a signal that initiates events leading to replication of the virus and lysis of the host cell (http://www.portlandpress.com/pp/books/online/glick/searchresdet.cfm? Term=bacteriophage).

Phagocyte/phagocytic cell: A cell that is capable of phagocytosis, or the uptake of particulate material by a cell. The main mammalian phagocytes are neutrophils and macrophages (http://www.mblab.gla.ac.uk/~julian/Dict.html).

Phylogenetic profiling: Phylogenetic profiling is a well established method for predicting functional relations and physical interactions between proteins (Pagel, P., P. Wong, and D. Frishman. 2004. A domain interaction map based on phylogenetic profiling. *Journal of Molecular Biology* 344(5):1331-1346).

Phylogenetics: The study of the relationships between organisms based on how closely they are related to each other (http://www.medterms.com/script/main/art.asp?articlekey=39615).

Phylogenomics: The use of evolutionary information in the prediction of gene function (http://genome.cshlp.org/content/8/3/163.full.pdf+html).

Phylum: In taxonomy and systematics, the highest level of classification below the kingdom (http://ec.europa.eu/research/biosociety/library/glossarylist_en.cfm).

Pleiotropic: Producing more than one effect; having multiple phenotypic expressions (http://www.merriam-webster.com/dictionary/pleiotropic).

Ribosomal RNA: A class of RNA molecules, coded in the nucleolar organizer, that has an integral (but poorly understood) role in ribosome structure and function. RNA components of the subunits of the ribosomes (http://www.biochem.northwestern.edu/holmgren/Glossary/Definitions/Def-R/ribosomal_RNA.html).

Serotype: The characterization of a microorganism based on the kinds and combinations of constituent antigens present in that organism; a taxonomic subdivision of bacteria based on the above.

Species: The basic unit of taxonomy. A species is defined as a group of individuals that are genetically related and can interbreed to produce fertile young of the same kind (http://www.sedgwickmuseum.org/education/glossary.html).

Transposons: Small, mobile DNA sequences that can replicate and insert copies at random sites within chromosomes. They have nearly identical sequences at each end, oppositely oriented (inverted) repeats, and code for the enzyme, transposase, that catalyses their insertion (http://www.mblab.gla.ac.uk/~julian/Dict.html).

Vancomycin-intermediate or vancomycin-resistant *Staphylococcus aureus* (VISA/VRSA): VISA and VRSA are specific types of antimicrobial-resistant staph bacteria. While most staph bacteria are susceptible to the antimicrobial agent vancomycin some have developed resistance. VISA and VRSA cannot be treated successfully with vancomycin because these organisms are no longer susceptibile to vancomycin. However, to date, all VISA and VRSA isolates have been susceptible to other Food and Drug Administration (FDA)-approved drugs (http://www.cdc.gov/ncidod/dhqp/ar_visavrsa_FAQ.html).

Virulence: The degree of pathogenicity of an organism as evidenced by the severity of resulting disease and the organism's ability to invade the host tissues.

Virulence factors: The properties (i.e., gene products) that enable a microorganism to establish itself on or within a host of a particular species and enhance its potential to cause disease. Virulence factors include bacterial toxins, cell surface proteins that mediate bacterial attachment, cell surface carbohydrates and proteins that protect a bacterium, and hydrolytic enzymes that may contribute to the pathogenicity of the bacterium (http://www.mgc.ac.cn/VFs/main.htm).

Zoonotic: Infection that causes disease in human populations but can be perpetuated solely in nonhuman host animals (e.g., bubonic plague); may be enzootic or epizootic.

Appendix D

Forum Member Biographies

David A. Relman, M.D. (*Chair*), is professor of medicine (infectious diseases and geographic medicine) and of microbiology and immunology at Stanford University School of Medicine, and chief of the infectious disease section at the Veterans Affairs (VA) Palo Alto Health Care System. Dr. Relman received his B.S. in biology from the Massachusetts Institute of Technology and his M.D. from Harvard Medical School. He completed his residency in internal medicine and a clinical fellowship in infectious diseases at Massachusetts General Hospital, Boston, after which he moved to Stanford for a postdoctoral fellowship in 1986 and joined the faculty there in 1994. His research focus is on understanding the structure and role of the human indigenous microbial communities in health and disease. This work brings together approaches from ecology, population biology, environmental microbiology, genomics, and clinical medicine. A second area of investigation explores the classification structure of humans and nonhuman primates with systemic infectious diseases, based on patterns of genome-wide gene transcript abundance in blood and other tissues. The goals of this work are to understand mechanisms of host-pathogen interaction, as well as predict clinical outcome early in the disease process. His scientific achievements include the description of a novel approach for identifying previously unknown pathogens; the characterization of a number of new human microbial pathogens, including the agent of Whipple's disease; and some of the most in-depth analyses to date of human indigenous microbial communities. Among his other activities, Dr. Relman currently serves as chair of the Board of Scientific Counselors of the National Institutes of Health (NIH) National Institute of Dental and Craniofacial Research, is a member of the National Science Advisory Board for Biosecurity, and advises a number of U.S. government departments and agencies on matters related to pathogen diversity, the future life sciences landscape, and the nature of present and future biological threats. He was co-chair of the Committee on Advances in Technology and the Prevention of Their Application to Next Generation Biowarfare Threats for the National Academy of Sciences (NAS). He received the Squibb Award from the Infectious Diseases Society of America (IDSA) in 2001, the Senior Scholar Award in Global Infectious Diseases from the Ellison Medical Foundation in 2002, an NIH Director's Pioneer Award in 2006, and a Doris Duke Distinguished Clinical Scientist Award in 2006. He is also a fellow of the American Academy of Microbiology.

Margaret A. Hamburg, M.D. (*Vice Chair*), was the founding vice president, Biological Programs, at the Nuclear Threat Initiative, a charitable organization working to reduce the global threat from nuclear, biological, and chemical weapons, and ran the program for many years. She currently serves as senior scientist for the organization. She completed her internship and residency in internal medicine at the New York Hospital-Cornell University Medical Center and is certified by the American Board of Internal Medicine. Dr. Hamburg is a graduate of Harvard College and Harvard Medical School. Before taking on her current position, she was the assistant secretary for planning and evaluation, U.S. Department of Health and Human Services (HHS), serving as a principal policy adviser to the secretary of health and human services, with responsibilities including policy formulation and analysis, the development and review of regulations and legislation, budget analysis, strategic planning, and the conduct and coordination of policy research and program evaluation. Prior to this, she served for nearly six years as the commissioner of health for the City of New York. As chief health officer in the nation's largest city, her many accomplishments included the design and implementation of an internationally recognized tuberculosis control program that produced dramatic declines in tuberculosis cases, the development of initiatives that raised childhood immunization rates to record levels, and the creation of the first public health bioterrorism preparedness program in the nation. She currently serves on the Harvard University Board of Overseers. She has been elected to membership in the Institute of Medicine (IOM), the New York Academy of Medicine, and the Council on Foreign Relations and is a fellow of the American Association for the Advancement of Science (AAAS) and the American College of Physicians.

David W. K. Acheson, M.D., F.R.C.P., is assistant commissioner for food protection in the U.S. Food and Drug Administration (FDA). Dr. Acheson graduated from the University of London Medical School in 1980 and, following training in internal medicine and infectious diseases in the United Kingdom, moved to the New England Medical Center and Tufts University in Boston in 1987. As an associate professor at Tufts University, he undertook basic molecular pathogenesis research on foodborne pathogens, especially Shiga toxin-producing *Escherichia* *coli*. In 2001, Dr. Acheson moved his laboratory to the University of Maryland Medical School in Baltimore to continue research on foodborne pathogens. In September 2002, Dr. Acheson accepted a position as chief medical officer at the FDA Center for Food Safety and Applied Nutrition (CFSAN). In January 2004, he also became the director of CFSAN's Food Safety and Security Staff, and in January 2005, the staff was expanded to become the Office of Food Safety, Defense and Outreach. In January 2007, the office was further expanded to become the Office of Food Defense, Communication and Emergency Response. On May 1, 2007, Dr. Acheson assumed the position of FDA assistant commissioner for food protection to provide advice and counsel to the commissioner on strategic and substantive food safety and food defense matters. Dr. Acheson has published extensively and is internationally recognized both for his public health expertise in food safety and for his research in infectious diseases. Additionally, Dr. Acheson is a fellow of both the Royal College of Physicians (London) and the Infectious Diseases Society of America.

Ruth L. Berkelman, M.D., is the Rollins Professor and director of the Center for Public Health Preparedness and Research at the Rollins School of Public Health, Emory University, in Atlanta. She received her A.B. from Princeton University and her M.D. from Harvard Medical School. Board certified in pediatrics and internal medicine, she began her career at the Centers for Disease Control and Prevention (CDC) in 1980 and later became deputy director of the National Center for Infectious Diseases (NCID). She also served as a senior adviser to the director of CDC and as assistant surgeon general in the U.S. Public Health Service. In 2001 she came to her current position at Emory University, directing a center focused on emerging infectious diseases and other urgent threats to health, including terrorism. She has also consulted with the biologic program of the Nuclear Threat Initiative and is most recognized for her work in infectious diseases and disease surveillance. She was elected to the IOM in 2004. Currently a member of the Board on Life Sciences of the National Academies, she also chairs the Board of Public and Scientific Affairs at the American Society of Microbiology (ASM).

Enriqueta C. Bond, Ph.D., is president of the Burroughs Wellcome Fund. She received her undergraduate degree from Wellesley College, her M.A. from the University of Virginia, and her Ph.D. in molecular biology and biochemical genetics from Georgetown University. She is a member of the Institute of Medicine, the American Association for the Advancement of Science, the American Society for Microbiology, and the American Public Health Association. Dr. Bond chairs the Academies' Board on African Science Academy Development and serves on the Report Review Committee for the Academies. She serves on the board and executive committee of the Research Triangle Park Foundation, the board of the National Institute for Statistical Sciences, the board of the Northeast Biodefense

Center and the New England Center of Excellence in Biodefense and Emerging Infectious Diseases, and the council of the National Institute of Child Health and Human Development. Prior to being named president of the Burroughs Wellcome Fund in 1994, Dr. Bond served on the staff of the IOM beginning in 1979, becoming its executive officer in 1989.

Roger G. Breeze, Ph.D., received his veterinary degree in 1968 and his Ph.D. in veterinary pathology in 1973, both from the University of Glasgow, Scotland. He was engaged in teaching, diagnostic pathology, and research on respiratory and cardiovascular diseases at the University of Glasgow Veterinary School from 1968 to 1977 and at Washington State University College of Veterinary Medicine from 1977 to 1987, where he was professor and chair of the Department of Microbiology and Pathology. From 1984 to 1987 he was deputy director of the Washington Technology Center, the state's high-technology sciences initiative, based in the College of Engineering at the University of Washington. In 1987, he was appointed director of the U.S. Department of Agriculture's (USDA's) Plum Island Animal Disease Center, a Biosafety Level 3 facility for research and diagnosis of the world's most dangerous livestock diseases. In that role he initiated research into the genomic and functional genomic basis of disease pathogenesis, diagnosis, and control of livestock RNA and DNA virus infections. This work became the basis of U.S. defense against natural and deliberate infection with these agents and led to his involvement in the early 1990s in biological weapons defense and proliferation prevention. From 1995 to 1998, he directed research programs in 20 laboratories in the Southeast for the USDA Agricultural Research Service before going to Washington, DC, to establish biological weapons defense research programs for USDA. He received the Distinguished Executive Award from President Clinton in 1998 for his work at Plum Island and in biodefense. Since 2004 he has been chief executive officer of Centaur Science Group, which provides consulting services in biodefense. His main commitment is to the Defense Threat Reduction Agency's Biological Weapons Proliferation Prevention Program in Europe, the Caucasus, and Central Asia.

Steven J. Brickner, Ph.D., is a research fellow in antibacterials chemistry at Pfizer Global Research and Development in Groton, Connecticut. He graduated from Miami University (Ohio) with a B.S. in chemistry with honors and received his M.S. and Ph.D. degrees in organic chemistry from Cornell University. He was an NIH postdoctoral research fellow at the University of Wisconsin, Madison. Dr. Brickner is a medicinal chemist with 25 years of research experience in the pharmaceutical industry, all focused on the discovery of novel antibacterial agents. He is an inventor or co-inventor on 21 U.S. patents and has published numerous scientific papers in the areas of oxazolidinones and novel azetidinones. Dr. Brickner has been a member of the IOM Forum on Microbial Threats since 1997 and is a member of the editorial advisory board for *Current Pharmaceutical Design*. Dr.

Brickner initiated the oxazolidinone research program at Upjohn, led the team that discovered Zyvox® (linezolid), and is a co-inventor of this antibiotic used to treat multidrug-resistant gram-positive infections. Zyvox is the first member of *any* entirely new class of antibiotics to reach the market in more than 35 years since the quinolones. He is a co-recipient of the 2007 American Chemical Society's Team Innovation Award and the Pharmaceutical Research and Manufacturers of America's 2007 Discoverers Award. He was named the 2002-2003 Outstanding Alumni Lecturer, College of Arts and Science, Miami University (Ohio).

Gail H. Cassell, Ph.D., is currently vice president, Scientific Affairs, and Distinguished Lilly Research Scholar for Infectious Diseases, Eli Lilly and Company, in Indianapolis, Indiana. She is the former Charles H. McCauley Professor and chairman of the Department of Microbiology at the University of Alabama Schools of Medicine and Dentistry at Birmingham, a department that ranked first in research funding from NIH during her decade of leadership. She obtained her B.S. from the University of Alabama in Tuscaloosa and in 1993 was selected as one of the top 31 female graduates of the twentieth century. She obtained her Ph.D. in microbiology from the University of Alabama at Birmingham and was selected as its 2003 Distinguished Alumnus. She is a past president of the American Society for Microbiology (the oldest and single-largest life sciences organization, with a membership of more than 42,000). She was a member of the NIH Director's Advisory Committee and a member of the Advisory Council of the National Institute of Allergy and Infectious Diseases (NIAID) of NIH. She was named to the original Board of Scientific Councilors of the CDC Center for Infectious Diseases and served as chair of the board. She recently served a three-year term on the Advisory Board of the director of the CDC and as a member of the HHS secretary's Advisory Council of Public Health Preparedness. Currently she is a member of the Science Board of the FDA Advisory Committee to the Commissioner. Since 1996 she has been a member of the U.S.-Japan Cooperative Medical Science Program responsible for advising the respective governments on joint research agendas (U.S. State Department-Japan Ministry of Foreign Affairs). She has served on several editorial boards of scientific journals and has authored more than 250 articles and book chapters. Dr. Cassell has received national and international awards and an honorary degree for her research in infectious diseases. She is a member of the IOM and is currently serving a three-year term on the IOM Council, its governing board. Dr. Cassell has been intimately involved in the establishment of science policy and legislation related to biomedical research and public health. For nine years she was chairman of the Public and Scientific Affairs Board of the American Society for Microbiology; she has served as an adviser on infectious diseases and indirect costs of research to the White House Office of Science and Technology Policy (OSTP); and she has been an invited participant in numerous congressional hearings and briefings related to infectious diseases, antimicrobial resistance, and biomedical research. She has served two terms on the Liaison Committee for Medical Education (LCME), the accrediting body for U.S. medical schools, as well as other national committees involved in establishing policies for training in the biomedical sciences. She has just completed a term on the Leadership Council of the School of Public Health of Harvard University. Currently she is a member of the Executive Committee of the Board of Visitors of Columbia University School of Medicine, the Board of Directors of the Burroughs Wellcome Fund, and the Advisory Council of the School of Nursing of Johns Hopkins.

Bill Colston, Ph.D., is division leader for the Chemical and Biological Countermeasures (CB) Division for the Global Security (GS) Principal Directorate at Lawrence Livermore National Laboratory. The newly formed CB Division is comprised of about 190 scientists from a variety of disciplines. The mission of this division is to provide national policy support, threat characterization, biological detection, chemical and explosives detection, instrumentation and systems development, decontamination and restoration, forensics and attribution, Biodefense Knowledge Center products, and incident response support operations. Prior to this assignment he held the positions of founding director of the Department of Homeland Security (DHS) Biodefense Knowledge Center (BKC) and deputy program leader for the Chemical and Biological Security Program. Dr. Colston holds a Ph.D. from the University of California, Davis, in biomedical engineering. He has published more than 40 publications in the scientific literature, holds more than 15 patents related to medical diagnostics and imaging devices, and has received three different research and development (R&D) 100 Awards. His research interests are focused mainly on molecular characterization of infectious disease, with direct relevance to new diagnostic devices.

Col. Ralph (Loren) Erickson, M.D., M.P.H., Dr.P.H., is the director of the Department of Defense Global Emerging Infections Surveillance and Response System (DOD-GEIS) headquartered in Silver Spring, Maryland. He holds a B.S. degree in chemistry from the University of Washington, an M.D. from the Uniformed Services University of the Health Sciences, an M.P.H. from Harvard University, and a Dr.P.H. from Johns Hopkins University. Residency trained and board certified in preventive medicine, Dr. Erickson has held a number of leadership positions within the Army Medical Department, including director of the General Preventive Medicine Residency Program, Walter Reed Army Institute of Research; director, Epidemiology and Disease Surveillance, U.S. Army Center for Health Promotion and Preventive Medicine; commander of the U.S. Army Center for Health Promotion and Preventive Medicine (Europe); and specialty leader for all U.S. Army preventive medicine physicians.

Mark Feinberg, M.D., Ph.D., is vice president for medical affairs and policy in global vaccine and infectious diseases at Merck & Co., Inc., and is responsible for global efforts to implement vaccines to achieve the greatest health benefits, including efforts to expand access to new vaccines in the developing world. Dr. Feinberg received a bachelor's degree magna cum laude from the University of Pennsylvania in 1978 and his M.D. and Ph.D. degrees from Stanford University School of Medicine in 1987. His Ph.D. research at Stanford was supervised by Dr. Irving Weissman and included time spent studying the molecular biology of the human retroviruses—HTLV-I (human T-cell lymphotrophic virus, type I) and HIV—as a visiting scientist in the laboratory of Dr. Robert Gallo at the National Cancer Institute. From 1985 to 1986, Dr. Feinberg served as a project officer for the IOM Committee on a National Strategy for AIDS. After receiving his M.D. and Ph.D. degrees, Dr. Feinberg pursued postgraduate residency training in internal medicine at the Brigham and Women's Hospital of Harvard Medical School and postdoctoral fellowship research in the laboratory of Dr. David Baltimore at the Whitehead Institute for Biomedical Research. From 1991 to 1995, Dr. Feinberg was an assistant professor of medicine and microbiology and immunology at the University of California, San Francisco (UCSF), where he also served as an attending physician in the AIDS-oncology division and as director of the virology research laboratory at San Francisco General Hospital. From 1995 to 1997, Dr. Feinberg was a medical officer in the Office of AIDS Research in the Office of the Director of NIH, the chair of the NIH Coordinating Committee on AIDS Etiology and Pathogenesis Research, and an attending physician at the NIH Clinical Center. During this period, he also served as executive secretary of the NIH Panel to Define Principles of Therapy of HIV Infection. Prior to joining Merck in 2004, Dr. Feinberg served as professor of medicine and microbiology and immunology at the Emory University School of Medicine, as an investigator at the Emory Vaccine Center, and as an attending physician at Grady Memorial Hospital. At UCSF and Emory, Dr. Feinberg and colleagues were engaged in the preclinical development and evaluation of novel vaccines for HIV and other infectious diseases and in basic research studies focused on revealing fundamental aspects of the pathogenesis of AIDS. Dr. Feinberg also founded and served as the medical director of the Hope Clinic of the Emory Vaccine Center—a clinical research facility devoted to the clinical evaluation of novel vaccines and to translational research studies of human immune system biology. In addition to his other professional roles, Dr. Feinberg has also served as a consultant to, and a member of, several IOM and NAS committees. Dr. Feinberg currently serves as a member of the National Vaccine Advisory Committee (NVAC) and is a member of the Board of Trustees of the National Foundation for Infectious Diseases (NFID). Dr. Feinberg has earned board certification in internal medicine; he is a fellow of the American College of Physicians, a member of the Association of American Physicians, and the recipient of an Elizabeth Glaser Scientist Award from the Pediatric AIDS Foundation and an Innovation in Clinical Research Award from the Doris Duke Charitable Foundation.

J. Patrick Fitch, Ph.D., is laboratory director for the National Biodefense Analysis and Countermeasures Center (NBACC) and president of Battelle National Biodefense Institute, LLC (BNBI). BNBI manages and operates the NBACC national laboratory for the Department of Homeland Security as a Federally Funded Research and Development Center (FFRDC) established in 2006. NBACC's mission is to provide the nation with the scientific basis for awareness of biological threats and attribution of their use against the American public. Dr. Fitch joined Battelle in 2006 as vice president for Biodefense Programs after more than 20 years of experience leading multidisciplinary applied science teams at the University of California's Lawrence Livermore National Laboratory (LLNL). From 2001 to 2006, he led the LLNL Chemical and Biological National Security Program (CBNP), with applied science programs from pathogen biology to deployed systems. CBNP accomplishments include performing more than 1 million assays on national security samples; setting up and operating 24/7 reach-back capabilities; setting up a nationwide bioalert system; receiving three R&D 100 awards; designing signatures for validated assays in the CDC Laboratory Response Network and the National Animal Health Laboratory Network; and designing. His advisory board activities have included the U.S. Animal Health Association, Texas A&M University DHS Center of Excellence, Central Florida University (College of Engineering), Colorado State University (College of Engineering), California State Breast Cancer Research Program, and *Biomolecular Engineering*. Dr. Fitch was a fellow of the American Society for Laser Medicine and Surgery and an associate editor of *Circuits, Systems and Signal Processing*. He has received two national awards for medical devices, a technical writing award for an article in *Science*, and an international best paper award from the Institute of Electrical and Electronics Engineers. He also co-invented the technology, developed the initial business plan, and successfully raised venture investments for a medical device start-up company. Dr. Fitch received his Ph.D. from Purdue University and B.S. from Loyola College of Maryland.

Capt. Darrell R. Galloway, M.S.C., Ph.D., is chief of the Medical Science and Technology Division for the Chemical and Biological Defense Directorate at the Defense Threat Reduction Agency. He received his baccalaureate degree in microbiology from California State University in Los Angeles in 1973. After completing military service in the U.S. Army as a medical corpsman from 1969 to 1972, Captain Galloway entered graduate school and completed a doctoral degree in biochemistry in 1978 from the University of California, followed by two years of postgraduate training in immunochemistry as a fellow of the National Cancer Institute (NCI) at the Scripps Clinic and Research Foundation in La Jolla, California. Captain Galloway began his Navy career at the Naval Medical Research Institute in Bethesda, Maryland, where he served as a research scientist working on vaccine development from 1980 to 1984. In late 1984,

Captain Galloway left active service to pursue an academic appointment at Ohio State University, where he is now a tenured faculty member in the Department of Microbiology. He also holds appointments at the University of Maryland Biotechnology Institute and the Uniformed Services University of the Health Sciences. He has an international reputation in the area of bacterial toxin research and has published more than 50 research papers on various studies of bacterial toxins. In recent years, Captain Galloway's research has concentrated on anthrax and the development of DNA-based vaccine technology. His laboratory has contributed substantially to the development of a new DNA-based vaccine against anthrax that has completed the first phase of clinical trials. Captain Galloway is a member of the ASM and has served as president of the Ohio branch of that organization. He received an NIH Research Career Development Award. In 2005, Captain Galloway was awarded the Joel M. Dalrymple Award for significant contributions to biodefense vaccine development.

S. Elizabeth George, Ph.D., is deputy director, Biological Countermeasures Portfolio Science and Technology Directorate, Department of Homeland Security. Until it merged into the new department in 2003, she was program manager of the Chemical and Biological National Security Program in the Department of Energy's National Nuclear Security Administration's Office of Nonproliferation Research and Engineering. Significant accomplishments include the design and deployment of BioWatch, the nation's first civilian biological threat agent monitoring system, and PROTECT, the first civilian operational chemical detection and response capability deployed in the Washington, DC area subway system. Previously, she spent 16 years at the U.S. Environmental Protection Agency (EPA), Office of Research and Development, National Health and Ecological Effects Research Laboratory, Environmental Carcinogenesis Division, where she was branch chief of the Molecular and Cellular Toxicology Branch. She received her B.S. in biology in 1977 from Virginia Polytechnic Institute and State University and her M.S. and Ph.D. in microbiology in 1979 and 1984, respectively, from North Carolina State University. From 1984 to 1986, she was a National Research Council (NRC) fellow in the laboratory of Dr. Larry Claxton at EPA. Dr. George is the 2005 chair of the Chemical and Biological Terrorism Defense Gordon Research Conference. She has served as councillor for the Environmental Mutagen Society and president and secretary of the Genotoxicity and Environmental Mutagen Society. She holds memberships in the ASM and the AAAS and is an adjunct faculty member in the School of Rural Public Health, Texas A&M University. She is a recipient of the EPA Bronze Medal and Scientific and Technological Achievement Awards and the DHS Under Secretary's Award for Science and Technology. She is author of numerous journal articles and has presented her research at national and international meetings.

Jesse L. Goodman, M.D., M.P.H., is director of the FDA's Center for Biologics Evaluation and Research (CBER), which oversees medical, public health, and policy activities concerning the development and assessment of vaccines, blood products, tissues, and related devices and novel therapeutics, including cellular and gene therapies. He moved to the FDA full-time in 2001 from the University of Minnesota, where he was professor of medicine and director of the Division of Infectious Diseases. A graduate of Harvard College, he received his M.D. from the Albert Einstein College of Medicine; did residency and fellowship training at the Hospital of the University of Pennsylvania and at the University of California, Los Angeles (UCLA), where he was also chief medical resident; and is board certified in internal medicine, oncology, and infectious diseases. He trained in the virology laboratory of Jack Stevens at UCLA and has had an active laboratory program in the molecular pathogenesis of infectious diseases. In 1995, his laboratory isolated the etiologic agent of human granulocytic ehrlichiosis (HGE) and subsequently characterized fundamental events involved in the infection of leukocytes, including their cellular receptors. He is editor of the book *Tick Borne Diseases of Humans* published by ASM Press in 2005 and is a staff physician and infectious diseases consultant at the NIH Clinical Center and the National Naval Medical Center-Walter Reed Army Medical Center, as well as adjunct professor of medicine at the University of Minnesota. He is active in a wide variety of clinical, public health, and product development issues, including pandemic and emerging infectious disease threats; bioterrorism preparedness and response; and blood, tissue, and vaccine safety and availability. In these activities, he has worked closely with CDC, NIH, and other HHS components, academia, and the private sector, and he has put into place an interactive team approach to emerging threats. This model was used in the collaborative development and rapid implementation of nationwide donor screening of the U.S. blood supply for West Nile virus. He has been elected to the American Society for Clinical Investigation (ASCI) and to the IOM.

Eduardo Gotuzzo, M.D., is principal professor and director at the Instituto de Medicina Tropical Alexander von Humbolt, Universidad Peruana Cayetan Heredia in Lima, Peru, as well as chief of the Department of Infectious and Tropical Diseases at the Cayetano Heredia Hospital. He is also an adjunct professor of medicine at the University of Alabama, Birmingham, School of Medicine. Dr. Gotuzzo is an active member of numerous international societies and has been president of the Latin America Society of Tropical Disease (2000-2003), the IDSA Scientific Program (2000-2003), the International Organizing Committee of the International Congress of Infectious Diseases (1994 to present), president-elect of the International Society for Infectious Diseases (1996-1998), and president of the Peruvian Society of Internal Medicine (1991-1992). He has published more than 230 articles and chapters as well as six manuals and one book. Recent honors and awards include being named an honorary member of the American

Society of Tropical Medicine and Hygiene in 2002, an associate member of the National Academy of Medicine in 2002, an honorary member of the Society of Internal Medicine in 2000, and a distinguished visitor at the Faculty of Medical Sciences, University of Cordoba, Argentina, in 1999. In 1988 he received the Golden Medal for Outstanding Contribution in the Field of Infectious Diseases awarded by Trnava University, Slovakia.

Jo Handelsman, Ph.D., is a Howard Hughes Medical Institute professor in the Departments of Bacteriology and Plant Pathology and chair of the Department of Bacteriology at the University of Wisconsin, Madison (UW, Madison). She received her Ph.D. in molecular biology from the UW, Madison, in 1984 and joined the faculty of UW, Madison, in 1985. Her research focuses on the genetic and functional diversity of microorganisms in soil and insect gut communities. She is one of the pioneers of functional metagenomics, an approach to accessing the genetic potential of unculturable bacteria in environmental samples. In addition to her research program, Dr. Handelsman is nationally known for her efforts to improve science education and increase the participation of women and minorities in science at the university level. She co-founded the Women in Science and Engineering Leadership Institute at UW, Madison, which has designed and evaluated interventions intended to enhance the participation of women in science. Her leadership in women in science led to her appointment as the first president of the Rosalind Franklin Society and her service on the National Academies' panel that wrote the 2006 report *Beyond Bias and Barriers: Fulfilling the Potential of Women in Academic Science and Engineering*, which documented the issues of women in science and recommended changes to universities and federal funding agencies. In addition to more than 100 scientific research publications, Dr. Handelsman is co-author of two books about teaching: *Entering Mentoring* and *Scientific Teaching*. Dr. Handelsman is the editor-in-chief of *DNA and Cell Biology*, and the series *Controversies in Science and Technology*, and a member of the National Academy of Sciences' Board on Life Sciences and the Institute of Medicine's Forum on Microbial Threats. She is a National Academies mentor in the life sciences, a fellow in the American Academy of Microbiology and AAAS, director of the Wisconsin Program for Scientific Teaching, and co-director of The National Academies' Summer Institutes on Undergraduate Education in Biology. In 2008, she received the Alice Evans Award from the American Society for Microbiology in recognition of her mentoring. In 2009, she received the Carski Award from the American Society for Microbiology in recognition of her teaching contributions, and she was named "A Revolutionary Mind" by *Seed Magazine* in recognition of her unorthodox ideas.

Carole A. Heilman, Ph.D., is the director of the Division of Microbiology and Infectious Diseases (DMID), at the National Institute of Allergy and Infectious Diseases, a component of NIH-HHS. As director of DMID she has responsibility for scientific direction, oversight, and management of all extramural research programs on infectious diseases (except AIDS) within NIH. In addition, since 2001 Dr. Heilman has played a critical role in launching and directing NIAID's extramural biodefense research program. Previously, Dr. Heilman served as deputy director of NIAID's Division of AIDS for three years. Dr. Heilman has a Ph.D. in microbiology from Rutgers University. She did her postdoctoral work in molecular virology at the National Cancer Institute and continued at the NCI as a senior staff fellow in molecular oncology. She moved into health science administration in 1986, focusing on respiratory pathogens, particularly vaccine development. She has received numerous awards for scientific management and leadership, including three HHS Secretary's Awards for Distinguished Service for her contributions to developing pertussis, biodefense, and AIDS vaccines.

David L. Heymann, M.D., is currently assistant director-general for Health Security Environment and representative of the director-general for Polio Eradication at the World Health Organization (WHO). Prior to that, from July 1998, Dr. Heymann was executive director and then assistant director-general of the WHO Communicable Diseases Cluster, which includes WHO's programmes on infectious and tropical diseases, and from which the public health response to severe acute respiratory syndrome (SARS) was mounted in 2003. From October 1995 to July 1998, Dr. Heymann was director of the WHO Programme on Emerging and other Communicable Diseases, and prior to that was the chief of research activities in the WHO Global Programme on AIDS. Before joining WHO, Dr. Heymann worked for 13 years as a medical epidemiologist in sub-Saharan Africa (Cameroon, Côte d'Ivoire, Malawi, and the Democratic Republic of Congo—formerly Zaire) on assignment from the U.S. Centers for Disease Control and Prevention (CDC) in CDC-supported activities. These activities aimed at strengthening capacity in surveillance of infectious diseases and their control, with special emphasis on the childhood immunizable diseases including measles and polio, African hemorrhagic fevers, poxviruses, and malaria. While based in Africa, Dr. Heymann participated in the investigation of the first outbreak of Ebola in Yambuku (former Zaire) in 1976, then again investigated the second outbreak of Ebola in 1977 in Tandala, and in 1995 directed the international response to the Ebola outbreak in Kikwit. Prior to these 13 years in Africa, Dr. Heymann worked 2 years in India as a medical epidemiologist in the WHO Smallpox Eradication Programme. Dr. Heymann holds a B.A. from the Pennsylvania State University, an M.D. from Wake Forest University, a diploma in tropical medicine and hygiene from the London School of Hygiene and Tropical Medicine, and has completed practical epidemiology training in the two-year Epidemic Intelligence Service (EIS) of CDC. He is a recipient of the Heinz Award for the Human Condition in Global Public Health, of the American Public Health Association Award for Excellence, and of the American Society of Tropical Medicine and Hygiene Donald MacKay medal; he is a member of the U.S. Institute of Medicine. He received the Outstanding Award of the American Public Health Association and the Outstanding Science Alumni Award from the Eberly College of Science of Pennsylvania State University. Dr. Heymann has published more than 149 scientific articles on infectious diseases and related issues in medical and scientific journals, and authored several chapters on infectious diseases in medical textbooks. He is currently editor of the nineteenth edition of the *Control of Communicable Diseases Manual*, a joint publication of WHO and the American Public Health Association.

Phil Hosbach is vice president, New Products and Immunization Policy, at sanofi pasteur. The areas under his supervision are new product marketing, state and federal government policy, business intelligence, bids and contracts, medical communications, public health sales, and public health marketing. His current responsibilities include oversight of immunization policy development. He acts as sanofi pasteur's principal liaison with CDC. Mr. Hosbach graduated from Lafayette College in 1984 with a degree in biology. He has 20 years of pharmaceutical industry experience, including the past 17 years focused solely on vaccines. He began his career at American Home Products in clinical research in 1984. He joined Aventis Pasteur (then Connaught Labs) in 1987 as clinical research coordinator and has held research and development positions of increasing responsibility, including clinical research manager and director of clinical operations. Mr. Hosbach also served as project manager for the development and licensure of Tripedia, the first diphtheria, tetanus, and acellular pertussis (DTaP) vaccine approved by the FDA for use in U.S. infants. During his clinical research career at Aventis Pasteur, he contributed to the development and licensure of seven vaccines, and he has authored or co-authored several clinical research articles. From 2000 through 2002, Mr. Hosbach served on the board of directors for Pocono Medical Center in East Stroudsburg, Pennsylvania. Since 2003 he has served on the board of directors of Pocono Health Systems, which includes Pocono Medical Center.

James M. Hughes, M.D., is professor of medicine and public health at Emory University's School of Medicine and Rollins School of Public Health, serving as director of the Emory Program in Global Infectious Diseases, associate director of the Southeastern Center for Emerging Biological Threats, and senior adviser to the Emory Center for Global Safe Water. He also serves as senior scientific adviser for infectious diseases to the International Association of National Public Health Institutes funded by the Bill and Melinda Gates Foundation. Prior to joining Emory in June 2005, Dr. Hughes served as director of the National Center for Infectious Diseases at CDC. Dr. Hughes received his B.A. and M.D. degrees from Stanford University and completed postgraduate training in internal medicine at the University of Washington, infectious diseases at the University of Virginia, and preventive medicine at CDC. After joining CDC as an EIS officer in 1973, Dr.

Hughes worked initially on foodborne and waterborne diseases and subsequently on infection control in healthcare settings. He served as director of CDC's Hospital Infections Program from 1983 to 1988, as deputy director of NCID from 1988 to 1992, and as director of NCID from 1992 to 2005. A major focus of Dr. Hughes' career has been on building partnerships among the clinical, research, public health, and veterinary communities to prevent and respond to infectious diseases at the national and global levels. His research interests include emerging and reemerging infectious diseases; antimicrobial resistance; foodborne diseases; healthcare-associated infections; vectorborne and zoonotic diseases; rapid detection of and response to infectious diseases and bioterrorism; strengthening public health capacity at the local, national, and global levels; and prevention of water-related diseases in the developing world. Dr. Hughes is a fellow of the American Association for the Advancement of Science, the American College of Physicians, and the Infectious Diseases Society of America, a member of IOM, and a councillor of the American Society of Tropical Medicine and Hygiene.

Stephen A. Johnston, Ph.D., is currently director of the Center for Innovations in Medicine in the Biodesign Institute at Arizona State University. His center focuses on formulating and implementing disruptive technologies for basic problems in health care. The center has three divisions: Genomes to Vaccines, Cancer Eradication, and DocInBox. Genomes to Vaccines has developed high-throughput systems to screen for vaccine candidates and is applying them to predict and produce chemical vaccines. The Cancer Eradication group is working on formulating a universal prophylactic vaccine for cancer. DocInBox is developing technologies to facilitate presymptomatic diagnosis. Dr. Johnston founded the Center for Biomedical Inventions (also known as the Center for Translation Research) at the University of Texas-Southwestern, the first center of its kind in the medical arena. He and his colleagues have developed numerous inventions and innovations, including the gene gun, genetic immunization, TEV (tobacco etch virus) protease system, organelle transformation, digital optical chemistry arrays, expression library immunization, linear expression elements, and others. He also was involved in transcription research for years, first cloning *Gal4*, then later discovering functional domains in transcription factors and the connection of the proteasome to transcription. He has been professor at the University of Texas Southwestern Medical Center at Dallas and associate and assistant professor at Duke University. He has been involved in several capacities as an adviser on biosecurity since 1996 and is a member of the WRCE SAB and a founding member of BioChem 20/20.

Gerald T. Keusch, M.D., is associate provost and associate dean for global health at Boston University and Boston University School of Public Health. He is a graduate of Columbia College (1958) and Harvard Medical School (1963). After completing a residency in internal medicine, fellowship training in infectious diseases, and two years as an NIH research associate at the Southeast Asia Treaty Organization (SEATO) Medical Research Laboratory in Bangkok, Thailand, Dr. Keusch joined the faculty of the Mt. Sinai School of Medicine in 1970, where he established a laboratory to study the pathogenesis of bacillary dysentery and the biology and biochemistry of Shiga toxin. In 1979 he moved to Tufts Medical School and New England Medical Center in Boston to found the Division of Geographic Medicine, which focused on the molecular and cellular biology of tropical infectious diseases. In 1986 he integrated the clinical infectious diseases program into the Division of Geographic Medicine and Infectious Diseases, continuing as division chief until 1998. He has worked in the laboratory and in the field in Latin America, Africa, and Asia on basic and clinical infectious diseases and HIV/AIDS research. From 1998 to 2003, he was associate director for international research and director of the Fogarty International Center at NIH. Dr. Keusch is a member of ASCI, the Association of American Physicians, the ASM, and the IDSA. He has received the Squibb (1981), Finland (1997), and Bristol (2002) awards of the IDSA. In 2002 he was elected to the IOM.

Rima F. Khabbaz, M.D., is director of the National Center for Preparedness, Detection, and Control of Infectious Diseases at CDC. She became director of the National Center for Infectious Diseases at CDC in December 2005 and led its transition to the current centers. She is a graduate of the American University of Beirut, Lebanon, where she obtained both her bachelor's degree in science and her medical doctorate degree. She trained in internal medicine and completed a fellowship in infectious diseases at the University of Maryland in Baltimore. She is also a clinical associate professor of medicine (infectious diseases) at Emory University. She began her CDC career in 1980 as an epidemic intelligence service officer in the Hospital Infections Program. She later served as a medical epidemiologist in CDC's Retrovirus Diseases Branch, where she made major contributions to defining the epidemiology of non-HIV retroviruses (HTLV-I and HTLV-II) in the United States and developing guidance for counseling HTLV-infected persons. Following the hantavirus pulmonary syndrome outbreak in the southwestern United States in 1993, she led CDC's efforts to set up national surveillance for the syndrome. Prior to becoming director of NCID, she was acting deputy director and, before that, associate director for epidemiologic science, NCID. Additional positions held at CDC include associate director for science and deputy director of the Division of Viral and Rickettsial Diseases. She played a leading role in developing CDC's blood safety programs and its food safety programs related to viral diseases. She also had a key role in CDC's responses to outbreaks of new and/or reemerging viral infections including Nipah, Ebola, West Nile, SARS, and monkeypox. She led CDC's field team to the nation's capital during the public health response to the anthrax attack of 2001. She is a fellow of the Infectious Diseases Society of America, a member of the American Epidemiologic Society, the American Society for Microbiology, and the Council of State and Territorial

Epidemiologists. She served on FDA's Blood Product Advisory Committee and on its Transmissible Spongiform Encephalopathy Advisory Committee. She also served on IDSA's Annual Meeting Scientific Program Committee and serves on the society's National and Global Public Health Committee. She is a graduate of the National Preparedness Leadership Initiative at Harvard University and of the Public Health Leadership Institute at the University of North Carolina.

Lonnie J. King, D.V.M., is currently director of CDC's new National Center for Zoonotic, Vector-Borne, and Enteric Diseases (NCZVED). Dr. King leads the center's activities for surveillance, diagnostics, disease investigations, epidemiology, research, public education, policy development, and disease prevention and control programs. NCZVED also focuses on waterborne, foodborne, vectorborne, and zoonotic diseases of public health concern, which also include most of CDC's select and bioterrorism agents, neglected tropical diseases, and emerging zoonoses. Before serving as director, he was the first chief of the agency's Office of Strategy and Innovation. In 1996, Dr. King was appointed dean of the College of Veterinary Medicine, Michigan State University. He served for 10 years as dean of the college. As dean, he was the chief executive officer for academic programs, research, the teaching hospital, diagnostic center for population and animal health, basic and clinical science departments, and outreach and continuing education programs. As dean and professor of large animal clinical sciences, Dr. King was instrumental in obtaining funds for construction of the $60 million Diagnostic Center for Population and Animal Health, initiated the Center for Emerging Infectious Diseases in the college, served as the campus leader in food safety, and had oversight for the National Food Safety and Toxicology Center. He brought the Center for Integrative Toxicology to the college and was the university's designated leader for counterbioterrorism activities for his college. Prior to this, Dr. King was administrator for USDA's Animal and Plant Health Inspection Service (APHIS). Dr. King served as the country's chief veterinary officer for five years and worked extensively in global trade agreements within the North American Free Trade Agreement and the World Trade Organization. Before beginning his government career in 1977, he was in private veterinary practice for seven years in Dayton, Ohio, and in Atlanta, Georgia. He received his B.S. and D.V.M. from Ohio State University in 1966 and 1970, respectively. He earned his M.S. in epidemiology from the University of Minnesota while on special assignment with the U.S. Department of Agriculture in 1980. He received his master's in public administration from the American University in Washington, DC in 1991. Dr. King has a broad knowledge of animal agriculture and the veterinary profession through his work with other government agencies, universities, major livestock and poultry groups, and private practitioners. Dr. King is a board-certified member of the American College of Veterinary Preventive Medicine and has completed the senior executive fellowship program at Harvard University. He served as president of the Association of American Veterinary Medical Colleges from 1999 to 2000 and was vice chair for the National Commission on Veterinary Economic Issues from 2000 to 2004. Dr. King helped start the National Alliance for Food Safety, served on the Governor's Task Force on Chronic Wasting Disease for the State of Michigan, and was a member of four NAS committees; most recently, he chaired the National Academies Committee on Assessing the Nation's Framework for Addressing Animal Diseases. Dr. King is one of the developers of the Science, Politics, and Animal Health Policy Fellowship Program, and he lectures extensively on the future of animal health, emerging zoonoses, and veterinary medicine. He served as a consultant and member of the Board of Scientific Counselors to CDC's National Center for Infectious Diseases and is a member of the IOM's Forum on Microbial Threats. Dr. King was an editor for the OIE (World Organisation for Animal Health) *Scientific Review on Emerging Zoonoses*, is a current member of FDA's Board of Scientific Advisors, and is president of the American Veterinary Epidemiology Society. Dr. King was elected to the IOM in 2004.

Col. George W. Korch, Ph.D., is commander, U.S. Army Medical Research Institute for Infectious Diseases, Ft. Detrick, Maryland. Dr. Korch attended Boston University and earned a B.S. in biology in 1974, followed by postgraduate study in mammalian ecology at the University of Kansas from 1975 to 1978. He earned his Ph.D. from the Johns Hopkins School of Hygiene and Public Health in immunology and infectious diseases in 1985, followed by postdoctoral experience at Johns Hopkins from 1985 to 1986. His areas of training and specialty are the epidemiology of zoonotic viral pathogens and medical entomology. For the past 15 years, he has also been engaged in research and program management for medical defense against biological pathogens used in terrorism or warfare.

Stanley M. Lemon, M.D., is the John Sealy Distinguished University Chair and director of the Institute for Human Infections and Immunity at the University of Texas Medical Branch (UTMB) at Galveston. He received his undergraduate A.B. degree in biochemical sciences from Princeton University *summa cum laude* and his M.D. with honors from the University of Rochester. He completed postgraduate training in internal medicine and infectious diseases at the University of North Carolina at Chapel Hill and is board certified in both. From 1977 to 1983 he served with the U.S. Army Medical Research and Development Command, followed by a 14-year period on the faculty of the University of North Carolina School of Medicine. He moved to UTMB in 1997, serving first as chair of the Department of Microbiology and Immunology, then as dean of the School of Medicine from 1999 to 2004. Dr. Lemon's research interests relate to the molecular virology and pathogenesis of the positive-stranded RNA viruses responsible for hepatitis. He has had a long-standing interest in antiviral and vaccine development and has served as chair of FDA's Anti-Infective Drugs Advisory Committee. He is the past chair of the Steering Committee on Hepatitis and Poliomyelitis of the WHO Programme on Vaccine Development. He is past chair of the NCID-CDC Board of Scientific Counselors and currently serves as a member of the U.S. Delegation to the U.S.-Japan Cooperative Medical Sciences Program. He was co-chair of the NAS Committee on Advances in Technology and the Prevention of Their Application to Next Generation Biowarfare Threats, and he recently chaired an IOM study committee related to vaccines for the protection of the military against naturally occurring infectious disease threats.

Edward McSweegan, Ph.D., is a program officer at the National Institute of Allergy and Infectious Diseases. He graduated from Boston College with a B.S. in biology in 1978. He has an M.S. in microbiology from the University of New Hampshire and a Ph.D. in microbiology from the University of Rhode Island. He was an NRC associate from 1984 to 1986 and did postdoctoral research at the Naval Medical Research Institute in Bethesda, Maryland. Dr. McSweegan served as an AAAS diplomacy fellow in the U.S. State Department from 1986 to 1988 where he helped to negotiate science and technology agreements with Poland, Hungary, and the former Soviet Union. After moving to NIH, he continued to work on international health and infectious disease projects in Egypt, Israel, India, and Russia. Currently, he manages NIAID's bilateral program with India, the Indo-U.S. Vaccine Action Program, and he represents NIAID in the HHS Biotechnology Engagement Program with Russia and related countries. He is a member of AAAS, the ASM, and the National Association of Science Writers. He is the author of numerous journal and freelance articles.

Stephen S. Morse, Ph.D., is professor of epidemiology and founding director of the Center for Public Health Preparedness at the Mailman School of Public Health of Columbia University. He returned to Columbia in 2000 after four years in government service as program manager at the Defense Advanced Research Projects Agency (DARPA), where he co-directed the Pathogen Countermeasures Program and subsequently directed the Advanced Diagnostics Program. Before coming to Columbia, he was assistant professor of virology at the Rockefeller University in New York, where he remains an adjunct faculty member. He is the editor of two books, *Emerging Viruses* (Oxford University Press, 1993; paperback, 1996), which was selected by American Scientist for its list of 100 Top Science Books of the 20th Century, and *The Evolutionary Biology of Viruses* (Raven Press, 1994). He was a founding section editor of the CDC journal *Emerging Infectious Diseases* and was formerly an editor-in-chief of the Pasteur Institute's journal *Research in Virology*. Dr. Morse was chair and principal organizer of the 1989 NIAID-NIH Conference on Emerging Viruses, for which he originated the term and concept of emerging viruses-infections. He has served as a member of the IOM-NAS Committee on Emerging Microbial Threats to Health, chaired its Task Force on Viruses, and was a contributor to the resulting report *Emerging Infections* (1992). He was a member of the IOM's Committee on Xenograft

Transplantation, and he currently serves on the Steering Committee of the IOM's Forum on Emerging Infections (now the Forum on Microbial Threats). Dr. Morse also served as an adviser to WHO and several government agencies. He is a fellow of the New York Academy of Sciences and a past chair of its microbiology section, a fellow of the American Academy of Microbiology of the American College of Epidemiology, and an elected life member of the Council on Foreign Relations. He was the founding chair of ProMED, the nonprofit international Program to Monitor Emerging Diseases, and was one of the originators of ProMED-mail, an international network inaugurated by ProMED in 1994 for outbreak reporting and disease monitoring using the Internet. Dr. Morse received his Ph.D. from the University of Wisconsin, Madison.

Michael T. Osterholm, Ph.D., M.P.H., is director of the Center for Infectious Disease Research and Policy and director of the NIH-sponsored Minnesota Center for Excellence in Influenza Research and Surveillance at the University of Minnesota. He is also professor at the School of Public Health and adjunct professor at the Medical School. Previously, Dr. Osterholm was the state epidemiologist and chief of the acute disease epidemiology section for the Minnesota Department of Health. He has received numerous research awards from NIAID and CDC. He served as principal investigator for the CDC-sponsored Emerging Infections Program in Minnesota. He has published more than 300 articles and abstracts on various emerging infectious disease problems and is the author of the best-selling book *Living Terrors: What America Needs to Know to Survive the Coming Bioterrorist Catastrophe*. He is past president of the Council of State and Territorial Epidemiologists. He currently serves on the IOM Forum on Microbial Threats. He has also served on the IOM Committee to Ensure Safe Food from Production to Consumption, and on the IOM Committee on the Department of Defense Persian Gulf Syndrome Comprehensive Clinical Evaluation Program, and as a reviewer for the IOM report *Chemical and Biological Terrorism: Research and Development to Improve Civilian Medical Response*.

George Poste, Ph.D., D.V.M., is director of the Biodesign Institute and Del E. Webb Distinguished Professor of Biology at Arizona State University. From 1992 to 1999, he was chief science and technology officer and president, Research and Development, of SmithKline Beecham (SB). During his tenure at SB, he was associated with the successful registration of 29 drug, vaccine, and diagnostic products. He is chairman of Orchid Cellmark. He serves on the board of directors of Monsanto and Exelixis. He is a distinguished fellow at the Hoover Institution at Stanford University. He is a member of the Defense Science Board of the U.S. Department of Defense and of the IOM Forum on Microbial Threats. Dr. Poste is a board-certified pathologist, a fellow of the Royal Society, and a fellow of the Academy of Medical Sciences. He was awarded the rank of Commander of the British Empire by Queen Elizabeth II in 1999 for services to medicine and for the advancement of biotechnology. He has published more than 350 scientific papers; has co-edited 15 books on cancer, biotechnology, and infectious diseases; and serves on the editorial board of several technical journals.

John C. Pottage, Jr., M.D., has been vice president for Global Clinical Development in the Infectious Disease Medicine Development Center at GlaxoSmithKline since 2007. Previously he was senior vice president and chief medical officer at Achillion Pharmaceuticals in New Haven, Connecticut. Achillion is a small biotechnology company devoted to the discovery and development of medicines for HIV, hepatitis C virus (HCV), and resistant antibiotics. Dr. Pottage initially joined Achillion in May 2002. Prior to Achillion, Dr. Pottage was medical director of antivirals at Vertex Pharmaceuticals. During this time he also served as an associate attending physician at the Tufts New England Medical Center in Boston. From 1984 to 1998, Dr. Pottage was a faculty member at Rush Medical College in Chicago, where he held the position of associate professor, and also served as the medical director of the Outpatient HIV Clinic at Rush-Presbyterian-St. Luke's Medical Center. While at Rush, Dr. Pottage was the recipient of several teaching awards and is a member of the Mark Lepper Society. Dr. Pottage is a graduate of St. Louis University School of Medicine and Colgate University.

Gary A. Roselle, M.D., received his medical degree from the Ohio State University School of Medicine in 1973. He served his residency at the Northwestern University School of Medicine and his infectious diseases fellowship at the University of Cincinnati School of Medicine. He is program director for infectious diseases for the Department of Veterans Affairs Central Office in Washington, DC, as well as the chief of the medical service at the Cincinnati VA Medical Center. He is a professor of medicine in the Department of Internal Medicine, Division of Infectious Diseases, at the University of Cincinnati College of Medicine. Dr. Roselle serves on several national advisory committees. In addition, he is currently heading the Emerging Pathogens Initiative for the VA. He has received commendations from the under secretary for health for the VA and the secretary of veterans affairs for his work in the Infectious Diseases Program for the VA. He has been an invited speaker at several national and international meetings and has published more than 90 papers and several book chapters.

Janet Shoemaker is director of the ASM's Public Affairs Office, a position she has held since 1989. She is responsible for managing the legislative and regulatory affairs of this 42,000-member organization, the largest single biological science society in the world. She has served as principal investigator for a project funded by the National Science Foundation (NSF) to collect and disseminate data on the job market for recent doctorates in microbiology and has played a key role in ASM projects, including production of the ASM *Employment Outlook in the Microbiological Sciences* and *The Impact of Managed Care and Health System*

Change on Clinical Microbiology. Previously, she held positions as assistant director of public affairs for ASM; as ASM coordinator of the U.S.-U.S.S.R. Exchange Program in Microbiology, a program sponsored and coordinated by the NSF and the U.S. Department of State; and as a freelance editor and writer. She received her baccalaureate, cum laude, from the University of Massachusetts and is a graduate of the George Washington University programs in public policy and in editing and publications. She has served as commissioner to the Commission on Professionals in Science and Technology and as ASM representative to the ad hoc Group for Medical Research Funding, and she is a member of Women in Government Relations, the American Society of Association Executives, and AAAS. She has co-authored published articles on research funding, biotechnology, biological weapons control, and public policy issues related to microbiology.

P. Frederick Sparling, M.D., is the J. Herbert Bate Professor Emeritus of Medicine, Microbiology, and Immunology at the University of North Carolina (UNC) at Chapel Hill, and professor of medicine, Duke University. He is director of the North Carolina Sexually Transmitted Infections Research Center and also the Southeast Regional Centers of Excellence in Biodefense and Emerging Infections. Previously he served as chair of the Department of Medicine and chair of the Department of Microbiology and Immunology at UNC. He was president of the Infectious Diseases Society of America from 1996 to 1997. He was also a member of the IOM Committee on Microbial Threats to Health (1990-1992) and the IOM Committee on Emerging Microbial Threats to Health in the 21st Century (2001-2003). Dr. Sparling's laboratory research has been on the molecular biology of bacterial outer membrane proteins involved in pathogenesis, with a major emphasis on gonococci and meningococci. His work helped to define the genetics of antibiotic resistance in gonococci and the role of iron-scavenging systems in the pathogenesis of human gonorrhea.

Brian Staskawicz, Ph.D., is professor and chair, Department of Plant and Microbial Biology, University of California, Berkeley. Dr. Staskawicz received his B.A. in biology from Bates College in 1974 and his Ph.D. from the University of California, Berkeley, in 1980. Dr. Staskawicz's work has contributed greatly to understanding the molecular interactions between plants and their pathogens. He was elected to the NAS in 1998 for elucidating the mechanisms of disease resistance, as his lab was the first to clone a bacterial effector gene from a pathogen and among the first to clone and characterize plant disease resistance genes. Dr. Staskawicz's research focuses on the interaction of the bacteria *Pseudomonas* and *Xanthomonas* with *Arabidopsis*, tomato, and pepper. He has published extensively in this area and is one of the leading scientists in the world working on elucidating the molecular basis of plant innate immunity.

Terence Taylor is director of the Global Health and Security Initiative and president and director of the International Council for the Life Sciences (ICLS). He is responsible for the overall direction of the ICLS and its programs, which have the goal of enhancing global biosafety and biosecurity. From 1995 to 2005, he was assistant director of the International Institute for Strategic Studies (IISS), a leading independent international institute, and president and executive director of its U.S. office (2001-2005). He studies international security policy, risk analysis, and scientific and technological developments and their impact on political and economic stability worldwide. He was one of IISS's leading experts on issues associated with nuclear, biological, and chemical weapons and their means of delivery. In his previous appointments, he has had particular responsibilities for issues affecting public safety and security in relation to biological risks and advances in the life sciences. He was one of the commissioners to the United Nations Special Commission on Iraq, for which he also conducted missions as a chief inspector. He was a science fellow at the Center for International Security and Cooperation at Stanford University, where he carried out, among other subjects, studies of the implications for government and industry of the weapons of mass destruction treaties and agreements. He has also carried out consultancy work for the International Committee of the Red Cross (ICRC) on the implementation and development of the laws of armed conflict and serves as a member of the editorial board of the *ICRC Review*. He has served as chairman of the World Federation of Scientists' Permanent Monitoring Panel on Risk Analysis. He was a career officer in the British Army on operations in many parts of the world, including counterterrorist operations and UN peacekeeping. His publications include monographs, book chapters, and articles for, among others, Stanford University, the World Economic Forum, Stockholm International Peace Research Institute (SIPRI), the Crimes of War Project, the *International Herald Tribune*, the *Wall Street Journal*, the *International Defence Review*, the *Independent* (London), *Tiempo* (Madrid), the *International and Comparative Law Quarterly*, the *Washington Quarterly*, and other scholarly journals, including unsigned contributions to IISS publications.

Murray Trostle, Dr.P.H., is a foreign service officer with the U.S. Agency for International Development (USAID), presently serving as the deputy director of the Avian and Pandemic Influenza Preparedness and Response Unit. Dr. Trostle attended Yale University where he received a master's in public health in 1978, focusing on health services administration. In 1990, he received his doctorate in public health from UCLA. His research involved household survival strategies during famine in Kenya. Dr. Trostle has worked in international health and development for approximately 38 years. He first worked overseas in the Malaysian national malaria eradication program in 1968 and has since focused on health development efforts in the former Soviet Union, Africa, and Southeast Asia. He began his career with USAID in 1992 as a postdoctoral fellow with AAAS.

During his career he has worked with a number of development organizations such as the American Red Cross, Project Concern International, and the Center for Development and Population Activities. With USAID, Dr. Trostle has served as director of the child immunization cluster, where he was chairman of the European Immunization Interagency Coordinating Committee and the USAID representative to the Global Alliance on Vaccines and Immunization. Currently, Dr. Trostle leads the USAID Infectious Disease Surveillance Initiative as well as the Avian Influenza Unit.